WHY WE BELIEVE
IN CREATION
NOT IN EVOLUTION

By Fred John Meldau,

Editor, **Christian Victory Magazine,**

Denver 11, Colo.

CHRISTIAN VICTORY PUBLISHING COMPANY

Denver 11, Colo.

Contents

Chapter	Page
FOREWORD, By Prof. Leroy Victor Cleveland	5
INTRODUCTION	7
1. The Case Presented	12
2. Strange Creatures that Witness against Evolution	20
3. OUR PLANET: A Witness to Creative Design	25
4. The Witness of the UNIVERSE and the Witness of the ATOM to the Fact of Divine Creation	54
5. The Witness of MICROSCOPIC FORMS OF LIFE to the Fact of Divine Creation	80
6. The Witness of "DESIGN" and "ADAPTATIONS" to the Fact of Divine Creation	99
7. The Perfect "BALANCE" and the Universal "INTERDEPENDENCE" of all Life on Earth Witness to the Superintendence of a MASTER MIND	109
8. The ENDLESS VARIETIES in Nature and the "PERSISTENCE OF SPECIES" and the many Strange and Odd Specimens of Life are Witnesses to the Fact of Divine Creation	122
9. BIRDS: "Winged Wonders"—Witnesses Par Excellence for God and Creation	158
10. BEES AND ANTS: The "Social Insects"—The Phenomenon of "Community Instinct"	201
11. THE MARVELOUS MYSTERY OF MAN	225
12. MYSTERIES, MIRACLES AND MISSING LINKS	273
13. "EYES," "SEX" and "SPECIALIZED ORGANS"	293
14. FOSSILS—FRAUDS—FABLES	324
Bibliography	343
Index	345

Why We Believe in Creation
Not in Evolution
Copyright 1959
Christian Victory Publishing Company
Denver, Colo.

Printed in the United States of America

FOREWORD

READING THE MANUSCRIPT of this book was a pleasant chore, involving a surprise. It was soon apparent that here was a "must" book for this troubled, confused hour—a volume seemingly specifically "come to the kingdom for such a time as this."

This volume is of strategic importance to millions who seek sound knowledge, and **evidence** honestly interpreted. Evolution is given a head-on challenge, and should indeed be challenged ere millions of students and laymen accept the fraudulent speculation. The dangerously antichrist hypothesis should be most searchingly evaluated before it is accepted. It is being accepted by many as a basis for scientific socialism, secularism, atheism, communism, moral relativism, collectivism, materialism, scientific humanism, and related "isms." For evolution and naturalism to supplant supernaturalism and creative Omniscience will be at a bitter cost.

Thinkers who would like to locate, quickly, ammunition against what they feel is a totally untenable theory, will find this book a comprehensive, well stocked arsenal. (And, incidentally, one strikingly free of such errors as are sometimes found in writings by superficial students of so profound and vast a subject).

The widespread discovery of TRUTH in the area in which this book deals has a vital bearing on the number one problem of our century: What is man, and whence came he? And the even more important question, Is the Nazarene a "made-over ape" or is He the Son of God as He said— the Saviour of the world, God manifest in the flesh—the One who can lead us into God's kingdom of immortality?

References to supernaturalism (to creationism or "creatology") now approach the state of being one hundred per cent blacked out in our schools, while most science teachers and writers jet-swoosh ahead on all fronts with a teaching directly opposite to the teaching of Scripture. It is no more than fair—basically right—that "the other side" of so vital an issue also should be given to our young people!

Do algae, amebas, worms, and so on up to apes and men "evolve higher" or do they not? Could it be true, as we sincerely believe, that **each living thing stays in its kind—or dies?** This book gives a convincing answer; and we may well believe that its position and material will be up-to-date for years to come. The reader will judge which "side" uses the best **reason**. Let us decide with truth—and so avoid possible **treason** before God!

Signed,
Leroy Victor Cleveland, Th.B., Ed.M., Ed.D. (Hon.),
Secretary, USA Division, **Evolution Protest Movement,**
Canterbury, Conn.

ACKNOWLEDGEMENTS

In a work of this kind, touching on scores of highly specialized fields of science, the author has had to seek the advice, criticism and suggestions of biologists and other scientists, and teachers of biology. He wishes to acknowledge his special indebtedness to Prof. Leroy Victor Cleveland, Canterbury, Conn., Willard L. Henning, Dep't of Biology, Bryan College, Dayton, Tenn., and Wayne Frair, Dep't of Biology, The King's College, Briarcliff Manor, New York. Others gave unstintingly of their time and resources of knowledge to assist in the preparation of this work. The author, however, must take full responsibility for the final form of all statements in the book.

INTRODUCTION
(To be read)

The famous Rufus Choate once engaged in a legal battle with the more famous Daniel Webster. The case depended on whether or not two wagon wheels belonged to the same axle on the same wagon. Choate advanced a brilliant argument, based on the theory of the "fixation of points" that the wheels came from the same axle. He had the jury almost convinced. Then Daniel Webster took the stand. He asked that the wheels and the axle he brought forward. It was evident they did not come from the same axle, for they were not the same size. To the honest and sensible jury, Mr. Webster simply said, "Look at those wheels, gentlemen, just **look** at them, and see for yourself that they did not, they **could** not, come from the same axle and wagon." That was all the argument he advanced. The fact was evident; and the jurymen were moved by the facts—and he won the case. If judged by the FACTS in the case, evolution hasn't a chance!

Our American jury system is based on the premise that the average man or woman, though not himself an expert, can decide an issue when evidence is presented to him. We will report **facts in this book, knowing that the average person can come to the right decision when facts are presented.**

Many professionals are easily fooled. Consider the case of the practical jokers in Paris who tied a brush to a donkey's tail and made him swish it on a canvas within reach. A clever and ambiguous title was given it—and the "picture" was duly accepted by an art committee for exhibition! Or that of Brian Hughes, a practical joker of New York, who took an alley cat with its spine so injured that its head was set at an angle, beribboned it and put it on a silk pillow, and entered it in a cat show as "Nicodemus, a Dublin Brindle." **"Nicodemus" got a blue ribbon!**

For forty years the world of scientists was fooled by the so-called "Piltdown Man," "discovered" by Charles Dawson in the south of England and long called "Eoanthropus" (dawn man), and reputed to be from 100,000 to 500,000 years old! The Smithsonian Institution of Washington gives details of the deception in "The Great Piltdown Hoax."

Careful "detective" work done by Dr. J. S. Weiner, and others, revealed that "the lower jaw and the canine tooth are actually those of a modern anthropoid ape, deliberately altered (filed down) so as to resemble fossil specimens." The faker had cunningly "fossilized" the jaw and teeth by staining them a mahogany color with an iron salt and bichromate!

If "experts" and "scientists" can be so easily fooled by a cock-headed alley cat or by a faked surrealistic painting, or a fraudulent fossil, will they not do as badly in trying to interpret the whole history of creation from bones, fossils and unproven theories?

Many of us prefer to believe the record of the divinely inspired Scriptures* that assures us that "God created" the heavens and the earth, man, and all living things. All the evidence supports the Bible record; so we are intellectually compelled to stand on the Scriptural position.

Evolution or Creation

Evolutionists and Creationists both realize that the theory of Evolution and the teaching that God created all things are mutually exclusive. Years ago Sir Arthur Keith said,

"Evolution is unproved and unprovable. We believe it because the only alternative is special creation, and that is unthinkable." Others have voiced the same opinion.

Huxley declared, "It is clear that the doctrine of evolution is directly antagonistic to that of Creation. . . . Evolution, if consistently accepted, makes it impossible to believe the Bible."

If Evolution Is True the Bible Is False

If evolution is true, not only is the Bible mistaken in its teachings that GOD CREATED all things, but also the doctrines of the Bible rest on a foundation of sand, and must collapse.

H. G. Wells sums up the situation in these pointed words: "If all animals and man evolved . . . then there were no first parents, no Eden, no Fall. And if there had been no Fall, then the entire historic fabric of Christianity—the story of the first sin and the reason for an atonement—collapsed like a house of cards."

One Smith, erstwhile president of the American Association for the Advancement of Atheism, bluntly said, "The descendants of apes don't need a Saviour."

But if the Bible is true—and we are absolutely certain it

---

*There is full evidence that the Bible is the Inspired Word of God. See the author's booklet, "57 Reasons Why We Know the Bible is the Word of God" (20c; Christian Victory Pub. Co., Denver 11, Colo.)

is—then evolution is merely the vain imaginings of biased men, men determined they will **not** believe in a Supreme Being, but ready to believe any kind of theory that might be a possible substitute for the evident fact of creation.

Evolution Is Now Gloating

With characteristic cocksureness, the authors (Linville, Kelly and Cleave) of the textbook "General Zoology," write, "All scientists at the present time agree that evolution is a fact."

Julian Huxley (grandson of Thomas Huxley, famous naturalist), English biologist, boasts in his new book, **Religion Without Revelation,** that God has nothing left to do—all belief in His intervention in nature or human history having been debunked. Huxley infers that since sin and forgiveness are no longer real, God, he declares, has been forced to abdicate, "evacuating section after section of His kingdom." Huxley's actual words are: "Operationally, God is beginning to resemble, not a Ruler, but the last fading smile of a cosmic Cheshire cat."*

If those blasphemous words of Huxley are a challenge to the Almighty, he will yet hear from Him; and so will all who reject His truth and seek to drive Him out of His own universe.

What Evolution Is Aiming For

Thousands of American scientists are confirmed "scientific socialists," bent on bringing to eventual fruition their ideal of "scientific socialism." Communism knows it does not have a chance to make America communist by means of a frontal attack on our ideology. But they also know they have a very good chance of bringing in communism through the back door of state socialism, and that is what they are trying to do.**

Believing they have won the battle of the mind in intellectual circles, because so many have accepted the philosophy of

---

* A reference to the grinning cat in Dodgson's "Alice's Adventures in Wonderland," in which a grinning cat, in withdrawing from Alice's view, disappears so gradually that last of all to vanish is its grin.—Editor.

**Kruschev made a prediction not long ago that "the U. S. will go socialistic." Many students of political trends think he is right. The only thing that can save America from such a national calamity is the presence of millions of BIBLE-BELIEVING CHRISTIANS in our good land, who believe God and believe His Word. It is our duty to WARN people of the falsity of evolution and of the subtle plans to make America and the world over into a godless socialistic state.

evolution, evolutionists are looking forward to "the next step in the evolution of man," the introduction of "societal organism." In his book on **"Evolution and Human Destiny,"** Fred Kohler says that evolution is leading us on to a "scientific" world state. We quote:

"The individual . . . must always suffer death. For the species as a whole (the race) . . . it would appear probable (that) a new integration step will take place leading to the formation of an entirely new organic entity—namely the societal organism." This, he adds, "is proceeding at a surprisingly fast pace."

Then he injects this horrifying thought: "The further evolution of human society would be greatly affected by the development of a reproductive system operating on a societal level. . . . An entirely different situation would prevail were it possible to sire future humanity from the best fraction of a percent of the human race" (Pp. 107-109).

So, many of the intellectual leaders of our nation are seeking to browbeat and brainwash "the common herd" into acceptance of the theories of evolution, jelled into the political formula of state socialism. And they are willing to prostitute our "rugged individualism," rape the human mind, banish marriage, and force the public into a goose-step mentality that can be led into the totalitarian setup of their predicted "societal organism." This eventually will demolish the home and set up a materialistic state based on "scientific breeding" that will produce a loveless, godless, Christless race, each individual being a mere cog in the state machine and Huxley's "scientific humanism" will replace God and Christ.

Could it be that these disciples of evolution, living in a Gospel-enlightened land, have closed their minds to the truth to the point where God has sent them strong delusion "that they should believe a lie" because "they received not the love of the truth, that they might be saved"? (see 2 Thess. 2:12, 11). One thing we know: tens of thousands of scientists and intellectuals in our country have turned from the Living God to faith in the preposterous theory of evolution.

Prof. Fleischmann, modern zoologist of Erlangen, after repudiating Darwinism, said, "The Darwinian theory of descent has not a single fact to confirm it in the realm of nature. It is not the result of scientific research, **but purely the product of imagination."**

The late Sir William Dawson, Canada's great geologist, said of evolution, "It is one of the strangest phenomena of humanity; it is utterly destitute of proof."

Such a great thinker as Dr. Robert A. Millikan, famous physicist and Nobel prize winner, said, "The pathetic thing is that we have scientists who are trying to prove evolution which no scientist can ever prove." (Dr. Millikan is an evolutionist; but he is honest enough to admit it is a theory that can NOT be proved).

Our Method of Presenting Proof

We plan to prove in this book that (1) the evidence shows that nature—inorganic and organic: the world around us—must be the work of an Almighty, All-Wise Creator; and (2) "Evolution" is an utterly inadequate answer to the facts of nature. By disproving "evolution" we help to establish the fact of creation; by establishing the fact of the creative work of God, we disprove evolution. So, we have a double-pronged approach, an approach using only the sword of TRUTH that will demolish this vain theory to all willing to listen to facts and vindicate the teaching of the Bible that

"Whatsoever the Lord pleased, that did He in heaven, and in earth, in the sea" (Ps. 135:6).

We will literally call "on heaven and earth"—even the sea— to bear witness for GOD and CREATION and against Evolution. The reader will find that the proof against evolution and in favor of creationism is overwhelming, positive, absolute. In fact, after spending much time in extensive research, we can say that **we do not know of a single plausible argument or fact that favors evolution;** but ALL facts unite to give their conclusive witness that GOD created all things.

In our pursuit of the truth we will present intriguing facts of life and the world, yea even the universe, that would seem to be incredible, were they not so well known, so well established by reputable scientists! THE WHOLE UNIVERSE AND ALL THAT IS IN IT IS AN ENDLESS CHAIN OF "MIRACLES"—UNBELIEVABLE WONDERS. We shall look into the soul of the universe and the heart of the atom; we shall thrill to unaccountable wonders in nature—as the inexplicable miracles of chromosomes and genes, metamorphosis, migration, commensalism, and cross-pollination. We will marvel at strange creatures like the platypus, the sea horse and other astonishing creatures that bear witness to divine plan and creative ingenuity.

We will recount the queer habits of ants and bees, hunting wasps, crabs and oysters! It is our belief that "every creature . . . on the earth, and under the earth, and such as are in the sea" bear witness to the wisdom and power of their Creator.

We trust many will be turned from the fable of evolution to the facts of life and eternity, as revealed in the Bible.

Chapter 1
THE CASE PRESENTED

In this chapter we will give definitions—and a summary of the entire problem of evolution and its solution. We also will give an important **four-word statement** that is a key to the correct understanding of the problem.

"Development" and normal "growth" are **not** evolution. "Evolution" is often used to include the development and progress in inventions and industry, in such phrases as "the evolution of the telephone" or "the evolution of the automobile." The proper word to express such thoughts is "development."

Equally improper is the use of the word evolution to describe the growth of a plant from a seed or a hen from an egg.

Evolution Defined

Darwin defines organic evolution (p. 523, **Origin of Species**) as "the belief that all animals and plants are descended from some one . . . primordial form."

Commenting on the views of Lamarck, Darwin approvingly said, "He upholds the doctrine that all species, including man, are descended from other species . . . all change in the organic world being the result of (natural) law and not of miraculous interposition."

The LINE OF DESCENT from the lower to the higher forms of life is often given (with some variations) as follows: Protozoa — primitive metazoa — worms — fish — amphibians — reptiles — birds and mammals — man. Some recent Zoology textbooks (as Storer and Usinger) no longer refer to "a line of descent" but they speak of "specialized forms" that descended from some supposed ancestral lines (now non-existent) from which all present forms of animal life arose. This is a meaningless evasion— an alternate approach that solves no problems. But always, the transmutation from the lower form of life to the higher presupposes the gradual change by natural, resident forces, unaided by any external, supernatural intervention.*

The Bible Teaching Set Forth

The Bible clearly teaches that God **created** the heavens and

---
*Many today hold to a modified theory of evolution—"Theistic evolution." It is based on the assumption that the higher plants and animals developed from lower forms of life, and that this was God's way of creating all higher forms of life, including man. This we are convinced is not in accord with either the facts of nature or Scripture, hence must be rejected.

the earth, and all forms of life on earth, including man. He created plants and animals in various "kinds" (families and genera) and gave each "kind" the power to reproduce, but only "after its kind"* (see Genesis 1:1, 11, 21, 25, 26-27).

The statement that life as created by God should "bring forth after its kind" does not preclude the bringing forth of a great variety of that kind. For example, we have the **canine** "kind" in which are the related dogs (of many varieties), foxes, wolves and hyenas; the **feline** "kind" bringing forth lions, leopards, tigers, and cats (many varieties).

Unfortunately, much confusion has resulted from the use of the word "evolution" to denote mere improvement of a species, or the development of new "varieties" within the species. Obviously, there are many "varieties" within each species—but to develop new varieties is definitely NOT evolution. Evolution teaches the change, or transmutation, via a generally slow, gradual process of mutation, of one genus into another, the lower into the higher. It does NOT refer merely to the **improvement** of a species. The controversy then is NOT over the improvement of a species by interbreeding, nor the de-development of different "varieties" within the species, but over the evolution of a NEW genus or "kind," the new developing from the old, the higher from the lower.

"Species" Defined

Although there are some exceptions to the rule, the usual definition for "species" is that it is a population (a closely related group of animals or plants) which interbreed and produce fertile offspring."**

According to this definition of "species" given by many scientists, the fertility of the offspring is proof that the parents were of the same species. Without change of species there can be no evolution—and without fertility there can be no descendants! Populations of different "kinds" (using the

---

*The Hebrew word used in Genesis 1 for "create" is **bara** and infers Divine power. The Hebrew word for "kind" is **min** and obviously refers to a related group capable of interbreeding and producing fertile offspring. It corresponds more to our word "genera" than "species" for some "species" according to recent classifications, do interbreed, with fertile offspring.

** See p. 4, "Evolution in the Genus Drosophila," Patterson and Stone; P. 120, "Systematics and the Origin of Species," Mayr; P. 122, "Readings in Evolution, Genetics and Eugenics," Prof. H. H. Newman, University of Chicago Press.

Bible word) will NOT interbreed—hence there are no offspring; and sometimes there are offspring of distinctly related "species" in which the offspring, like the mule, are sterile. The Creator has so made life that it interbreeds only in closely related species or genera; and as soon as interbreeding is attempted between more distantly related species or genera, the offspring is sterile and an impassable roadblock is put up so that it becomes impossible for one genus to transmute into another! Interbreeding between "families" or "phyla" is impossible.

We now are ready to give the four-word statement which is a key phrase that explains the operations and limitations of the divinely-given law of life, first revealed in Genesis 1. In nature we find endless variety within the species or genera; but absolutely NO CHANGE from one genus to another. Summed up in four words the laws governing all life prove there are:

MUTATIONS BUT NO TRANSMUTATIONS

which means that there are many varieties within any group, but there can never be one "kind" of life (genus) mutating (changing) into another.

We now call attention to three fundamentally important facts: (1) Practically all species exist in great variety. (2) The generally recognized phyla (major groups) of life are static; there is no evidence whatever of change from one phylum to another by evolutionary processes. (3) Practically all so-called "proofs" of evolution offered by evolutionists are merely "mutants," variants, minor changes within the same species.

Let us now examine evidence for these three facts:

(1) Practically all Species exist in great variety. Variety is the law of the Creator. In trees, no two leaves are exactly alike; in humans, no two fingerprints are identical; not even two snowflakes, out of the trillions that have fallen, have ever been found that are identical.

As we all know, species are divided into sub-species, varieties, strains, races, breeds. In the dog family there are such mutants (varieties, breeds) as the St. Bernard, the Collie, the Bulldog, the Spitz, the Fox Terrier, the Pomeranian, the Cocker Spaniel, the Greyhound, and a hundred more—but they are all dogs!* But who ever heard of a dog changing into a cat? We see then many "horizontal" differences, but no "vertical" changes from one genus into another.

In fact "all animals and plants mutate," so scientific breeders can produce cattle without horns, white turkeys, seedless grapefruit, and many other varieties seemingly superior to the original stock, but all within the limits of the original "kind."**

In every realm of life the story is the same. More than 20,000 **new** species of protozoa have been discovered with the aid of the compound optical microscope.

One is astounded to learn that there are "80,000 species of snails"—and they are all snails!

In plant Division I, containing the Algae and Fungi, "there are more than 20,000 species of algae alone." In the Fungi series there are over 100,000 species! Among these are innumerable bacteria, molds, mildews, yeasts, smuts, rusts, mushrooms, etc. Division II contains over 20,000 species of mosses—and so the story goes.

Even in such obvious forms as the tiger, many variations occur in nature. It is large and long-haired in some sections; smaller and shorter-haired in India; and very small in Sumatra.

In mankind we see the same phenomena: one species, **Homo sapiens,** with many races: and with no two individuals—not even identical twins—exactly alike! One writer points out there are usually "48 (or 46) chromosomes in each adult cell, and these (not counting the variations possible through the interchange of the many genes in each chromosome) make possible 17 million combinations of human characteristics!"

---

* 108 distinct breeds of dogs are now recognized but so "expert has the dog fancier become in developing variants that there is no reason why the number should not be almost indefinitely increased."

** In nature too we find the development of "varieties," though the process is usually much slower than when man does it. Flies in this country have wings; but the only flies to be found on the storm-swept island of Kerguelen, in the southern Indian Ocean, creep around without wings, or little stubby vestiges of wings—but they are still flies. Scores of similar phenomena can be recounted—but such variants induced by environment always stay within the confines of their own "kind."

We should note also that man has been able to hasten, in some instances, the process of breeding new varieties of animals and developing new varieties of plants by artificially producing mutants through the use of radiation, heat or chemicals. Mutations in barley and corn have been produced by X-raying seeds. But all such mutations **remain in the same genus** as the seeds that were treated. Mutations produced by X-ray exposure are almost always harmful.

(2) **There has never—there can never—be any change by evolutionary processes, from one "kind" (genus) to another.** In the countless billions of living organisms and dead fossils there has never been seen the slightest tendency to advance out of the confines of the original "kind" to which each organism belongs. On the contrary, there is found in every living creature the most stubborn determination **not** to evolve. This is called "fixity of species" and is a commonly observed phenomenon. Here are a few of the many hundreds of competent scientists who could be called on to bear witness to this fact.

Prof. Coultre, University of Chicago, said, "The most fundamental objection to the theory of natural selection is that it cannot originate characters; it only selects among characters already existing."

Sir William Bateson, F.R.S., British naturalist who died in 1925, said, "We cannot see how the differentiation into species came about. 'Variations' of many kinds, often considerable, we daily witness, but no origin of species."

Georges Cuvier (1769-1832), French "dictator of biology" in Napoleonic times firmly maintained the doctrine of the fixity of species. **Since his death, no facts have been discovered that in any wise militate against the fundamental principle he stood for.** Incidentally, France, to this day, has NOT accepted the theory of evolution with the zest this country has.

In recent times, Dr. Austin H. Clark, F.R.G.S., said: "The greatest groups of animals in life do not merge into another. They are and have been fixed from the beginning. . . . No animals are known even from the earliest rocks, which cannot at once be assigned to their proper phylum or major group. . . . A backboned animal is always unmistakably a backboned animal, a star-fish is always a star-fish, and an insect is always an insect, no matter whether we find it as a fossil or catch it alive at the present day. . . . If we are willing to accept the facts, we must believe that there were never such intermediates, . . . that these major groups, from the very first, bore the same relation to each other that they do at the present."

At a later date, when Dr. Clark (recognized as one of the world's greatest biologists) was biologist of the United States national Museum, he stated bluntly that Darwin, Lamarck and all their followers were wrong "on almost all vital points." "So far as concerns the major groups of animals, the creationists seem to have the better of the argument. There is not the slightest evidence that any of the major groups arose from any other. Each is a special animal-complex . . . appearing as a special and distinct creation."

Richard Goldschmidt, Ph.D., D.Sc., Professor of Zoology, University of

California, says, "Geographic variation as a model of species formation" will not stand under thorough scientific investigation. "Darwin's theory of natural selection has never had any proof . . . yet it has been universally accepted" (p. 211). "There may be wide diversification **within the species** . . . but the gaps (between species) cannot be bridged. . . . Subspecies do not merge into the species either actually or ideally" (see pp. 138, 183. The Material Basis of Evolution; Yale University Press). He says, further, "Nowhere have the limits of (any) species been transgressed, and these limits are separated from the limits of the next good species by the unbridged gap, which also includes sterility" (P. 168, ibid.).

Dr. Etheride, speaking of the British Museum, said, "In all this great museum there is not a particle of evidence of transmutation of species."

Prof. T. H. Morgan said, "Within the period of human history we do not know of a single instance of the transformation of one species into another." (P. 43, "Evolution and Adaptation.")

Yves Delage, renowned biologist, said, "If one take his stand on the exclusive ground of facts . . . the formation of one species from another has not been demonstrated at all."

Darwin himself confessed, "Not one change of species into another is on record." "We cannot prove that a single species has been changed (into another)." (Vol. 1, p. 210; "My Life and Letters").

Without "transmutation" the theory of evolution is as devoid of proof as any other fairy tale.

(3) Practically all so-called "proofs" of "transmutation" offered by evolutionists are merely mutants, variants, minor changes within the same species.

We have before us a half dozen articles in current magazines that seek to offer "proof" of evolution; everyone of them merely cites mutations made within the species. All such mutations are as commonplace as varieties of chickens and as meaningless, as far as proving evolution is concerned. We quote from a few of them:

In the May, 1957 "Scientific American" is "A Study in the Evolution of Birds" by H. N. Southern. He calls attention to the fact that some guillemots have heads that are all black, and others, having white rings around their eyes, are called "bridled." He takes several pages to describe this phenomenon: "The frequency of the bridled character varied consistently with the latitude: at the southern end of the range, in Portugal, not a single spectacled (bridled) guillemot was seen, but northward the proportion of bridled birds increased fairly regularly until it reached more than 50% in Iceland. . . . It was obvious that in some way the bridled trait, or something associated with it, conferred a considerable advantage in the northern part of the range" (P. 130).

We all know that "variations" in birds are as commonplace as the different breeds of pigeons.

In another article in the "Scientific American" on "Evolution Observed," by Francis J. Ryan, he states that though "almost nowhere in nature can we see evolution in action" . . . "we are now beginning to realize that objective in the laboratory. . . . With bacteria as subjects we have actually been able to observe evolution in progress."*

After admitting that "although bacteria will mutate, they are really remarkably stable," he said that they "obtained successively fitter and fitter types through 7,000 generations." They developed strains resistant to penicillin when the environment contained penicillin and strains resistant to streptomycin when the environment contained streptomycin! Every doctor in the land knows that bacteria soon become penicillin-resistant, when penicillin is used repeatedly but they still remain the same genus of bacteria as they were before!

Actually, all they demonstrated is the well-known fact that bacteria as well as other forms of plant life will mutate under differing environments. But after all is said and done, after those experiments which lasted through "7,000 generations," "the equivalent of 1,000,000 years of human history," these scientists still had BACTERIA! Such mutations, produced in bacteria, whether by man, or by nature, are no more proof of evolution than to assume that because one can breed yellow dogs and black dogs he can eventually breed tigers from those yellow dogs and black dogs!

Darwin himself was deceived by a similar phenomenon: variations in 14 species of finches, and other animals, on the Galapagos Islands. "He reached the conclusion that since variation in individual characteristics existed among the members of any species, selection of some individuals and elimination of others must be the key to organic change." (See "Charles Darwin," Scientific American; also "Darwin's Finches," Scientific American). Since Darwin's time it has been abundantly proven that mutations and variants are confined to their "kind" and do NOT lead to transmutation.

Life magazine not long ago had a series of articles on "Evolution." They too sought to demonstrate "evolution" by examples of what they termed "evolution through isolation" and "evolution through adaptation," which were nothing more than natural mutations. They gave the example of "five species of birds of paradise" found in different locations

* He explains that though a human generation is 20 years, for bacteria a generation is only 20 minutes; so in two years bacteria can grow through "more generations than man has in a million years."

in New Guinea. These five species had minor variations of color—but all were birds of Paradise! In seeking to prove "evolution through adaptation" they gave as an illustration the gray and the brown lizards that live in the White Sands Desert. True, they were different colors—but both were still lizards.

Further illustrations would be pointless. Let us remember that the law of life is: MUTATIONS (varieties in the species), but NO TRANSMUTATION FROM ONE "KIND" (genus) TO ANOTHER—**hence, there is no evolution!**

The evidence proves the Bible teaching to be correct: though there are many varieties in each species yet each genus **persists in breeding "after its kind,"** stubbornly refusing to do otherwise.

Both "natural selection" and "chance mutations" have been ruled out as possible explanations of the so-called "evolutionary process." Charles H. Hapgood in a recent article in The Saturday Evening Post, presents these facts:

"It is still widely supposed that the principle of NATURAL SELECTION explains the origin of new forms of life. The truth is, on the contrary, that the impossibility of explaining evolution through natural selection, without the assistance of some other factor, became obvious to geneticists about the year 1900. Statistical studies by J. B. S. Haldane and others showed that the amount of time that would be required for new traits to become established in a species by natural selection alone was so immense that even whole geological periods would not suffice to produce new species. As a way out of the difficulty it was suggested that mutations might account for more rapid changes in life forms. It soom became evident, however, that the very great majority of all mutations, since they are random, must be harmful and will be eliminated, in due course, by the process of natural selection itself. **The net result** of mutations, therefore, must be to SLOW DOWN, rather than to accelerate, the process of evolution. The time element is by no means the only problem left unsolved by evolutionary theory. . . ." (The Earth's Shifting Crust, by Charles H. Hapgood, Saturday Evening Post, 1-10-'59).

So modern science has eliminated both Charles Darwin's theory of evolutionary changes through "natural selection," and the more recent theory of comparatively rapid evolutionary advances through MUTATIONS. With both "NATURAL SELECTION" and "RANDOM MUTATIONS" eliminated as possible factors in the "evolutionary process" there is nothing left for evolution to lean on. Long ago the Lamarckian theory of "acquired characteristics" was knocked out by Weismann and others. The fact is, every feature of the theory of evolution that has been advanced has been demolished by scientific investigation and careful research.

Chapter 2
STRANGE CREATURES
That Witness Against Evolution

THIS IS A STRANGE WORLD, with many creatures in it that seem to come out of some unreal, imaginary land; in fact, some are so incredible that one would not accept them as real did he not know they exist. It reminds one of the man who seeing an elephant for the first time said, "Why, there ain't no such animal!"

These strange creatures, who live on land and in the sea, are a most powerful witness for the Creator. Let us call up some of these unusual "witnesses" and hear their intriguing story.

(1) **The Portuguese "Man-of-War" (Physalia).** We quote from a recent magazine article.

"The ruthless **Physalia** employs a lethal combination of ingenious tackle, murderous chemical warfare and a remarkable working partnership with a small fish, the **Nomeus.**

"The deadly sequence starts as this innocent-looking little **Nomeus** swims about, apparently aimlessly, in the vicinity of the man-of-war. A bigger fish, seeing this tempting and seemingly defenseless morsel, makes a grab for it. The **Nomeus** darts away with an unexpected burst of speed, straight toward a tangle of seaweed-like tentacles that hang down from the man-of-war. The larger fish plunges recklessly after it, into these harmless-looking streamers. In a fraction of a second he is paralyzed and the fiendish tentacles are drawing him in to be consumed by the hungry man-of-war.

"What happened to him as he plunged into the streamers is a process that astounds and mystifies scientists who consider it one of nature's deadliest mechanisms. The tentacles are studded with tiny pear-shaped capsules. Sheathed in each capsule is a compressed hair. The instant any creature touches one of the capsules, this hair shoots out like a harpoon, its sharp point penetrating the body of the victim. Through the hair flows a powerful acid which has the power to paralyze the creature into which it is thus injected. . . . So great is the force of the driven threads that they can easily puncture human skin. When they do, the victim may be doubled up in violent spasms and enter a state of shock. . . .

"But why do these deadly hairs, which respond so instantly and savagely to the slightest touch by the other creature, let **Nomeus** go unscathed? Actually, the **Nomeus** can be harmed by the **Physalia's** sting. . . . But the **Nomeus** seems to be especially adapted for dodging the tentacles of the **Physalia.**

"When the man-of-war captures and eats a fish which the **Nomeus** has lured into his trap, the little decoy fish gets the scraps from the man-of-war's meal." (Coronet).

Such highly technical equipment that is so involved, can not be duplicated by man, and such an ingenious strategem of luring other fish to kill them for food could NOT be developed by mere chance. The whole involved system, depending on the cooperation of a fish and including "one of nature's most deadly mechanisms," is the work of an intelligent Master Mind who apparently delights to create numberless insoluble mysteries in nature and distribute strange abilities to equally strange creatures—all the work of His Mind and Hand. Unguided evolution could work around in the dark for a billion years and not come up with such an amazing mechanism!

(2) **Sea Cucumbers.** Sea cucumbers are fleshly "echinoderms" that creep about on ocean floors. They have queer rows of "tube feet" by which they attach themselves to rocks.

Two most unusual means of self preservation granted to these humble creatures by the Creator are given here (reproduced from a scientific article).

"When attacked, some sea cucumbers throw out their viscera, leaving them for the enemy, meanwhile escaping and regenerating a new set. Others throw out slime threads which entangle the enemy. The animals creep about on the ocean floor by muscular movements of the body wall. The tube feet are little used. The animals swallow sand or mud and digest the organic material. . . . They are used in China in making a soup."

Not evolution, but God gifted the lowly sea cucumber with the ability to disgorge its insides—and then manufacture new ones at its leisure.

We would like to know, if evolution made this startling feat possible for the sea cucumber, why do not all other animals in the same category, and higher, have the same ability? Obviously, such miracles in nature are the work of a Supreme, intelligent Creator.

(3) **THE CASE OF THE MOVING LEFT EYE**

There are so many marvels in sea life that we will have to give a separate chapter to "Fish Wonders." However, we want now to tell of the miracle of the "moving eye." Let us take the "plaice" as an example—though any flat fish will do. We refer to the **Pleuronectidae,** i.e., fishes that swim on their sides.

"On emerging from the egg the young plaice, almost microscopic and quite transparent except for its black pinpoint of eyes, is an ordinary round fish with eyes in the position of those of other round fishes, swimming around like a round fish. But after a month a strange thing happens; the left eye begins to move. Meanwhile, the body slowly flattens sideways and the baby fish, a surface swimmer so far, begins to sink slowly towards the bottom. The left eye is still gradually moving, and by six weeks has reached the top of the head. A week later it has gone right around and has almost reached the right eye. By now the young plaice has sunk to the bottom and is lying on what was its left side, but which from now on will be its underpart—the white side—and the two eyes are close together on what is now the top of the head" (The Living Sea, pp. 153, 154).

With plaice, soles, dabs, flounders and halibuts it is always the **left** side that goes down and the left eye that moves; these are called "dextral fishes." But with other species (like the turbot and brill), called "sinistral fishes," the reverse flattening process takes place, and in these fishes it is the **right** eye that travels toward the left eye and away from the right side on which they lie.

No one knows WHY, but it is always the same eye in the same genus that does the travelling.

Moreover, many of these fish possess "homochromic mimicry"—the ability to change their color to suit their surroundings, as a protective measure—to an amazing degree.

They become the same color as their surroundings; and if the color of their surroundings is changed, the fish soon takes on that new color!

Evolution has no adequate explanation of this phenomenon. WHY aren't the fish born flat? WHY in some does the right eye move and in others the left? If as evolutionists say, the round fish "evolved" into flat fish that they might have more safety on the bottom of the sea, then the question arises, why did not **all** round fish "evolve" into flat fish? And if they evolved, why did not they "evolve all the way and have their eyes on the flat side of the head when they were born, instead of having the seemingly awkward arrangement of having the one eye move from one side of the head to the other **after** it was born? The truth is, this is one of those many cases in nature in which the new-born of the species has an entirely different environment from the adult form, and so God performed a miracle when He made these fish so that they could live both as round fish in infancy, near the surface, and as flat fish, in adult life, at the bottom of the sea. Common sense tells us, the only answer to such an unusual arrangement in nature is, **God made it so.**

Actually, there are myriads of STRANGE PLANTS and ANIMALS having characteristics that to the superficial observer seem to be "without rhyme or reason," that can not be accounted for by blind evolution, but show the handiwork of an intelligent Designer and Creator. We mention but a few more of the innumerable oddities:

The "RAILROAD WORM" of South America has a red light and two rows of yellow lights, which make it look like a train. WHY this strange arrangement? The only possible answer is, God made it so!

The CHINA-MARK MOTH, exquisitely decorated, spends its entire caterpillar stage **under water.** This is so drastically contrary to usual experience that there can be no possible "evolutionary chain" leading up to it or departing therefrom. WHY does this creature have this strange life cycle? No one knows; but the answer is, God made it so!

An ALGERIAN LOCUST is able to use its own blood as a weapon. It can shoot, like an accomplished Texas gunman, literally "from the hip." There is a pore between the first and second joints at the base of the leg. This pore can be opened when danger threatens, and a blistering stream of locust blood ejected to a distance of 20 inches! Why do other locusts not have this strange power—if evolution did the job? Obviously, the creature was designed, made that way; God made it so!

Not to be outdone, the BAMBARDIER BEETLE "squirts from its hind end a reddish acid fluid which explodes with a pop. As the 'shot' comes into contact with the air it dissolves into a cloud of bluish smoke which, hovering like a gas barrage, covers the beetle's retreat. The gas has irritant properties and generally succeeds in putting the enemy to flight" (Nature Parade, p. 122). This is ingenuity so involved, using knowledge of engineering and chemistry so advanced, that man can not duplicate this miracle! It demands creation as the only reasonable explanation.

There is a species of BLIND TERMITES that shoot to kill. They have a bi-lobed gland on the head which contains a fluid that solidifies on being exposed to the air. Although this termite is blind, it possesses a mysterious sense of direction which enables it to fire its lethal syringe as accurately as if it could see. "This termite discharges his 'jet' right in the face of an invading ant—and the ants that receive the fatal douche run about as if demented, . . . and they usually die" (Sir J. Arthur Thompson). Certainly such uncanny ability can not be attributed to mere chance.

The bird that gets a NEW STOMACH. The curlew has the incredible ability of coughing away its old stomach (usually in the Fall of the year) and growing a new one! (How handy that would be, if men, having worn-out stomachs, could do that!) The curlew lives on snails and their shells, and so God gave it the ability to get a new stomach when the old one was worn out by the shell diet. This is one of those unaccountable oddities that demand as the simple, yet satisfactory, explanation—God made it so!

Then there is the miracle of complexity, the SEA WORM **"Eunice gigantea"** that has "what may be regarded as 300 separate brains, 250 stomachs, 600 hearts and 30,000 muscles!" Moreover, the **"Eunicidae"** "have a proboscis with seven to nine jaws articulated together." No wonder it is classed as "a rapacious worm." What did it evolve from? Why is its proboscis **more complex,** far more complex, than many higher forms of life?

In future chapters we will tell more about other ODD INSECTS, QUEER PLANTS, INCREDIBLE FISH—especially the unbelievable world of DEEP SEA FISH—that defy all laws of continuity and that are so utterly different from other forms of life that no possible explanation of "natural evolution" can account for these living miracles. Certainly such wonders were NOT the work of a phantom abstraction. Let us give credit to whom credit is due: GOD DESIGNED AND CREATED ALL LIFE IN NATURE!

Chapter 3

OUR PLANET: A Witness to Creative Design

"The Lord by wisdom hath founded the earth" (Prov. 3:19)

"The earth was fitted for man, and man was fitted for the earth." This is so evident, when one considers the facts, as to be axiomatic. The question arises, could a world that shows such marvelous evidence of design "just happen"—or does the fact of "design for an intended purpose" prove Divine creation. We say unhesitatingly, "it would be easier for a blind monkey in a dark room to pick up individual letters from a jumbled mass of type" and assemble Shakespeare's "Hamlet" than for the innumerable miracles in life—such as Man, with his marvelously intricate and well-functioning body and brain, and the Earth on which man lives, with its myriads of miracles and its innumerable adaptations to man's needs—just to happen.

So overwhelming is the evidence for Intelligent Design and Creative Wisdom, in the make-up of our world, that we believe all who examine the evidence we here present will agree, "**The world is the work of a Divine Creator.**" **If one footprint** on the sand convinced Robinson Crusoe that a person was on his island, then by the same logic we know that a Creator made the world, because He left, as it were, the countless footprints of His activities.

When Moses wrote the Pentateuch, the Hindus held this view about the earth in relationship to the Universe: "the earth is borne by four elephants standing on a tortoise's back" (quoted from Flammarion's Astronomical Myths). How different from such childish nonsense is the simple declaration of Scripture:

"In the beginning God created the heavens and the earth" (Gen. 1:1).

The same Bible tells us in simple, non-technical language, that "He (God) . . . hangeth the earth upon nothing" (Job 26:7). We all know this is exactly the truth! How can one explain the vast difference between the puerile theories of the nations that were contemporaries of ancient Israel, and the simple yet profound teachings of Moses and Job, save on the ground that they spoke "by inspiration of God."

Scientists have advanced many theories about the origin of the earth—and not one of them is accepted as a logical explanation today! Against **every** theory advanced by man, some scientist has advanced facts to prove that theory wrong. So, they conclude, "We must honestly admit that we do not know how the earth and the other planets came into being." For example, the two best known theories of the origin of our solar system are: (1) Immanuel Kant's and Laplace's Nebular Hypothesis, and (2) The Planetesimal Theory, advanced by Buffon, Sir James Jeans and others. Scientists say concerning the first of these theories,

"We now know that a rotating mass of nebulous gas would not throw off rings" (Popular Science). "Since the days of Laplace, we have found out that if a ring of gas **were** thrown off from a spinning nebula, it would break up into many small pieces and these would never come together and join into making a planet—as the Nebular Hypothesis teaches. If the planets had been born that way, then the sun would be the fastest spinning object in the whole solar system. Actually, it is the slowest" (Adapted from E. V. McLoughlin).

Concerning the second theory, the same authority says,

"Unfortunately, astronomers and physicists have had to discard this Planetesimal Theory (that ages ago a large star came near our sun and pulled off from the sun 'tidal waves of hot gases' that followed the whirling motions of the sun and became the planets. The second star later went off at a tangent, leaving our sun with its planets). First of all, we find it impossible to explain by the Planetesimal Theory how the planets got so far away from the sun. They would all be much closer to the sun than the earth is, if the theory were true. But there is a more serious difficulty: we can not figure out how those 'tidal waves' of hot gases could ever have liquified and later become solid rock. These original gases would have been so hot they would almost immediately have run off into space, and would never have had a chance to turn into drops of liquid or solid rock." (Adapted from the Book of Knowledge. See Vol. 1, pp. 133-136.)

The Strange Phenomena of "Phoebe" and the Rings of Saturn

To remind the reader of the many difficulties involved in creating an acceptable theory as to the origin of the earth, we call attention to the "Rings" of Saturn and to Saturn's moon "Phoebe."

One of the most spectacular sights in the heavens is Saturn and its three broad, flat "rings," as seen through a large telescope. These "rings" are not flaming and fiery, as was formerly supposed, but it has been discovered with the aid of the spectroscope that they are composed of "countless millions of tiny cold, dark bodies," tiny moonlets if you please, each shining by the reflected light of the sun.

Saturn has nine moons. Phoebe (8,000,000 miles away), most remote of the nine, revolves around Saturn in retrograde motion, in the opposite direction to that of the other eight.*

These phenomena—that some of the moons of the planets revolve in the opposite direction from what one would expect, and that there seems to be no reason for this change in direction—have had the effect of keeping sober men who know the facts from concocting reckless "theories" about the origin of the planets, knowing they have to account for such strange phenomena.

So the little moon Phoebe, revolving backwards off in space, 8,000,000 miles from Saturn, and almost a billion miles from the sun, is a WITNESS FOR GOD! Who but a versatile and Almighty and Sovereign Creator could make a system like our solar system that **works perfectly** despite the fact that part of its "machinery" is apparently going the wrong direction?

Here is another question that theorists find hard to answer: Why has not Saturn drawn these trillions of tiny pieces of matter to its surface? In three tiers they surround Saturn with a tiara of glory—but WHY they are there and how they got there, no man on earth knows. And they are there despite the fact that some theorists say they shouldn't be there! And they most certainly would NOT be there if the Nebular Hypothesis or the Planetesimal Theory were correct.

Is There Life on the Other Planets?

All of the planets, except Mars, are immediately ruled out as possibilities because they are either too far away—hence too cold—or too close to the sun—hence too hot. Then too, they are either too small or too large, and so have either too little or no atmosphere, or have a poisonous atmosphere of the heavier gases.

Some believe that life on Mars, however, is a distinct possi-

---

* All of the five moons of Uranus revolve around their planet with retrograde motion. Neptune has two moons, one of which is retrograde in motion. Strangest of all: Jupiter has twelve moons, and FOUR OF THEM (those numbered 8, 9, 11 and 12) TRAVEL BACKWARD, or with retrograde motion, in their orbits, contrary to the usual direction of rotation and revolution of the planets and their moons. Why eight of Jupiter's moons should revolve one way and four in the opposite direction is "one of the unsolved mysteries of astronomy."

bility. But note these facts: Mars is 142,000,000 miles from the sun. Astronomers took a close-up look at Mars in 1954, when it approached to within 63,000,000 miles of our earth. They were convinced no human life exists there. Dr. Gerard P. Kuiper, director of McDonald observatory (Mt. Locke, Tex.), said:

"Human life on Mars is entirely out of the question because of the severe night temperature and because there is not enough oxygen in the air." Editor's note: The amount of oxygen on Mars is infinitesimally small—only about one-fourth of one per cent per unit of volume as is in the air on our earth.

"The temperature on Mars ranges from a little above freezing at the equator during the day to about -90 degrees F. at night. Hence there is no liquid water on Mars; hence, no oceans or lakes."

Dean B. McLaughlin, University of Michigan astronomy professor, says "the dark markings on Mars' red surface are drifts of volcanic ash, and not vegetation," as has been long conjectured by many. He says this volcanic ash "in the dry, oxygen-poor atmosphere of Mars, is green rather than brown as on earth."

The Bible says, "God created the heavens and the EARTH." As we have seen, the earth is unique in our solar system—and as far as we know from science, it is unique in the entire universe. This can be accounted for on no other ground than that it was CREATED by an Intelligent, all-powerful God. And it was created for a specific purpose: that it might be a suitable dwelling place for mankind.

"Extraordinary Combinations" on Earth

When one considers "the extraordinary combinations of important characteristics that made the earth exactly what it is," one marvels at the ingenuity and carefulness of the Architect.

The MASS and SIZE of the earth are just right. If the earth were 9,500 miles in diameter instead of the 8,000 that it is, it would double the weight of the air. With twice as much oxygen, the amount of water would be greatly increased: so much so that the entire surface of the planet would be covered with an ocean. The conditions on earth would be similar to what now prevail on Jupiter, "which is always surrounded

---

* The facts about Mars are so well established, the question is no longer subject to controversy. Patrick Moore, in his book, A Guide to the Planets, says, "The earth is unique because it is the only world in the solar system upon which we could survive."

by very thick clouds; (scientists) estimate that the ocean that envelopes Jupiter must be 1,500 miles deep."

If the earth were much lighter than it is, its gravitational pull would be less, so that it would not be able to hold as much air as we now have. The lighter gases would escape first and heavier gases, like carbon dioxide, would remain, so the combination of gases in the air would be affected as well as its volume and density—and life as we know it would no longer be possible on earth. Conditions on earth would approximate those on the moon.

Dr. Wallace estimated that if the mass of our earth were only one-fourth less than it is (i.e., if its diameter were 7,200 miles instead of 8,000), almost the whole earth, due to a lessening of its atmospheric mantle, would be reduced to "a snow and ice-clad waste." As a matter of fact, Dr. Wallace more than a half century ago concluded "there was evidence of design" in the size of our earth. He proved that if there were a variation of only 10 per cent, either in the increase or decrease of the size of our world, no life on earth would be possible!

The earth is just THE RIGHT DISTANCE from the sun, relative to the amount of heat and light the sun pours into space. If the earth were much closer, it would be too hot; if it were much farther away, it would be too cold.

So vital is this that Dr. William J. Humphreys, formerly with the United States weather bureau, once lectured on it. He told the American Meteorological Society that if the average temperature of the earth were raised but two or three degrees "you could bid good-bye to all the big cities of the earth," for the glaciers would melt and that in turn would raise the mean level of the oceans 150 feet. This would also inundate thousands of square miles of our most fertile lands.

If the average mean annual temperature were only a few degrees colder, vast amounts of moisture would be put in cold storage in the arctic regions, so robbing the oceans of much of their water. The result would be greatly increased desert areas of the world.

The Earth's Tilt and Rotation Are Just Right

The earth's axis, which now points toward the North Star, is tilted just right—at the strange angle of 23 degrees from the perpendicular, that is, in relation to the plane of its orbit. Because of this tilt the sun appears to go north in the summer and south in the winter, giving us four seasons in the temperate zone. For the same reason, there is "twice as much of the land area of the earth that can be cultivated and inhabited as there would be if the sun were always over the equator, with no change of seasons."

Think what would happen if the earth were tilted any other way than it is.

If the earth had been tilted as much as 45 degrees instead of what it is, temperate zones would have torrid zone heat in the summer and frigid zone cold in the winter. On the other hand, if the axis of the earth were vertical to the plane of its orbit, January and July would have the same climate and ice would accumulate until much of the continents would be ice-covered six months and flooded the other six months. If the axis of the earth were horizontal to the plane of its orbit there would result "a crazy jumble of fierce heat and deadly cold," with prolonged nights on half the earth and prolonged days on the other.

The earth ROTATES at just the right speed, making a complete revolution every twenty-four hours in its trip around the sun. The result is, the earth's crust is evenly heated like a chicken on a turning spit. Were our day a year long, as it is on Mercury,* there would be scorching heat on one side, and bitter cold on the other.

If our day of 24 hours were longer or shorter, all present balanced adjustments would be upset, and life on earth would become intolerable, if not utterly impossible.

The speed of rotation of the earth and its inclination on its axis of 23 degrees do NOT come under any law of the universe—they are just made that way, because it was best for God's intended purpose for the earth. The other planets and satellites do not have the same speed, nor do any of them have the same inclination as the earth.

The Sun in Relation to the Earth

The SUN, 93,000,000 miles away, is a MIRACLE of the first magnitude. It is exactly big enough, and exactly the right distance from us to do as God intended for it to do: provide us with light, heat, power, and many other blessings.

---

* Mercury's period of rotation on its axis is about the same as its period of revolution around the sun, eighty-eight days, making the day the same length as the year! Consequently, the same side of Mercury is always turned toward the sun. That side is extremely hot, with a calculated temperature of about 770 degrees F. The side which is always turned away from the sun is forever dark and cold, with a temperature approaching absolute zero. If SUCH a condition prevailed on earth, there could of course be no life on this planet. Clearly, an all-wise Designer made the perfect adjustments in all things to make the earth a habitable place.

It is the power of the sun that lifts the water from the oceans in the form of vapor, and so makes possible the rainfall over the continents. It is the power and heat of the sun that drives the winds of the world that in turn carry the rain clouds over the lands. It is the mysterious power of photosynthesis—a process not fully understood to this day—that enables plants to manufacture organic food from carbon dioxide, water, mineral salts and SUNSHINE. Animals live on this organic food, manufactured by the plants.

Herschel told us what would happen if the sun were suddenly extinguished.

"In three days from the extinction of the sun there would, in all probability, not be a vestige of animal or vegetable life on the globe. . . . The first forty-eight hours would suffice to precipitate every atom of moisture from the air in deluges of rain and piles of snow, and from that moment would set in a universal frost such as Siberia or the highest peak of the Himalayas never felt—a temperature of between two and three hundred degrees below zero."

The mystery of the sun's unfailing stability. Despite the fact that the sun gives forth such vast amounts of power, of which we on earth get but a comparatively infinitesmal amount, the sun does NOT burn out or lessen or change its output. Who feeds the sun? What mysterious power keeps it going? The recent explanation is, "nuclear transmutation," but even nuclear fission or nuclear fusion, or a combination of both, consume POWER, and matter—and so the sun should run down. Fred Hoyle in his book "The Nature of the Universe" discusses this problem. He says,

"But is nuclear transmutation taking place in the Sun nevertheless? . . . Energy generation must indeed be taking place inside the Sun, and at such a rate as to compensate for what is being radiated into surrounding space. . . . How then can the astrophysicist explain why the Sun does not collapse and also why it has remained pretty much its present size, as the geologists have shown, over at least the last 500,000,000 years?" (Pp. 37,34.)

Hoyle suggests (following Eddington) that the sun keeps its same approximate size and a uniform temperature by "nuclear transmutation," despite the fact it is slowly running down—a sort of self-balancing system that keeps the heat and light and power radiating from the sun at a more or less uniform level. He suggests that in a long time—possibly ten billion years—the sun will burn itself out and explode.

This is extremely interesting. It confronts us with this marvelous fact: the Sun IS a self-regulating furnace that consumes unbelievable quantities of fuel and radiates in all directions unbelievable amounts of ENERGY—and yet it stays stable year in and year out. It is like the burning bush that was not consumed that attracted Moses. It is both mystifying and marvelous.

The Christian finds his answer to this problem in the Bible. God created all things in, by and through Christ. It is Christ who sustains the Sun!*

"Who (i.e., Christ) is the image of the invisible God. . . . For by Him were all things created . . . all things were created by Him and for Him: and He is before all things, and by Him ALL THINGS CONSIST (are held together)" (Col. 1:15-17).

More Intricate Analysis of the Sun's Rays

The sun's rays are MIRACLES that science seeks to explain. When we ask what light is, we face a problem that has baffled the wisest, most learned scientists for generations—and still does. The ancient Greeks thought that light consisted of minute particles radiating from the sun. The physicists of fifty years ago theorized that light was "a series of waves in an all-pervasive material called ether." Today, the "electro-magnetic theory" is advocated, and light is regarded as "waves, or bunches of particles." So the current theory combines "particles" and "waves." "Photons (the energy released when an electron jumps from one orbit into another) get together in bunches and form 'electro-magnetic waves,' which are light." (Article in "Science Illustrated," on "What Is Light?"). Understand the theory? Don't worry; no one else does either. That light from the sun could travel 93,000,000 miles at 186,280 miles per second, and reach us in a perfect state, as true light, is a miracle that none less than an infinite Creator can achieve!

But light is but one of many MYSTERIOUS FORCES coming from the sun! In recent years the theory that light, heat

---

*This answer will not, of course, satisfy the scientist who leaves God out and accepts only a natural explanation for the facts of a physical universe. We have no controversy with honest researchers who seek to discover the physical laws and reasons behind all natural phenomena; nevertheless, the universe is so full of MYSTERIES and unsolved problems that through sheer force of FACTS we are driven to the conclusion that a Supreme Being, whom the Bible reveals, not only created the Universe, but also keeps it going. True faith is not inconsistent with the facts of science. On the other hand, the sheer logic of FACTS forces us to acknowledge the Creator.

and radio waves are all "electromagnetic radiations" has won acceptance. We know that there are also other kinds of "electromagnetic radiations."

"The sum total of these radiations forms what is called a complete electromagnetic spectrum. Arranged in the order of increasing wave lengths, the radiations in this spectrum include gamma rays, X-rays, ultra-violet rays, visible light (from violet to red), infrared (heat) rays and radio waves. They all have the velocity of light—186,280 miles per second; they differ only in wave length, and accordingly, in frequency."

Another authority says, "Ether wave vibrations of a frequency of 3,000 billion to 800,000 billion per second cause us to experience the sensation of radiant heat. Vibrations ranging from 400,000 billion to 800,000 billion per second, falling on our retina, cause us to experience light and color. Certain electric waves below 3,000 billion vibrations per second are not sensed, nor are the ultraviolet and X-rays, which are caused by the inconceivably rapid vibrations in ether, amounting to 800,000 billion to 6,000,000 billion per second!"

Confirmation of the "electromagnetic radiations" theory has come in the last few years.

"Between us and the sun there is a daily traffic of radiation as familiar and predictable as tomorrow's sunrise. Less familiar, and much harder to explain, is the fact that across that vast distance the sun bombards us not only with light and heat, but also with streams of **electrified particles of** its own substance. Although we cannot see or feel them, they have astonishing effects upon the atmosphere of the earth." (Condensed from "Corpuscles from the Sun," by Walter Orr Roberts, Scientific American).

To try to explain SUCH UNBELIEVABLE PHENOMENA as coming to pass as the result of "fortuitous circumstances" or "mere chance" is the height of absurdity; such wonders are the work of a supreme Creator.

Because of these "electromagnetic waves" coming from the sun, we not only have Light, and Heat, and Power, but all the modern miracles of radio, television, radar, X-ray, and a hundred other achievements of modern science that have revolutionized our lives since the turn of the century.

The Moon and the Tides Are Just Right

If the moon were half as far away, or twice its present diameter, great tides would wreck most of our harbors, periodically submerge low-laying islands and coastal plains, and drive inland a hundred miles on some rivers. If the moon were smaller and farther away, it would not have sufficient pull on our tides to cleanse our harbors or adequately rejuvenate (with oxygen) the waters of our oceans.

Obviously, Divine wisdom is shown in the creation of the Moon as a servant of the earth.

Those who visit our seashores are fascinated by the rythmic pounding of the breakers against the rocky coast and the resulting splash and spray with each incoming breaker. This rythmic "breathing"—for such it truly is—is very important to the life of the sea. All animal life in the oceans must have oxygen. The breakers, activated in part by the pull of the tides, aerate the water, giving it a new supply of oxygen.

Both the sun and the moon affect the tides. As a matter of fact, both the sun and the moon create "tides" in the atmosphere as well as in the oceans! And this ebb and flow of gravitation pull helps circulate the air in our atmosphere.

"Far over our heads daily moon tides heave and billow on the bosom of our ocean of air" (Sydnew Chapman, in "Tides in the Atmosphere," Scientific American).

The Miracle of Our Atmosphere

To make the world habitable for man, it needs (1) a regular and sufficient supply of light and heat (which we get from the sun); (2) an abundant and general distribution of water; (3) an atmosphere of proper density and composition.

We have exactly the right kind of an atmosphere—and the fact that we do have it is a MIRACLE! Our atmosphere is made up largely of nitrogen (78%) and oxygen (21%). In addition, our atmosphere contains an essential amount of argon, carbon dioxide and water vapor, beside traces of these gases: neon, helium, methane, krypton, xenon, hydrogen, ozone and others. Now here is the miracle:

"The signal fact is," says Helmut E. Landsberg, writing in the August, 1953, Scientific American, "that the rare gasses are present in our atmosphere in only small amounts, MUCH SMALLER THAN THOSE KNOWN ELSEWHERE IN THE UNIVERSE. At the same time, oxygen, nitrogen, carbon dioxide and water vapor are present on earth IN MUCH GREATER ABUNDANCE than elsewhere." Mr. Landsberg reminds us of the source of his information. "The relative distribution of the elements in the universe has been determined by spectroscopic analysis of solar and stellar matter and by chemical analysis of meteorites" (P. 83).

If the ingredients of our atmosphere followed the proportions prevalent in the Universe **life on earth would be impos-**

---

* Sir James Jeans called attention to the fact that if the moon were as big as our earth and 8,000 miles away we would be afflicted with tides a mile high! And that of course would create chaos on earth.

sible. It is clear, Divine Intelligence designed our atmosphere and Divine power executed the plan.

The Miracle of Self-Adjustment

Mr. Landsberg further states,

"During the past billion years this atmosphere has probably been essentially in a state of equilibrium. Production and consumption of the various gases balance. The major producer in the process is the volcanoes; the big flywheels are plants and the oceans" (Ibid.).

The omniscient God, knowing what conditions would prevail on the earth when man entered the industrial era, created the world so that its atmosphere would be self-regulating. Here is one problem and its solution:

Although 180 billion tons of carbon dioxide gas have been released into the atmosphere by the burning of mined fuel during the last fifty years ... and during the next 100 years the increasing use of fossil fuels in our world-wide industrial civilization should result in the production of about 1,700 billion more tons of carbon dioxide, yet the end result will not be injurious to man because "well over ninety per cent of any excess carbon dioxide introduced into the atmosphere eventually finds its way into the ocean, being readily dissolved by the ocean water" (Dr. Robert E. Wilson).

There are other means in nature to keep the "balance" of the atmosphere as God created it. For example, plants use carbon dioxide gas in the process of photosynthesis—and that helps keep down an excess of carbon dioxide gas.

The Miraculous "Nitrogen Cycle"

About 78% of the atmosphere is nitrogen and 21% oxygen. Nitrogen is an important constituent of plant and animal life—but they are unable directly to extract nitrogen from the atmostphere. Scientists call the bacteria that capture and fix nitrogen in the soil, so that plants can absorb it, "nitrogen fixers." These minute organisms then act as "middlemen" between free nitrogen in the air and the plants and animals who are unable to absorb directly from the air the nitrogen they need. It is good that the Beneficent Creator made nitrogen an inactive (inert) gas.* For had it been otherwise, the world would long ago have ended in catastrophe.

"It is fortunate that nitrogen is chemically inert. If it were less reluc-

---

* "It is amazing to know that nitrogen SHOULD be much more active, chemically, than it is. Nitrogen's thermodynamic relations indicate that it has the potentiality of being MUCH MORE ACTIVE than it ordinarily is" (Martin D. Kamen, Scientific American).

tant to combine with other elements IT MIGHT READILY COMBINE WITH WATER TO FORM NITRIC ACID . . . and then the oceans would turn into diluted nitric acid—a catastrophe certainly as horrible as any visualized in speculations about atomic warfare" (Martin D. Kamen, "Discoveries in Nitrogen Fixation," Scientific American).

An error in making even **one gas** would have made the world uninhabitable! The Creator is the perfect Chemist; all His works are perfect. The evidences for Divine Design in the world are overwhelming.

The vital place that nitrogen fixation holds in the world is seen in this striking quotation from Mr. Kamen:

"Nature's nitrogen cycle is as important to us as the carbon cycle of photosynthesis, by which plants recapture carbon dioxide from the air and convert the carbon into organic compounds. The nitrogen fixers and the photosynthetic organisms, linked in a majestic partnership, keep the living economy of the world solvent."

Lightning has a vital role in replacing nitrogen in the soil. "One hundred million tons of usable plant food a year are supplied to the soil by lightning. Nitrogen and oxygen are combined by lightning into plant food." (Farmer's Digest).

This is another testimony to this great fact: God not only created the world in wisdom, He made it self-supporting; nature is so adjusted that it keeps itself in perfect BALANCE.

The fact that the earth is "self-adjusting" is an awe-inspiring wonder.

The Miracle of Oxygen

Oxygen, which constitutes 21% of the atmosphere is essential to all life. It is in the air in exactly the right proportion.*

"We may almost say that the more chemistry and mineralogy we know, the more puzzled we are about the oxygen in the air. For, the more common rocks of the world contain so large a proportion of oxygen, that it makes up nearly half their weight. This is true of granite, sandstone, limestone . . . sand and clay. The wonder is that any free oxygen is left over for the air" (Forethought in Creation, P. 14).

Nitrogen is inert; oxygen is overly active: and yet the Creator designed the world in creation so that there was just the right amount of oxygen as well as nitrogen.

---

* Evidence that oxygen is in the air in exactly the right amount is seen in the tragic experiences of many hospitals a few years ago. In giving premature babies extra oxygen, they gave too much—and hundreds of babies were blinded. For several months doctors did not know the cause. Eventually they discovered the blindness was due to too much oxygen given the babies in the incubators.

The Atmosphere Has Just the Right Density

Obviously, the Creator, knowing all laws of His Universe, made the atmosphere at JUST THE RIGHT DENSITY at the surface of the earth. How do we know? It has been discovered recently that while

"the temperature eight miles up is a frigid 67 degrees below zero, it becomes a scorching 170 above at an altitude of 30 miles up" (Joseph Kaplan, in "The Air Above Us").

It is clear that if the atmosphere were rarified, life on earth would be impossible because of the extreme cold; if it were **very** rarified, life would be impossible because of the extreme heat.

What the Atmosphere Does for Us

In this brief treatise it would be manifestly impossible to list one thousandth of all things that the atmosphere does for us, but we give a few to show Purpose in Creation. We are living at the bottom of a great ocean of air which not only provides us with life-giving oxygen and water, but also much more.

(1) The atmosphere shields us from the constant barrage of meteors that reach us from outer space. Hitting our atmosphere, most of them are burned up before they reach the earth.

(2) The atmosphere gives us a protective covering from the harmful effects of the ultra-violet rays from the sun. Dr. Florence E. Miller, scientist connected with the Smithsonian Institute, says that we live "miraculously" on this planet, protected from eight "killer rays" from the sun, by a thin layer of ozone high up in our atmosphere.

"If that little belt of ozone, approximately forty miles up and only one-eighth of an inch thick (if compressed), should suddenly drift into space, all life on earth would perish."

The whole subject of "ozone" in the air is so unusual, it seems almost incredible.

"There are two kinds of ultra-violet rays—long and short. The long are deadly, and are absorbed and neutralized by that ozone belt. If permitted to come through, they would blind and blister the human race, and would soon destroy all life on earth. The short violet rays, on the other hand, are necessary to life. If that belt of protective ozone were too thick, so that the short rays could not come through, we would all die of rickets.

"The most deadly of the 'eight killers' comes through in quantities just sufficient to render great service, without destroying us. It keeps down the growth of green algae, one-celled plants that grow in streams. Unrestricted, they would multiply so rapidly that they would clog all the streams of the world, and lead to endless flooding of the world. On the other hand, it is fortunate for the earth as we know it that the algae were allowed to develop in moderation."

Ozone itself is poisonous to human beings; but it is present in such a small amount and is so high up in the atmosphere (20 to 40 miles up) that it acts as a protective shield. Who can read such wonders in creation and not give glory to the God who made all things?

(3) The atmosphere is, moreover, a perfect blanket, an unfailing insulation against both heat and cold. It prevents the rapid escape of heat. Were it not for our atmosphere "we should be frozen as hard as a board every night" (Fred Hoyle in, "The Nature of the Universe"). It also acts as "a great reservoir and distributor" of heat. Dr. Wallace said,

"The atmosphere has the 'peculiar' property (most fortunate for us) of allowing the sun's rays to pass freely through it to the earth which it warms, but acting like a blanket in preventing the rapid escape of the non-luminous heat so produced." Dr. Wallace points out the proof of this: When you get high enough in the tropics, you find snow lying on the ground all year round! (Quoted in "The Bible and Modern Science," P. 84).

(4) The atmosphere also provides us with a wonderful "night air glow." This is the light reflected to the surface of the earth from the upper atmosphere. "It amounts to about five times the total starlight falling on the earth" (Science Digest, Jan. 1953). While the sun is the great source of light, it needs the cooperation of the atmosphere to properly diffuse the light, both by day and by night; this is essential for the proper illumination of the earth.

"Dust"—a Witness for the Creator

The "witchery of the soft blue sky," to borrow a phrase from Wordsworth, has charmed people for ages past. Were it not for DUST in the atmosphere the heavens would be black. "At 17 miles up the sky is always black, with only a luminous glow from below." To the presence of dust in the atmosphere we owe also our sunrise splendors, gorgeous sunsets and radiant afterglow.

The dust also is necessary to help form the rain drops. "Scientific investigation has shown that if there were no dust in the air, not a drop of rain or a snowflake would fall on the earth, and no clouds or fog would form." In view of this, it is interesting to read in the Bible:

"The clouds are the dust of His feet" (Nah. 1:3); and Proverbs 8:26 speaks of "the highest part of the dust of the world."

Conservative estimates show that star dust to the amount of 100,000 tons is filtering onto the earth yearly; and each day billions of tiny meteoric bodies invade the earth's atmosphere and are destroyed by friction.

So we see that every little part of creation is essential to the whole and has its necessary function in the Divine Plan.

Water: The Miracle of Nature

We all know that a world without water would be lifeless; but a world in which water followed the customary laws of physics **would also soon become lifeless!** Water has been called "the most uncommon of the common substances." What a marvelously versatile substance water is! Its "molecules lock together in flinty embrace as ICE"—and think of all the uses of ice in our economy. In another form water covers the earth with a dry, protective blanket of SNOW in the winter, and piles up ton after ton of stored water in the valleys of the mountains. It falls as RAIN to quench the thirst of the dry earth in the spring and summer; it feeds our rivers and fills our oceans and lakes. It shades us from the heat of the sun as vapor in the CLOUDS. As STEAM it drives powerful machinery. Truly, water is one of the greatest gifts of God to man. Without water there could be no inhabited earth.

Water, unlike any other substance (except bismuth) is heaviest at 4 degrees Centigrade—slightly above freezing. Above that and below that it is lighter. Because of this, **ice floats** and freezes from the top on down, and it also causes cold water lakes to have two annual "turn-overs," thereby making a life-saving exchange of oxygen for poisonous carbon dioxide at the bottom of the lake. If ice were heavier than water, instead of lighter, when a lake froze over the ice would sink and the water would soon be solid ice from the bottom up. All rivers and lakes would freeze solid! This would kill all fish, prevent thawing in the spring and so upset the scheme of things as to make life on earth impossible. Who changed the law that cold contracts and heat expands, when it applies to water? The Creator designed it so.

Water has many other virtues that make it indispensable

for a habitable earth. Water is nature's best "air-conditioner." Water has an immense capacity for storing heat energy. Thus, the oceans can absorb enough heat during summer to cool the air; and during winter this heat is given off, thus moderating the cold weather.

Then too, water is "the closest approach to a true universal solvent" that we know. It dissolves with ease a fantastic variety of materials; and yet it does NOT dissolve the rocks of the seashore, otherwise the continents would melt away and disappear in the oceans.

Though sea water weighs 800 times more than air, when it is vaporized by the heat of the sun it is lighter than air and is lifted into the clouds! This remarkabe miracle makes rain possible. Did all these strange and wonderful characteristics of water "just happen"? Hardly. Water with all its marvelous characteristics is one of God's creations.

"And God said, Let there be a firmament in the midst of the waters, and let it divide the waters from the waters. And God made the firmament, and divided the waters which were under the firmament from the waters which were above the firmament: and it was so" (Gen. 1:6-7).

Ages before the modern science of meteorology, the Bible revealed the "rain cycle."

"The wind . . . whirleth about continually, and the wind returneth again according to his circuits. All the rivers run into the sea; yet the sea is not full; unto the place from whence the rivers come, thither they return again" (Ecc. 1:6-7).

Modern science describes this amazing rain cycle in these words:

"The energy of the sun received on a square mile of ocean will evaporate and raise about 5,435 tons of water vapor into the air in an hour. Water vapor rises to considerable heights and moves with the air currents (over the continents). When the air holding water vapor cools, the vapor condenses into billions of tiny droplets, so small it takes about 8,000,000 to make a fair-sized drop of rain! . . . Each of these tiny droplets must have a tiny particle of dust on which it clings. As this condensation takes place in the atmosphere, clouds are formed; when the air has been chilled sufficiently to form clouds these tiny droplets coalesce and form larger drops, and when the air can no longer support them, it rains." (Condensed from Science Digest.)

The Miracle of the "Lightning-Nitrogen" Cycle

Few people realize it, but there is an almost constant discharge of electricity from the earth.

This electrical discharge, almost exclusively in fair weather, is not very strong. Over the whole globe at any one moment the total is only about 1,500 amperes and represents a continuous expenditure of energy at the rate of around 50,000 kilowatts. This 1,500-ampere discharge was first discovered in 1917. . . . After three years of study of the atmosphere over separated points Dr. Vannevar Bush reported that the current comes from the still air lying above the clouds of thunderstorms. The current there is going downward, instead of upward. And the amount of it that reaches the earth, over all the world, is estimated to total—what do you think?— 1,500 amperes! This cycle goes on steadily. The discharge that rises from the earth HELPS PRODUCE LIGHTNING. There are about 16,000,000 thunderstorms over the world every year. These thunderstorms, with the accompanying lightning produce about 100,000 tons of nitrogen compounds annually, dropped into the soil, thereby helping greatly to fertilize it!

And so the Creator uses the powers of lightning and the imperceptible discharge of electricity from both the earth and the clouds to create the "Lightning-Nitrogen" Cycle that does so much good to the soil!

We Have Just the Right Number of Clouds

By the way—speaking of clouds, rain, thunderstorms and lightning, brings to mind this interesting fact: Our world has clouds—but not too many. The planet Venus "is covered entirely with a deep blanket of clouds that touch each other and seldom if ever break apart. Our world, on the other hand, averages one-half open sky where no clouds interfere with our sun's work" (George Leo Patterson). Blind unguided chance? Evolution? Oh no; the Living God made this adjustment in nature as well as all others that together make our world a habitable planet.

The Place of Rivers in the Rain Cycle

Every continent and all major islands are drained by rivers.

Africa has the Nile, Niger and Congo; in North America we have the St. Lawrence, Columbia, Colorado* and the Mississippi; South America has the Amazon and the Orinoco; Europe has the Rhine, the Rhone, the Danube and the Volga; Asia has the Yellow River, the Ganges, the ancient Tigris and Euphrates rivers, and many more.

---

*We can not help but mention how prone men are to suggest impossible theories and make preposterous statements. For example, in the article, "The Face of the Land" that appeared a few years ago in "Life" Magazine, the author said, "In the million years since man first evolved on the earth (?) . . . the Colorado River has carved the Grand Canyon a mile deep, eight miles wide at the rim, in about the same period of time." This is utterly impossible. Engineers, in building the Boulder dam, found a layer of large and small rocks—three to four feet in thickness—in the bottom of the Colorado River channel. This layer of loose rocks acts as a buffer

All of these rivers, with their tributaries, drain vast areas of the continents—and add greatly to the beauty and productivity and commerce of the world. Almost all the great cities of the world are on the seashore or on a large river.

It is not hard for us to realize that these mighty rivers as they flow continuously toward the sea are "part of a cycle which goes on eternally and without which we could not exist. For this cycle assures us of an adequate supply of fresh water; and water is necessary for all living things —both animals and plants—as well as for the far-flung industries upon which our modern civilization is based."

The Place of Mountains in the "Rain Cycle"

"It is hardly an exaggeration to say that rivers could not exist if there were no mountains. . . . For one thing, the chill sides of mountains offer vast condensing surfaces where the clouds give up their moisture in the form of rain or snow. . . Mountains also provide slopes and necessary elevation down which the water runs in its journey to the sea."

So the mountains become God's reservoirs to store moisture in the form of snow and ice until the warmth of spring thaws the ice and melts the snow and sends forth the water to the lowlands below—at the time of year when moisture is needed by the crops.

"The snow-laden summits of the Alps, the Himalayas and the Andes represent an enormous reserve, which is later made available for widespread areas of the earth's surface" (Popular Science).

One is led to ask, "Who (but God) hath measured the waters in the hollow of His hand, and meted out heaven with a span, and comprehended the dust of the earth in a measure, and weighed the mountains in scales, and the hills in a balance?" (Isa. 40:12).

The Sublime Beauty of Snow

"A snowflake is one of God's most beautiful architectural marvels. Snowflakes are infinite in variety and beauty—the great majority being six-sided crystals, each geometrically perfect and differing from all others in design" (Dr. Arthur I. Brown).

Who can watch the myriads of snowflakes filter down through the winter skies, pile up in fleckless beauty, and not know that this is God's world? Such miracles as snow could not "just happen." No one but an infinite God could create trillions upon trillions of delicate snow flakes each winter,

---

and protects the actual hard, rock bottom of the river channel FROM ANY ABRASION WHATEVER! Century after century could pass WITHOUT THE COLORADO RIVER WEARING DOWN THE CHANNEL ONE INCH —at least along many miles of its course where it is lined by this layer of rocks.

with NO TWO OF THEM IDENTICAL! That is a miracle of creative genius that should prove to the most skeptical that GOD is the Master Architect who made all things, including the beauty and endless variety of snowflakes.

"If, through the centuries, snowflakes accumulated only one-half of one per cent FASTER than they melted, the water of the earth would eventually pile up in mountains of ice until . . . the earth at last gathered all the water of the earth in frozen embrace" (B. H. Shadduck, Ph.D.).

If the sun were farther away, or if the earth were smaller, and the atmosphere lighter, if the earth were tilted other than it is, if the mean temperature of the earth were a few degrees lower than it is, or even if water froze at a temperature a few degrees higher than it does—our intricately well-balanced world WOULD SOON PERISH IN THE THROES OF IRREMEDIABLE MALADJUSTMENT.

William O. Field, writing on "GLACIERS" in the Scientific American, said:

"Of the earth's total water budget not much more than 1 per cent is in the solid form of ice or snow, and far less than that in the form of water vapor in the atmosphere. Yet these proportions make up a delicate balance which is immensely important to life on the earth. Any appreciable change in the ratios of water, ice and atmospheric moisture would have catastrophic consequences for man and his economy. The ice piled in glaciers on the lands, for instance, exercise a vital control over sea levels, climate and the continents' water supplies."

The Continuous Miracle of the "Soil Replenishing" Cycle

Pick up some loamy garden soil between the thumb and the forefinger. Within that thirtieth of an ounce of soil is a teeming world of living things—"perhaps as many microbes as there are human beings on earth," some two billion or more. These thousands of millions of micro-organisms in the soil are both plant and animal organisms. They are there as part of the Creator's ingenious system to keep the soil rejuvenated. One species will assist another in "breaking down complex compounds of dead organic matter to simpler substances," to become food for plant life.

The root systems of grasses and other plants help break down rock particles, the first step toward creating a **new** soil. As the roots spread through the soil, they deposit there nitrogen and other new substances built from air, sunlight and water. When the plant dies, it offers food to many organisms in the soil—and they then decompose the remains of roots, stems and leaves, so they can become food for succeeding generations of plants.

Earthworms, too, help in the manufacture of wholesome soil. They eat decomposing plants and soil, mixing and digesting the whole, casting it up on the surface. Earthworm castings not only bring soil up to the air and so help aerate the soil, and loosen it, but also the castings are a rich, rejuvenated soil.

Is it not wonderful that, through the use of micro-organisms, earthworms, the root system on plants, and certain fungi, God has established a self-perpetuating system of universal soil rejuvenation that keeps the soil fertile for its perennial crops? Such a wonderful and practical system could have been devised and put into operation only by an all-wise Creator.

The Blessing—and Menace—of Molds

Most of the 80,000 to 100,000 species of fungi are scavengers; they grow upon the remains of dead plants and animals, and convert these into rich soil.

And so, year in and year out, this ceaseless activity goes on —and man is able to plant seeds year after year, and get crops.

There are enemies and dangers of course. Some fungi destroy flour, wood, leather and innumerable other products. They are especially active when the relative humidity of the air is from 70 to 75%. In equatorial regions where the humidity is high fungi do much damage. Here is a warning:

"Fungi have done fairly well at converting a major portion of the world into mold. Given a slight but consistent increase in temperature and relative humidity over a large portion of the globe . . . and they probably would become a dominant form of plant life"—destroying the complex life as we now know it. (Amazing Appetites of Molds.)

Here again we see the perfect balance of God's world. IF THE RELATIVE HUMIDITY AND THE AVERAGE TEMPERATURE WERE INCREASED, THE WORLD AS WE KNOW IT WOULD SOON BE RUINED BY MOLDS!

The Miracle of the Stable Elements

God "doeth great things past finding out; yea, and wonders without number" (Job 9:9, 10).

The nearly 100 elements on earth—from hydrogen to uranium and beyond—are just the right kind of elements for a habitable world, and are here in the right amounts and in the right combinations. Comparatively small amounts of gold, silver and other precious metals, and the precious stones, are here for the use of man. Iron, copper, aluminum and other industrial metals are here in larger quantities, for they are needed by man in his industrial enterprises. Coal, oil and natural gas are hidden in the bowels of the earth (and yet they are near the surface), to be discovered and used by man when needed. Phosphorus, necessary to the organic world, would spread death and destruction, if it were here in too large quantity. The same is true of chlorine, fluorine, and other elements.

The continents must be made of rocks and mineral combinations not soluble by water (silicon, aluminum, magnesium and iron compounds)—and they are; otherwise all land would soon be carried into the sea.

Without oxygen, there could be no water; without calcium, there could be no lime, no bones; without nitrogen, there could be no plant life—and so the story goes. There is definite need for EVERY ONE of the nearly 100 elements found in nature, in the intricate economy of this world and in the economy of man.

The elements combine into innumerable combinations—all with the infinite precision of absolute perfection. "No druggist's prescriptions are made up with the thousandth part of the accuracy with which nature works," for all chemical compounds unite in strict conformity to atomic balances. The water molecule consists of two hydrogen atoms and one oxygen. It is staggering what chemical combinations are possible. Consider the various resins, plastics, alcohols, carbohydrates, nitrates and phosphates. Consider what possibilities are latent in common things. George Washington Carver developed over 200 useful items from peanuts; well over 200 medicines, dyes and synthetic products are made from coal; from the casein in skim milk are made such products as cloth, plastics, glue, buttons, paints, etc. Obviously, GOD made the elements for man, and as a challenge to man, that through research he might put to use all of the wealth of things God has put here for him.

Because these elements are constant and stable, the chemist can work with them with absolute assurance. Their properties are known and unchangeable. There is no "evolution" in the elements; and thank God there is not; all nature is reliable, stable.

"The properties of elements are to be regarded as fully determined from the earliest conceivable epoch, and are perfectly changeless in time" (L. J. Henderson, "The Order of Nature").

Could these approximately 100 elements create themselves? A man would be a fool to believe that. Creation demands a creator.

There are thousands of other amazing facts about the elements. We would like to discuss design exhibited in the "Periodic Table," and the orderliness and design in evidence in chemical combinations; but these subjects are rather involved for the average non-technical reader, so we content ourselves with calling attention to the more obvious evidences that all of us can see and understand.

The Miracle and Mystery of the Seas

"The sea is His and He made it" (Ps. 95:5).

"From the beginning men have recognized the sea as a supreme wonder and paradox of the natural world—at once a thing of beauty and terror . . . a source of life and a fearful and capricious destroyer" (The World We Live In—The Sea; Life Magazine).

We know of course that life on earth would be utterly impossible without an abundance of water. No oceans, no rain; no rain, no life. But when the Creator made the oceans He really did a magnificent job. Nearly three-fourths of the earth's surface is covered with water that has an average depth of two miles! These oceans contain some "300 million cubic miles of water" that form an immense, life-packed and life-giving reservoir.

The first miracle, in the light of what the rest of the universe is like, is **that there IS an ocean here!**

"In the universe as a whole, liquid water of any kind—sweet or salt— is an exotic rarity."

"Contrary to common belief, the liquid state is exceptional in nature; most matter in the universe seems to consist either of flaming gases, as in the stars, or frozen solids drifting in the abyss of space. Only within a hairline band of the immense temperature spectrum of the universe— ranging through millions of degrees—can water manifest itself as a liquid" (Life Magazine).

Scientists have wondered and theorized where all the water came from. Most scientists today tell you that the earth's water came from volcanoes—from water "sealed in the heart of the young planet from the beginning." (See, "The Earth Is Born,"

Life, Dec. 8, 1952). But how **could** it if the earth was originally a whirling mass of flame and fire, or a whirling mass of hydrogen gas, as many theorists claim?

"Most authorities agree that this first great flood (of water, coming from volcanoes) could NOT have filled the ocean basins as they are filled today, or indeed supplied much more than 20% of the water that now laps high on the continental ramparts" (Ibid.*).

WHO MADE THE OCEANS? The answer is, God made them. For an enlightening discussion of this problem, read Job 38:1-11.

Obviously, the very presence on our earth of such vast quantities of water is a special work of the Creator.

The second miracle, in view of the vast quantity of water there is on our earth, is **that there is any land area at all!** There is full evidence that at one or more times the entire earth **was** covered with water! (See Genesis 1:2; also Genesis 7). We know this because practically everywhere on earth may be found sedimentary rocks containing fossils from the sea. All land area of the earth was once "a part of the sea floor that happens now to be . . . sticking out."

All authorities admit,

"This planet is mostly sea, and there is nothing really to prevent it being ALL sea. A little natural levelling of the land would do so" (The Living Sea.)

Were we to level off completely the earth's present land surfaces (including the bottom of the oceans), the present continental masses would be about 1½ miles under water! Remember, only 29% of our earth is above water level—and if all the land now above sea were dumped into the depths of the oceans, it would fill only one-eighteenth of the present area of the oceans! Although the oceans average two miles in depth, the land area above water averages only ½ mile in height. How shall we account for this? The only possible ex-

---

* Editor's note: The article in Life Magazine claims that the rest of the water present on earth, after the original influx of 20%, "was squeezed to the earth's surface for thousands of centuries thereafter, and disgorged by volcanoes through fissures in the ocean floor," until it reached its present level. We disagree. We know that originally the entire world was covered with water. Furthermore, we know that oxygen is a VERY SCARCE ELEMENT IN THE UNIVERSE: so where did the earth get all the oxygen necessary to create so much water, in addition to the vast quantities of oxygen, in chemical compounds, in our rocks? God made it so.

planation is the MIRACLE OF CREATIVE DESIGN! **God made it so!**

One authority says,

"Every ocean bed has long, narrow chasms where the bottom falls away AS THOUGH SOME TITANIC FORCE HAD SUCKED THE CRUST INWARD TOWARD THE EARTH'S CORE. Curiously, these great oceanic trenches appear near the continental slopes or along the edge of island arcs rather than in mid-ocean" ("The Miracle of the Sea," in Reader's Digest).

The Bible tells us that "GOD (caused) the waters under the heaven to be gathered together unto one place, . . . and the dry land appeared" (Gen. 1:9).

One is reminded of the statement in Job 38:10-11, where the Lord told Job that He had established His decree "and said, Hitherto shalt thou come but no further, and here shall thy proud waves be stayed" (Margin, Job 38:10-11).

In view of these facts we today ask, as the Lord Himself asked Job, "WHO SHUT UP THE SEA WITH DOORS?" (Job 38:6).

The Amazing Wealth of the Seas

"Every cubic mile of sea water contains 100 million tons of common salt, six million tons of magnesium, and 4 million tons of potash." In addition to vast quantities of at least forty other elements, there are "7 tons of uranium and 5 grams of radium to the cubic mile (The World of Water) . . . "and in a cubic mile of sea water there are about $93,000,000 in gold and $8,500,000 in silver" (Wealth from the Salt Seas).

"All commercial iodine was formerly obtained from seaweeds; and this (iodine) is perhaps the most mysterious of all substances in the sea" (Wealth from the Seas).

"A monopoly of the world's bromine is held by the ocean, where 99% of it now occurs. The tiny fraction present in rocks was originally deposited there by the sea."* Today we extract thousand of tons of bromine from the sea and add it to our gasoline to make it "high test."

We also get vast quantities of MAGNESIUM from the oceans. It is a strong, light-weight metal, used in the manufacture of airplanes.

Of all the legacies given to the race by the sea, perhaps the most valuable is petroleum.

"The origin of petroleum is most likely to be found in the bodies of plants and animals buried under the sediments of former seas. . . . Wher-

---

*So BROMINE becomes another silent but effectual witness for the Divine Creator. The fact that it is present ONLY in the oceans argues for the Divine creation of the seas—otherwise, if the ocean waters originally came up out of the bowels of the earth through volcanoes, WHY IS THERE NOT BROMINE IN THE EARTH?

ever great oil fields are found, they are related to past or present seas" (Wealth from the Salt Seas; quoted from Science Digest).

To sum up: "The ocean is the earth's greatest storehouse of minerals. In a single cubic mile of water there are about 166 million tons of dissolved mineral salts, and in all the oceans there are about 50 quadrillion tons!" (Ibid).

"According to the Dow Chemical Company, which alone manufactures 500 preparations from substances found in the oceans, each cubic mile of sea water stores 175 million tons of dissolved chemicals worth FIVE BILLION DOLLARS!" (The Unknown Deep).

Obviously, God created the oceans TO BE OF USE TO MANKIND—as He planned and created everything else on earth.

Space forbids more than the mere mention of the enormous WEALTH mankind gets from fish and other edible creatures of the sea (crabs, lobsters, oysters, etc.). All combine to serve mankind, even as the Creator planned: for He made the earth and all that is in it **for the benefit of mankind!**

The Maintenance of Fresh Water Rivers and Lakes

Life as we know it must have FRESH WATER in large abundance, as well as the briny waters of the vast oceans. In the economy of God He solved this problem by keeping the water vapors that the sun lifts into the clouds FRESH, with no additional chemicals. So, when more than 24,000 cubic miles of rain descends each year over the continents it is all fresh water with no other chemicals added. If there had been that **one mistake** of having the laws of nature such that the ocean waters as they are would have been lifted in vapor, then soon the land areas of the earth would be salty marshes on which no crops could grow!

Further Miracles and Mysteries of the Seas

Great books have been written on this subject; so here again we must content ourselves with but a bare mention of the intriguing facts. The mysteries and miracles of the sea include such wonders as the fascinating Miracle of **"protective coloring"** for the fish of the sea. God is behind these wonders!

"Many surface fish, like the mackerel and the herring, are colored blue or mottled blue and green on top and silvery white beneath" because that is the color of the water as viewed from above—and the bottom corresponds to the color of the sands on the sea floor. "In the shallow waters near the shore fishes that live habitually among seaweed are striped and mottled"—to protect them from their enemies.

Another mystery is that of **Migration in the Sea.**

"The whale will travel thousands of miles between the food-rich waters of the polar seas and the warm breeding grounds near the tropics. Salmon travel hundreds of miles to return to the stream in which they were born. Eels from European rivers make a journey of 3,000 miles to the Sargasso Sea when they are ready to spawn. The tiny larvae, in their turn, set off on the long journey to the shores of Europe (where they have never been), taking several years for their marathon swim. There are no land marks to guide them, yet these little creatures find their way unerringly through the monotonous vastness of the sea"! (The World of Water, p. 85).

It would take many volumes to describe that vast array of **strange deep sea life** in the canyons and lower depths of the ocean: and we would be fascinated by the descriptions of the sea horses, sea cucumbers, sea spiders, quill worms, ribbon worms, glass-like sponges, and squid, odd fish with enormous heads, others with luminescent and electric light equipment and a host of other wonders of the deep that live in lightless depths of a mile or more below the last glimmer of the sun.

The Marvelous Circulation of the Seas

To provide oxygen, and phosphates, to help keep the temperatures of the earth more equal, the Creator has made the seas with a most intricate system providing proper CIRCULATION. This circulation is the result of

"the waves and currents that cause continual movement in the waters of the sea; these are created by the wind and the weather and the rotation of the earth (called the Coriolis Effect) . . . also by the rhythmic movements of the tides."

"The forces that unite the oceans and keep them in reciprocal motion, agitating the depths, impelling warm waters to the frozen ends of the earth and cold water in return to the sunny tropics, are intricate and interlocking, but essentially they are three in number: the wind, the rotation of the earth and the changing density of the water. Climate, gravity, and the varying density of the salt water . . . have smaller parts in perpetuating the motion" (Life Magazine).

All of us are familiar with the vast ocean currents. Both the Atlantic and the Pacific currents "form clockwise and counterclockwise patterns." Among the most important of these vast ocean currents are the famous GULF STREAM (which keeps moving a volume of water equal to a thousand Mississippis at flood tide), the POLAR STREAM, the JAPANESE CURRENT, the BRAZIL CURRENT, the PERU CURRENT, and a score of other lesser known ocean currents!

Surely, this gives full evidence that Someone PLANNED

it that way—to keep the oceans in proper balance as far as heat and cold are concerned, and to keep the minerals—especially phosphates—and oxygen in good supply.

The Three Amazing "Cycles" of the Oceans

In addition to the regular movements of water, there are at least three LIFE CYCLES in the oceans of more than passing interest. "In contemplating the intricate balance of these natural forces the mind is filled with deepest awe" (Miracle of the Sea).

The first is the amazing FOOD CYCLE, beginning with the "grass of the sea"—the plankton that grows in great abundance in the upper 250 feet of the sea waters.

> This "pasture of the sea" can produce nearly twenty tons per acre during the year—several times the yield of crops taken from the soil! This so-called "plankton" (the primary source of food for the living things in the sea) is made up in part of algae (single-celled plants) that grow profusely; these include the miscroscopic "diatoms"—minute plants. In the blanket of floating plankton are also myriads of tiny animals little bigger than the diatoms—animals such as radiolarians, foraminifera, tintinnids and minute animals called copepods, that FEED on the little plant cells.

This "plankton" is food for the fish. Herring, sardines, mackerel and other small fish feed on it continuously. Then larger fish, like salmon and tuna, feed on the herring and mackerel—and these in turn are devoured by sharks, seals and porpoises.

One miracle about plankton we must mention is:

> "Plankton . . . make daily migrations through depths of hundreds of feet as they adjust their environment to the light conditions they prefer" (The World of Water).

This is the Creator's very practical plan for serving fish that live at different levels their daily meal of plankton.

The second Cycle is the OXYGEN Cycle. All life in the sea breathes and lives on oxygen. ALL OXYGEN IN THE SEA MUST COME FROM THE SURFACE LAYER—about 250 feet deep. In this upper layer we find this miracle:

> "The microscopic plants (in plankton, mentioned above) are busy at their work every moment of taking in carbon dioxide and giving out oxygen, by means of the process called photosynthesis, while at the surface itself atmospheric oxygen is taken into solution direct. What remains then is merely a matter of distribution. This is effected by currents and rises and falls of the water, due to the action of winds and waves. In effect, the

sea is being continually stirred like soup in a saucepan" (The Living Sea; p. 214).

This is nothing less than a MIRACLE OF DIVINE DESIGN.

The third life Cycle is the astonishing PHOSPHATE CYCLE. Phosphorus is of great biological significance because it is vital to life. Most of earth's phosphorus is distributed as simple or complex phosphates. The cycle is stated for us in an article on "Phosphorus and Life," by D. O. Hopkins, printed in the 1952 "Annual Report of the Smithsonian Institution." We quote:

"All igneous rocks contain phosphorus, mainly as apatite, a complex form of calcium phosphate." Phosphates are now found in solution in the sea. "The initial assimilation of phosphate in the sea is largely made by algae (in the plankton)." This plankton is eaten by the fish. "A large proportion of sea life dies (a natural death), and their remains sink to lower depths; the eventual decomposition of this dead organic matter returns the phosphates to the (lower depths of) the sea. There is therefore in the lower water levels a steady building up of phosphates; (but) there are the REGULAR INVERSIONS OF THE UPPER AND LOWER LAYERS OF SEA WATER which result in the further utilization of deep-sea phosphate for plankton growth." "All the deep-sea fishing grounds are places where an exceptional uprising of bottom water takes place. With it, a supply of nutrients, particularly phosphate, is brought back to the sunlit zone of plankton growth."

And so again the Creator has planned things to keep the economy of the earth—and the sea—solvent.

And so the EARTH—and all things therein: atmosphere, soil, oceans, rain, clouds, mountains, elements—becomes a vast and unified witness for GOD. Such intricate and involved MIRACLES prove not only CREATION but also the fact that the Creator made the earth a well-balanced and self sustaining and self-rejuvenating system. The humble heart cries out with the Psalmist:

"O Lord, how manifold are Thy works. In wisdom hast Thou made them all; the earth is FULL of Thy riches" (Ps. 104:24).

We have given scores of facts; hundreds more could be given. Enough has been presented to show that the world is NOT the result of "fortuitous chance." But we say with Dr. Paul Francis Kerr, noted mineralogist:

"I cannot believe that the facts of science are mere accidents. The more we study the earth, the more sense it makes. What I have studied about the earth has made me no less a believer in a Supreme Power, but actually more so. . . . We have seen so much of God's handiwork we can say, GOD MUST BE.

"Honest thinkers must see, if they investigate, that only an infallible Mind could have adjusted our world and its life in its amazing intricacies."

It is as easy to believe that the **Mona Lisa** came into existence by stray bits of variously colored pigments being hurled through space that happened to hit a canvas now in the Louvre as to believe that this marvelous world came about by chance. Or, to change the illustration: tear apart the ten thousand parts of an intricate IBM business machine. Place all the parts in a large bowl, and let an ape stir them with a huge spoon—and see how long it will take until all these parts fit together to make the machine! And even if after ten billion years you DID get the machine (which you would not), **how could you account for those finely machined parts in the first place** if someone did not make them? An adjusted, self-sustaining creation DEMANDS an Intelligent, all-powerful Creator.

Chapter 4

The Witness of the UNIVERSE and the Witness of the ATOM to the Fact of Divine Creation

"In the beginning GOD CREATED the heavens and the earth" (Gen. 1:1).

I. THE WITNESS OF THE UNIVERSE

SINCE TIME IMMEMORIAL man has watched the night sky and wondered about the nature of the Universe of which his world is a part; but only in the last fifty years has he begun to understand the immense pattern of the heavens. We now know that our universe is not merely a universe of individual stars, but a universe of millions of "star systems" called "galaxies," similar to the Milky Way Galaxy of which our solar system is a part. As far as relative size is concerned, our earth now emerges as "a cosmic pebble circling a minor star in one of millions of enormous galaxies" rushing around in space that seems to be limitless.* Such inconceivable vastness has elicited from men such expressions as "the cold, awful depths of space"; "our Milky Way Galaxy leaves our earth, by comparison, as a tiny speck of dust in New York's Grand Central Station." **

As we contemplate this vast universe, we are impressed with three outstanding facts:

(1) The inconceivable SIZE of the universe

(2) The presence of LAW in the universe, coupled with the demonstration of limitless POWER

(3) The continuous display of a surpassing GLORY

Each of these factors bears witness to A SUPREME BEING OF UNIMAGINABLE POWER, WISDOM AND GLORY!

(1) The Inconceivable SIZE of the Universe

Our sun, 93,000,000 miles away, is 866,000 miles in diameter

---

* SIZE itself is no adequate criterion of importance. The fact that ON EARTH ALONE, as far as astronomers are able to determine, exists INTELLIGENT LIFE, makes this comparatively small earth of vastly greater importance than immense, lifeless galaxies!

** It is interesting to note that Isaiah uses a similar figure—comparing the nations of earth as "dust" to the God of the Universe. "Behold, the nations . . . are counted as the small dust of the balance. . . . All nations before Him are as nothing; and they are counted to Him less than nothing" (Isa. 40:15).

—1,300,000 times the volume of our earth. Betelguese, one of the stars in the constellation Orion, has a diameter of 215,000,000 miles—248 times the diameter of our sun; Arcturus, one of the super-giant stars in our galaxy has 25,600 times the volume of our sun! Antares, a double star in the constellation of Scorpio, is said to have a diameter of about 400,000,000 miles—over FOUR times the distance from earth to our sun! In other words, Antares is so large that if it were a hollow ball, and our earth and sun were placed inside it, our earth could follow its orbit around the sun, and not even come half way to the outer edge of Antares! The largest star known in our galaxy is one of the stars of the binary (double) star Epsilon Aurigae, which is said to have a diameter ten times that of Betelguese, or 2,150,000,000 miles! Other supergiants in other galaxies are presumed to be still larger!

Job stood aghast at the greatness of God when he contemplated His universe. He said, "He is wise in heart and mighty in strength . . . which alone S-P-R-E-A-D-E-T-H O-U-T the heavens . . . which maketh Arcturus, Orion, and Pleiades" (Job 9:4-9).

Astronomers, as you know, measure stellar distances by the yardstick called a "light year"—the distance light will travel, going at the speed of approximately 186,000 miles per second, in a year. A light year is about six trillion miles (6,000,000,000,000). Arcturus, the great star in the constellation Boötes, is 241 trillion miles away—a distance so great, it would take light from Arcturus 40 years to reach our earth, while light from the other giant mentioned above, Betelguese, takes about 100 years to reach us! Our closest star, Alpha Centauri (it is really a binary, a double star) is about 25,000,000,000,000 miles away; and it takes light 4 1/3 years to reach us.

Great as these distances are, they are almost insignificant in comparison with other distances astronomers speak of. We are told, "it takes light 100,000 years to travel from one edge of our galaxy to the other." Astronomers now estimate that with the 200-inch telescope on Mount Palomar they will be able (with certain refinements they are now working on) "to reach out into space **two billion light years**"! Such distances are of course utterly incomprehensible by the human mind; they leave one awestruck and stunned with amazement.

We are further told that "our galaxy has a hundred billion (100,000,000,000) stars" (Scientific American magazine;

Sep't., 1956); and that there are, beyond our own Milky Way Galaxy, "at least a billion more galaxies, each having approximately as many stars in it as our galaxy" (see article by Jan H. Ort, Scientific American).*

No man can even begin to give an accurate estimate of the number of stars in our universe. A noted astronomer (Jeans), when asked about how many stars there are in the universe, answered, "There must be as many stars in the universe as there are grains of sand on all the seashores of the world."

His statement brings to mind three Scriptures, written ages before the advent of modern telescopes, when men believed there were "about 3,000 stars." One of these Scriptures is Genesis 15:5:

"And He brought him (Abraham) forth abroad, and said, Look now toward heaven, and tell the stars, if thou be able to number them . . .".

Then, in Genesis 22:17, we read these astonishing words that connect the number of stars with the sands of the seashore! "I will multiply thy seed as the stars of heaven and as the sand which is upon the sea shore."

The other Scripture is Jeremiah 33:22: "As the host of heaven cannot be numbered, neither the sand of the sea measured . . .".

Modern astronomers and the inspired prophets of the Bible agree on this point: the stars of the universe are as innumerable as the sands of the seashore! The question naturally arises, How could the writers of the Bible have known that the stars are as innumerable as the sands of the seashore, except by Divine Revelation? **The same God who created the universe inspired the Bible.**

The Andromeda Nebula, the nearest of the great outside spiral galaxies, is "at least two million light years from us" (George W. Gray), and it appears "to be larger than our own Milky Way Galaxy." Twenty million light years is about as far as the 200-inch telescope is able to resolve a nebula (galaxy) into individual stars. More astonishing yet is the assertion that "in the average region of space the average distance

---

* Some authorities place the estimated number of galaxies (star systems similar to our own Milky Way Galaxy) that lie far beyond our galaxy, at from 500 billion (Hubble) to a trillion—instead of a billion. One begins to wonder if they are not swept off their feet and are misled by some features of the universe that they do not as yet fully understand and so have misinterpreted—such as perhaps the so-called "curvature of light." But of this we are confident: The greater the universe, the greater is the God who made it! Nothing is beyond the power of an infinite GOD!

between galaxies is about three million light years" (Rudolph Minkowski; in Scientific American).

Such a vast universe bears testimony to the fact that the God who made it is ALMIGHTY beyond all comprehension—truly OMNIPOTENT. If the creation of the "world" should impress men with the omnipotence of God, how much more the creation of the universe!

"For the invisible things of Him since the creation of the world are clearly seen, being perceived through the things that are made, even His eternal power and Godhead" (Rom. 1:20).

Isaiah tells us that GOD is far greater than His universe; for the universe is called His "tent."

"To whom then will ye liken God? . . . Have ye not known? have ye not heard? . . . It is He that sitteth upon the circle of the earth, and the inhabitants thereof are as grasshoppers; that S-T-R-E-T-C-H-E-T-H O-U-T the heavens as a curtain, and S-P-R-E-A-D-E-T-H them out as a tent to dwell in. . . . Lift up your eyes on high, and behold who hath created these things, that bringeth out their HOST by number; He calleth them all by names by the greatness of His might, for that HE IS STRONG IN POWER; not one faileth" (Isaiah 40:18-26).

The Bible does not start by giving an argument for the existence of God, it simply introduces Him as the "Creator of the Heavens and the earth" (Gen. 1:1). Nothing could be more grand or majestic—or simple. Later in Scripture we are, however, given the perfect argument for GOD in Hebrews 3:4:

"For every house is builded by some man; but He that built all things is God."

It is as simple as this: We all know when we see a house that some one had to build it; we all **should** know, from the same reasoning, that Someone who is Almighty had to build the greater "house," the universe.

One looks into the sky and asks, Where did all this vast system of stars come from? The atheist's answer is, matter is eternal, matter is indestructible. But the Christian's answer is: GOD alone is eternal, and He created matter. The creation of the atomic bomb gave proof to all that matter is NOT indestructible, but is really "congealed energy." When an atom bomb explodes, matter is turned into energy! If one insists that matter is eternal and was never created, he still faces the problem of where MOTION came from; who or what started the stars revolving around each other? What supplied the original terrific "push"? No matter how a thinker may try, eventually he is forced by sound logic back to GOD as the Original Cause of both matter, motion and **continuance** of motion.

(2) **The presence of LAW in the universe, coupled with the demonstration of unbelievable POWER**

"By understanding hath He established the Heavens" (Prov. 3:19).

The Universe is a marvelous, majestic, wellnigh infinite CLOCK, with "wheels within wheels"—to borrow a thought from the book of Ezekiel.

One writer speaks of "the great clock-work of the Universe." Another says,

"The heavens are orderly; the stars are not scattered helter-skelter for ever and ever. They are arranged in orderly systems called gallaxies. . . . Any system, of couse, has a shape. Our galaxy is shaped something like a pinwheel" (Astronomer, writing in the Denver Post).

Otto Struve, in "Surprising Facts About Stars," says,

"But in its very mysterious way, nature has created order in this disorder (what should be disorder)."

And a writer in "Popular Science," writes an article on **"Balancing the Heavens,"** in which he speaks of

"the apparently purposeful and deliberate movements of other heavenly bodies like the planets." He explains that "the movements of the stars and comets and meteors, as well as of the planets and the satellites of the planets, result from the uniform working of mechanical laws."

"BALANCING THE HEAVENS"—what a happy expression; and how true! The billions of stars, planets and their satellites are all BALANCED between the force of "gravitation"* and the law of "inertia," which states that "matter will persist in a state of rest, but when once it is put in motion, it tends to keep going at the same rate and in the same direction, unless acted upon by some external force." GRAVITATION pulls the stars toward their center or nucleus—and toward each other; INERTIA—after once the stars have been

---

* In the 17th century Sir Isaac Newton discovered the principle of universal gravitation. "Gravitation" is one of the most mysterious of all forces—for it acts at a distance **across vacant space**. It seems wholly unreasonable to think that a body can "reach out across space and put hooks on another distant body and pull it toward itself." This is an utter mystery—and will ever remain so. "We do not know, nor can we ever expect to know, the mechanism of gravitation" (Science, Nov. 23, 1923). It is most assuredly a LAW created by God. All LAW demands a LAW-GIVER.

put in motion—keeps them in motion, and in a regular orbit determined by the pull of gravity. And so, by the BALANCE between two great laws of nature, we have the explanation of the ceaseless motion in orbits of the stars and every other heavenly body!

But the serious thinker at once asks two pertinent questions:

(a) What or Who STARTED all this motion? Where did the original "push" come from? Science has no answer. The Bible Believer has the perfect answer: the Creator gave the universe its original motion—and He is the One who keeps it going! Remember, the universe is NOT an absolute vacuum; there is scattered through space a small quantity of hydrogen gas, as well as smaller amounts of dust particles. But be the matter ever so small, MATTER IN SPACE WILL IN TIME SLOW DOWN AND STOP ANY MOVING BODY. The finest pendulum ever made, in the most nearly perfect vacuum man can make, will neither START itself—nor **keep going** indefinitely.

(b) WHO keeps this vast machine going?

The man-made satellites require a tremendous "push" to get them 500 to 1,000 miles above the earth; but they gradually lose altitude and will eventually fall to earth. One of the fundamental maxims of the physical sciences is the Second Law of Thermodynamics: that is, there is a universal tendency to run down, a universal trend "toward randomness" and decline. On the average, things will get into disorder and run down if left to themselves. Obviously, Someone not only had to START this vast machine, but **Someone also has to keep it going.** Again, we are forced back to GOD. The fact is, if God should suddenly die—which is of course impossible—the universe would become chaotic in a very short time, even though it is running smoothly now, due to the truth of the Second Law of Thermodynamics: the universal tendency toward decay and decline.

The Bible presents the **Triune God** (Father, Son and Holy Spirit) as both the Creator and active Sustainer of the universe.

"For in Him (Christ) were all things created, in the heavens and upon the earth, things visible and things invisible . . . ALL things have been created through Him, and unto Him; and He is before all things, and in Him all things consist (are held together)" (Col. 1:16-17).

The heavens are not only "balanced" between the law of gravitation and the law of inertia, they also are balanced in a most amazing and vast and intricate array of revolving and interdependent systems, from our solar system on up to vast galaxies and systems of galaxies! *

Our earth revolves around its axis every twenty-four hours; once a year we spin around the sun, going over 500 million miles annually; our sun, carrying with it the entire planetary system,** goes in an enormous orbit around the center of our galaxy, a trip that takes 230 million years! And it travels at the rate of 175 miles per second! The journey our sun takes is around the center of our galaxy that is 100,000 light years from one edge to the other. And now astronomers tell us that "we have evidence that our Milky Way Galaxy and those relatively near it form a distinct 'Galaxy of Galaxies' that might be called a supergalaxy" (Gerard de Vaucouleurs).

Writing of this supergalaxy, Gerard de Vaucouleurs says, in the July, 1954 Scientific American:

"This supergalaxy—a gigantic system of galaxies—appears to be a strangely flattened cluster perhaps 40 million light-years across. Its uncounted population of galaxies may run into tens of thousands. Its CENTRAL NUCLEUS is roughly marked by the well-known cluster of galaxies beyond Virgo (one of the 89 Constellations), some 15 million light-years away" (Cornell University).

"Surveys showed that galaxies tend to cluster in groups, containing up to a thousand or more. The exploration has in fact suggested to some that our own galaxy may be an outrider in a supergalaxy, just as the sun is an outrider in our galaxy." (Harold P. Robertson, in The Universe; Scientific American, Sept., 1956).

As a matter of fact, one student of astronomy advanced the theory, based on mathematical calculations, "that you can find the center of creation in the motions of 108 great galaxies." Be that as it may, there is beyond doubt a CENTER

---

* The amazing accuracy and smoothness with which the Universe revolves—as a flawless, perfect machine—can be seen in the perfection that characterizes the journey of our earth around the sun: It takes the earth "365 days, 5 hours, 48 minutes and 48 seconds to make its journey around the sun; and in this circuit . . . the earth has never varied one second in a thousand years." None but an infinite GOD could achieve such flawless, continuous PERFECTION.

**In our planetary system, as we mentioned in our last chapter, our earth shows hundreds of evidences of DIVINE DESIGN—it was created to be a suitable habitation for mankind. In the rest of the universe both WATER and OXYGEN are very scarce; in fact, hydrogen and helium make up 99% of all matter in the universe. But on our earth we have a well-balanced variety of about 100 elements, with a large amount of oxygen, that combine to make a practical "world" for man to dwell in. GOD MADE IT SO!

of the entire Universe, around which all galaxies and all super-galactic systems revolve in an apparently endless procession of majestic grandeur!

From one vast center, one tremendous nucleus, the Almighty controls His universe! While many modern scientists ignore God, they do so at the expense of true logic and sound reasoning. A system as vast and as intricate and as involved and as orderly as our UNIVERSE demands not only a Supreme Architect, but also an Almighty Creator and an Omnipotent Superintendent to keep it going!

We have spoken of the law of GRAVITATION and the law of INERTIA. There are scores of other "laws" in the universe that are perfect and necessary to make up this vast universe. All students of astronomy are familiar with Kepler's three laws of planetary motion. There are "laws" of motion, laws of heat, laws of light, laws of sound—and all are PERFECT, never-changing, never-failing.

Rear Admiral D. V. Gallery (USN; writing in the Saturday Evening Post), said,

"The stars . . . in their orbits and velocities through the heavens faithfully obey a great code of LAW. Earth's scientists can quote and explain this code in great detail—until you ask, "Whence came these laws?"

And scientists have FAITH in the laws of the universe. They predict the coming of comets into our solar system years before we see the comet—then they predict its return at some future date—and they do so unfailingly, because the laws of the Universe are unfailing!

Prof. Einstein said in **The World as I See It:**

"The scientist's religious feeling takes the form of a rapturous amazement at the **Harmony of Natural Law,** which reveals an Intelligence of such superiority that, compared with it, all the systematic thinking and acting of human beings is an utterly insignificant reflection" (p. 29).

Edwin B. Frost, at one time astronomer with the Yerkes Observatory, wrote:

"Everything that we learn from the observational point of view in the study of astronomy seems to me to point precisely and always toward a purposeful operation in nature. . . .

"I cannot imagine planets getting together and deciding under what law they should operate. NOR DO WE FIND ANYWHERE IN THE SOLAR OR STELLAR SYSTEM THE DEBRIS THAT WOULD NECESSARILY ACCUMULATE IF THE UNIVERSE HAD BEEN OPERATING AT RANDOM. . . .

"You cannot fail to recognize that LAW has been long at work when you examine the wonderful structure of the aspirals (the spiral-shaped galaxies).

"In a purposeful creation I find it not at all inconsistent to believe that there must be a Mind developing the purpose. . . . If the universe is purposeful, then it is plain to me that man, who is the highest form of development on this earth, must himself be distinctly a result of purpose rather than accident." (Quoted by Dr. Graebner, from the Chicago Tribune, July 13, 1931).

(3) **The Continuous Display of a Surpassing GLORY**

"The heavens declare the glory of God" (Ps. 19:1).

A look into the night sky is a fascinating sight; this inherent glory of the skies is greatly enhanced by the discoveries of modern astronomy. When viewed in a telescope, the color of stars becomes decidedly more pronounced; as a matter of fact, many stars glow with a brilliance more dazzling than a cut and polished gem. The difference in color in stars is due to their temperature. Very hot stars are blue-white or white; cooler ones are orange or red. Through the telescope many stars can be seen as green, orange, violet, pink and many other colors! It is a most inspiring spectacle! A writer, describing the Constellations, says,

"We are apt to think that the stars which are so thickly scattered over the sky are all of one kind, all similar to one another. No impression could be more mistaken. Not only are they of many different kinds, but they show individual differences of extraordinary interest. . . . Let us take as an example one small region in the constellation Andromeda. . . . One star, Almaach, looks like any other star, but actually it consists of three stars; one of these is orange-colored; around this one revolve a pair of stars—one green, one blue. Not far from Almaach is the radiant point of the Andromed shooting stars. . . ."

Here are some more of the wonders of the heavens:

In the constellation Aquarius is "a magnificent globular cluster of stars like a swarm of glittering bees"; not far away (on a sky photograph) is "a lovely pale blue nebula."

The star Beta Orionis is a prodigious sun having a luminosity 13,000 times that of our sun. . . . The chief star in Canes Venatici is seen (under the telescope) to be actually two great suns, one yellow and the other a gorgeous lilac.

As you know, stars are classified according to their brightness: the brighter being those of the first magnitude or less. Those which are just visible, without the aid of a telescope, are the sixth magnitude. A first magnitude star is 100 times as bright as one of the sixth magnitude. Stars of the twenty-second magnitude have been photographed by the larger telescopes. Think of what marvels are to be seen in a star cluster of 100,000 stars, each a pinpoint of glory on the photograph!

Star clusters and the nebulae are "the spectacular showpieces of heaven." One writer speaks of "the gorgeous diadem of resplendent suns forming the Pleiades."

"The form of star clusters are often exceedingly beautiful and interesting. Lines, either straight or forming loops, arches, streamers, or more complicated figures, are traceable in many. (Cf. Job 38:31). . . . In Auriga there is a cluster in the form of a cross; on each arm there are two particularly bright stars, distinguished above all the rest. Similar devices of symmetry or effective placing are frequent. Well-defined geometrical forms, such as triangles and rectangles, are by no means uncommon" (Star Clusters and Nebulae).

"Very beautiful indeed are these wonderful **balls of suns** when seen in a great telescope. What appears to be one star to the naked eye, in Centarus, becomes transformed under the telescope into a vast sphere of brilliant, glorious suns, of which over 5,000 have been counted by photographic means!"

Another wonder of the heavens is the so-called PULSATING STARS. There are thousands of them.

"No more curious spectacle is afforded by the heavens than that of a throng of seeming signal lights waxing and waning every few hours under the sway, obviously, of some common law, yet with no trace of unanimity, some fading while their neighbors are on the rise."

A star which is marked on the sky chart as Delta Cephei alternately brightens and dims with remarkable regularity. It takes five days and eight hours to pass from its brightest phase down to its faintest and then back to its brightest. Astronomers are unable to account for this phenomenon. There are thousands of such stars, called "cepheids" (since they appear to be the same general type as Delta Cephei), each having its own characteristic change in luminosity.

One writer, describing the "Wonderful Milky Way," waxes enthusiastic, and says,

"Our knowledge of star varieties, of the giant stars and dwarf stars, of the violet-white Sirian, and of all other kinds, and our knowledge of the marvelous movements, variations and systematic relations of stars, expand and deepen the GLORY AND MYSTERY OF THE HEAVENS. The sense of some vast, undiscovered plan comprehending the movements and relations of all is altogether in keeping with the sublimity with which the night sky impresses everyone. But when we review the attempts to construct this scheme we are baffled by a sense of their inadequacy and artificiality."

How true! Human attempts, outside of the pale of Divine Revelation, are completely "inadequate" to explain this glorious Universe! When we contemplate the marvels of the universe, we agree with Dr. Fitchett:

"It were as easy to believe that Milton's 'Paradise' were set up in all its stately march of balanced syllables by an ape, or that the letters composing it had been blown together by a whirlwind, as to believe that the visible universe about us—built upon mathematical laws, knitted together by a million correspondences, and crowded thick with marks of purpose—is the work of mindless force."

MORE ABOUT "INADEQUATE THEORIES"

The three most popular theories today about the origin and nature and future of the universe are: (1) The Evolutionary, **"Expanding Universe"** theory. It postulates that "about five billion years ago the universe 'exploded' and began from a hard, concentrated 'primeval nucleus' of matter and radiant energy,"* and that the universe is "still expanding as a result of that original explosion: all galaxies are rushing away from the original nucleus at terrific speeds." But when astronomer Hubble's calculations suggested that the "more distant galaxies were rushing away from us at a 25,000-mile-per-second speed," he became skeptical of his own theory. Such fantastic speeds for vast bodies of matter seem absurd. Furthermore, his theory makes **our earth** the center of the expanding universe—and this is not consonant with other theories that put our solar system **near the edge** of our galaxy, with no one knowing exactly where our galaxy is relative to the universe.**

(2) **The "Steady-state" Expanding Universe** theory, advanced by Fred Hoyle (of Cambridge) and other astronomers. This theory postulates that the "expanding universe" is maintained in a "steady state" by the "continuous creation of new matter, from which is evolved new galaxies as the older galaxies rush out into limitless

---

*Neither Georges Lemaitre, Belgian astronomer, nor Edwin Powell Hubble and George Gamow, American astronomers (nor any other astronomer) who espoused this theory has ever deigned to tell us where this original "primeval nucleus of matter and radiant energy" came from!

The basis for belief in an "Expanding Universe"—the well known "Doppler effect"—is subject to other interpretations. We have no quarrel with those who accept this theory—but let us remember, it is but one of **many theories. To us it seems absurd to believe that "distant galaxies are rushing away from us at speeds of 25,000 miles PER SECOND" and more. Such speeds for vast masses of matter are fantastic and unreal.

space." **He doesn't tell us from what the new matter is being created.** So he predicates an absurdity, for matter cannot create itself. "Something" cannot be produced from "nothing," except by the power of the Creator.

(3) **The "Finite Universe of Curved Space"** theory, suggested by Albert Einstein—based on his theory of relativity. He suggested that space "may be curved into a non-Euclidean form" (i.e., not flat, like the geometry of Euclid that was limited to a plane), which would give us a closed but unbounded universe of finite volume, if the curvature is "positive."*

This theory approximates what the Bible teaches: a finite, though very large universe, the work of the Hands of the Almighty Creator.

"In the beginning GOD CREATED the heavens and the earth" (Gen. 1:1).

There are of course scores of other theories as to the nature of the universe; we have selected three that are widely discussed at the moment. **There is not a single theory of the origin and nature of the universe, that denies Divine creation, but that actually resolves itself into an absurdity, or else can be shown to be fallacious, by known facts.**

Is it not wonderful to be able to turn from man's vain thoughts and philosophies and rest on the fundamental fact that the ETERNAL, ALMIGHTY GOD made this vast universe as it is: a glorious display of His infinite wisdom and vast power! A great universe in no wise militates against creation: it simply proves the greatness of the Creator!

"The heavens declare the glory of God; and the firmament showeth His handiwork" (Ps. 19:1).

"When I consider Thy heavens, the work of Thy fingers, the moon and the stars, which Thou has ordained; what is man that Thou are mindful of him?" (Ps. 8:3, 4).

"Thou, even Thou, art Lord alone; Thou hast made heaven, the heaven of heavens, with all their host, the earth and all things that are therein" (Neb. 9:6).

"Worthy art Thou, our Lord and our God, to receive the glory and the honor and the power: for Thou didst create all things, and because of THY WILL they are, and were created" (Rev. 4:11).

---

*If the curvature is "negative"—as some say—(like the curvature from the inside of a tire that radiates **outward,** and not like the curvature from the outside of a tire that radiates inward) the lines of curvature would expand endlessly. The inferences from this "negative" curvature hypothesis are preposterous.

It should be unnecessary for us to have to refute the charge that "the Bible teaches that God created the heavens and the earth 4,000 B.C." The Bible clearly teaches that **"In the beginning"** God created the universe (Gen. 1:1; John 1:1-3)—and that faroff date may have been "five billion" years ago, more or less.

The Bible **does** teach that in comparatively recent times God "re-created" the earth and made it habitable for mankind, and that God created Adam and Eve comparatively recently.

The LIMITATIONS of Modern Astronomy

We are told that the 200-inch telescope "can peer out into space a distance of two billion light years"—which means that a diameter of FOUR BILLION LIGHT YEARS OF SPACE comes into the view of this modern telescope! This is most astonishing; but we must call attention to some limitations, and possibilities of error, that confront modern astronomy. ALL of these "arguments" and "objections" presented here are quoted from modern astronomers and scientists.

"The only way we can judge the distance (of distant galaxies) is by the faintness of their light. But we must also remember that we are looking far back in time. The intrinsic brightness of galaxies may change with time. CONSEQUENTLY WE CANNOT BE SURE THAT A DISTANT GALAXY IS FAINTER THAN ANOTHER THAT IS FARTHER AWAY" (Modern Cosmology, Scientific American).

Suggesting that the theories arising from the "red-shift" ("Doppler effect") on the spectrum may be wrong, H. P. Robertson, Professor of Mathematics, California Institute of Technology, says,

"Possibly the reddening of light from the distant galaxies is due TO SOME UNKNOWN SMALL INFLUENCE on it during its tremendous journey to us, rather than to a Doppler effect caused by the motion of the source. Then too, the distant nebulae may remain indefinitely where they are (rather than be rushing out into space) and the degradation (running down) of the universe may be caused by the frittering away of light rather than of loss of matter through escape (by rushing off into outer space)."

Commenting on Einstein's theory of the Curvature of Space, a writer in the Book of Knowledge says,

---

*Many Bible students believe there is a vast period of time—giving room for all geologic ages—between Genesis 1:1 and Genesis 1:3. Apparently a pre-Adamic judgment brought the original earth into the chaos and darkness described in Genesis 1:2. See Jerem. 4:23-26, Isa. 24:1 and Isa. 45:18 that clearly indicate that the earth underwent "a cataclysmic change as the result of divine judgment." (Scofield Reference Bible, note on Gen. 1:2).

"If space actually is so curved, then it would be reasonable to assume that rays of light from a star, which start on their way through the universe, will be curved and bent to fit the form of the universe . . . and we might conceivably be able to observe 'ghost images' of stars or nebulae or galaxies on the opposite side of our universe!"

It is interesting to note that there has been PROOF of Einstein's theory of the Curvature of Space.

"Einstein's idea of the gravitational curvature of space-time was triumphantly affirmed by the discovery of perturbations in the motion of Mercury at its closest approach to the sun and of the DEFLECTION OF LIGHT RAYS BY THE SUN'S GRAVITATIONAL FIELD" (P. Le Corbeiller, in "The Curvature of Space," Scientific American). Mr. Corbeiller also mentions the interesting fact that our own EARTH illustrates the "Finite Universe" idea. He says, "This is a most remarkable fact: the surface of the earth is boundless and yet it is finite." And this suggests to the Christian thinker the nature of the universe as a whole.

H. C. van de Hulst, writing on " 'Empty' Space" in the Nov., 1955, Scientific American, says,

"Another proof that interstellar space is not empty came . . . (when) about in 1930, astronomers discovered with some shock that as the light of stars passes through certain regions of interstellar space it is dimmed and scattered in various directions. . . . If there was indeed an interstellar haze which dimmed the light of distant stars or made them altogether invisible, then many of their calculations of star distances and their picture of our galaxy WERE WRONG. Further studies proved that the fear was justified. STARLIGHT PASSING THROUGH THE CROWDED REGIONS OF OUR GALAXY LOSES ROUGHLY HALF OF ITS ENERGY BY ABSORPTION AND SCATTERING IN EVERY 2,000 LIGHT YEARS OF ITS TRAVEL. As a result, even with our most powerful telescopes, we cannot see the center of our galaxy, some 25,000 light years away. Beyond about 6,000 light-years from our observing station most of our studies of the galaxy are literally lost in the fog." (Caps ours).

Many of the theories of modern astronomy are entirely untenable, or are at least challenged by other outstanding authorities. Reason and logic and the Bible assure us that this Universe was made and is upheld by an Almighty Being of Great Glory, Wisdom and Power!

We might add here the statement in the article on "Exploring the Depths and Heights" (Popular Science; p. 3611):

"Still other theories of the origin of the universe have been proposed; but none has been definitely proved or widely accepted. The origin of our universe (to science) REMAINS AS BAFFLING A MYSTERY AS EVER."

There is one other thought we call attention to:

The Innumerable MYSTERIES in the Universe

Not only is the "origin of the universe" an insoluble "mystery" to science (for some scientists wilfully reject the fact of GOD), but there are scores of other mysteries that continue to baffle the modern astronomer.

Dr. Ira S. Bowen, of Palomar Observatory, suggests the baffling problem that science has been unable to solve.

"Curiosity will never let man rest until he solves the riddle of the Universe. How old is the Universe? We now know that the creation date goes back two or three billion years. HOW IT ALL BEGAN is still a mystery, but maybe some time we'll get at least part of the answer."*

We suggest to Dr. Bowen that in Genesis 1:1 is the answer to HOW IT ALL BEGAN.

Of the thousands of baffling mysteries wrapped in the intricacies of the universe we mention but a few.

(1) The Mysteries of Radio-activity and Nuclear Fusion:
Proof that the Universe had a Beginning

"To assume that the universe had no beginning . . . fails to account for the CONTINUED EXISTENCE OF RADIOACTIVITY" (Scientific American). (Obviously, if the universe had no beginning, radioactivity—the degeneration of elements that are radio-active into baser elements, such as the degeneracy of radium into lead—would have run its course ages ago, and ALL elements would have degenerated into the baser metals).

ALSO, since hydrogen is the "mother" element of the universe, and since the stars are kept active by means of nuclear fusion—the slow transmutation of hydrogen into helium—and since the universe STILL IS MADE UP OF 98% hydrogen and only 1% helium, obviously, the universe had a beginning NOT TOO LONG AGO. Otherwise, the hydrogen of the universe long ago would have been turned into helium, by the processes of nuclear fusion going on all the time in the stars! To put the matter bluntly and plainly—THE STARS WOULD HAVE "BURNED OUT" LONG AGO

---

* Dr. Bowen speaks in the same article (National Geographic Magazine) of "The riddle of man's place in the Universe." This "riddle" also is solved in the Word of God. Man being an intelligent being, with a free will, and created in the image of God (Gen. 1:26, 27), and being ALIVE, is of vastly more importance than a lifeless mass of suns, stars and even galaxies! The EARTH, though comparatively small, is of such great importance in God's scheme of things, that He has plans eventually to MOVE HIS HEADQUARTERS DOWN TO THE "NEW EARTH" when He creates "a new heavens and a new earth" (see Rev. 21 and 22). The drama of human history became the very cynosure of the Universe when Christ, God's Son, came to earth, in the Incarnation, that He might redeem a fallen race by His atoning death on the cross. The entire intriguing story is told in the Bible. It is summed up in John 3:16.

IF THE UNIVERSE HAD NO BEGINNING. It does not take much thinking to see that the universe HAD to have a beginning, not too long ago. (Five billion years is not long for the lifetime of a vast universe). Everywhere you turn, one is forced back to the teaching of the Bible! GOD CREATED THE UNIVERSE!

A science writer in "SCIENCE DIGEST" speaks of this fact that explains the continuous "burning" of all the stars of the sky: "So when hydrogen is converted into helium, both heat and light are produced, and a certain amount of hydrogen is completely converted into helium."

(2) **The Mystery of the Exploding Stars.** About two dozen exploding stars show up every year in our nearest neighbor spiral galaxy, the Andromeda nebula. These exploding stars are called "novae" or "supernovae." No one fully understands why they do this. The "supernovae" flare up into a vastly greater display than the "novae."

"What sets off a nova's flare-up may be a true explosion of the star, or perhaps a nuclear chain reaction like that in the atomic bomb" (National Geographic Magazine). But then again, it may be from some other cause. No one knows.

(3) **The Mystery of the "Variables" or Pulsating Stars.** Among the stars are groups of stars that astronomers call "variables" or "pulsating stars." They grow brighter and then dimmer again with "much exactness—about as much, say, as Old Faithful geyser in Yellowstone National Park." Each of these stars has its own rhythm. A number pulsate in a few hours or a day; others may consume several months or a year or more for their particular cycle. The entire subject of "pulsating stars" is most fascinating.

(4) **The Mystery of Cosmic Rays.** Every minute of the day, "mysterious rays from some remote corner of space, possibly the stars, come hurtling through the atmosphere to bombard the earth with showers of particles. . . . In the time it takes to read this paragraph, you will be hit by more than 200 particles, which you can't hear, feel or see . . .". ("What's Behind Those Cosmic Rays?").

Moreover, there are in our solar system (and presumably in other parts of the universe),

"Vast aggregations of infinitesimally minute particles—electrons, protons and nuclei of atoms—which constitute a hitherto unsuspected element of the solar system. Hydrogen nuclei have been detected in them, and they may consist chiefly of this mother gas of all creation."

The more research that is done, THE MORE COMPLICATED THE PHYSICAL UNIVERSE PROVES TO BE! It is all so involved, so well-balanced and so wisely planned that the mind of man is incapable of grasping all its wonders!

There are many other "Mysteries" in the universe that would make interesting topics for discussion, but we have suggested enough to let us all realize that an INFINITE GOD is the Designer and Creator of this vast system!

In view of the fact that man is created in the image of God, we agree with the conclusion of Dr. Schilt. Dr. Schilt, an astronomer, was asked,

"Is man just a mite on a planet in a vast universe? Is man less significant than a flea on an ant's back?"

"No," the astronomer replied. "Finding all these star systems is just a game—an artificial game. The thing we really have to account for is the OBSERVER. Now if the earth is the only place where there is an observer, then IT CERTAINLY IS 'THE CENTER OF THE UNIVERSE.' . . . It is conceivable that no life exists in our universe except on our own earth. . . . So it makes more sense to me that man IS the 'center' of the universe, that he is the sole observer."

And God made the EARTH especially for this "sole observer."

"The heaven, even the heavens, are the Lord's: but the earth hath He given to the children of men (Ps. 115:16).

This reminds one of the Eighth Psalm:

"When I consider Thy heavens, the work of Thy fingers, the moon and the stars, which Thou hast ordained; what is man, that Thou art mindful of him, and the son of man that Thou visitest him. . . . Thou madest him to have dominion over the works of Thy hands; Thou hast put all things under his feet" (vs. 3-6).

>"Countless suns are ever circling
>Through the boundless realms of space,
>And the God whose hand has made them
>Keeps each orb in its true place.
>All revolve in perfect order
>Harmony complete we see,
>Yet the God whose will they follow
>Is the God who thinks of me."

II. THE WITNESS OF THE ATOM

Relatively speaking, "man stands somewhere midway between the stellar universe and the atom." The one is well nigh infinitely large—so large that it stretches out into space far beyond the reach of his most powerful telescopes. The other is almost infinitely small—so small that it diminishes into incredible minuteness, and then shaves off into the virtual nothingness of some of its particles, far beyond the reach of man's most powerful electron microscopes. Both the universe of the stars and the world of the atoms witness to the almighty power and infinite wisdom of our God.

The Molecule

Science defines the molecule as "the smallest particle of any chemical compound." Most all matter on earth is made up of these tiny particles called molecules. They are the "basic building blocks" of all chemical compounds, such as salt (a compound of sodium and chlorine) and water (a compound of hydrogen and oxygen). Over a million different kinds of molecules (hence that many chemical compounds) are known to modern science.

Yet molecules are small—very small. So small in fact that "a ¼-ounce teaspoon of water has in it 9,940,000,000,000,000,-000,000,000 molecules!"

The Atom

All molecules, even though so small, are divided into yet smaller units called atoms. If a substance is made up of atoms of one kind, it is called an element; but if a substance is made up of two or more kinds of atoms, it is called a chemical compound. While there are, as stated above, over a million known and classified chemical compounds, nature has only around 100 essential elements.*

Sixty years ago the atom (derived from the Greek word that means "that which cannot be cut or divided") was defined as "the smallest indivisible particle of a chemical element." In this century scientists have discovered "with mounting astonishment" that the atom itself is made up of three major particles! An atom is so small that a tungsten atom has been estimated to be only 5/1000ths of one-millionth of an inch in diameter.

Inside the Atom

Today we know that all atoms except the hydrogen** are com-

---

* In recent years scientists, through modern atom-smashing machines, have created some 10 or 12 more "elements" that are not found in nature. These are usually very unstable, and are beyond the atomic weight of uranium. To science, there are about 105 elements, with nearly 100 of them found in nature.

** The hydrogen atom has only a proton and an electron. All other elements, from helium on up the atomic scale, have protons, neutrons and electrons.

posed of (1) a nucleus, made up of (a) positively charged **protons** and (b) uncharged **neutrons;** and (2) **electrons,** negatively charged particles that revolve at high speed around the nucleus. Electrons are about 1840 times lighter than protons and neutrons.

The nucleus of an atom is so small that it is only "a millionth of a millionth of a millimeter" in diameter! And the electrons whirling around the nucleus are so small it would take roughly 500,000,000,000,000,000,000,000,000,000 of them to weigh one pound.

Sub-atomic Particles

Up to 1930 scientists had discovered only protons and electrons in the atom. In 1932 James Chadwick discovered the neutron. Now scientists have evidence that convinces them that in each atom there are not only protons, neutrons and electrons, but **an impressive list of about thirty or more** particles in each atom, so that it all becomes very confusing to the average untrained layman. They tell us not only of "photons" (the "quantity unit of radiation"), but also of "pions," the "positron," the "neutrino," the "meson" and a fantastic array of other particles—and even "anti-particles"—that theoretical and experimental physicists have found evidence for.*

It is clear that the minute "indivisible" atom is unbelievably complex—so much so that it is fair to say that scientists never will be able fully to fathom its depths and to know and understand all that constitutes the "sub-atomic world."

The Five "Miracles" of the Atom

As we consider the marvels of the atomic world, we are literally FORCED to see the Hand of the Almighty Creator—for outside of Divine creation there is absolutely no theory that can adequately account for the wonders of the atom. We list here but five of the many "marvels" of the atom.

(1) **The Miracle of the Minute SIZE of the Atom, and Its Constituent Particles**

If an atom is so unbelievably small that "it takes 2,500,000,000,000 protons in a row to make a line an inch long"—how small are some of its lesser particles? Who but an infinite God could create such wonders?

Look around you: see the dirt, the trees, the houses, the

---

*In the September '58, **Scientific American** Magazine are listed 32 sub-atomic particles of matter and energy presently known to physicists.

people; look above into the sky and see the stars. ALL things in all the universe are made up of these tiny atoms, each a miracle of creation! How marvelous is our God!

(2) **The Miracle of the Tremendous SPEEDS of the electrons in Revolution Around the Nucleus of Each Atom**

Each atom is a miniature "solar system," with the electrons whizzing around the nucleus (protons and neutrons) at astonishing speeds. These minute "electrons" dash around the nucleus "millions of times per second"! Moreover, the electrons revolve around the nucleus "in orbits whose diameters are about 10,000 times larger than the nucleus."

"Electrons revolve around the nucleus in an orbit less than one-millionth of an inch in diameter; and they make the revolutions several thousand million times every second."

(3) **The Miracle of the "Empty SPACE" in Each Atom**

An atom is built like our solar system. IT IS ALMOST ALL EMPTY SPACE. This seems incredible, but it is factual.

"Within the atom, electrons revolve around their nucleus several thousand million times a second. Each electron has as much room to move within the atom as a bee has to move around in a cathedral" (Sir Oliver Lodge).

Scientists say that "if you eliminated all the empty space in every atom in the body of a 200-pound man he would be no bigger than a particle of dust" (Arthur S. Eddington, in "The Nature of the Physical Universe"). And if the entire earth were likewise compacted "it would become a ball only one-half mile in diameter."

(4) **The Miracle of the ELECTRIC CHARGE IN EACH ATOM**

The proton in each atom has a positive charge of electricity, and each electron has a negative charge that exactly balances the positive charge in the proton. Who put the electricity in the atom, and balanced the normal atom so finely? None but GOD could work with such infinitesimal particles and make them perfect, as they are.

The remarkable thing is, in the normal, stable atom, there are just as many electrons outside the nucleus as there are protons inside.

(5) **The Miracle of the Immense COHESIVE FORCE in the Nucleus of the Atom**

Ordinarily, like charges of electricity in different objects that are close to each other REPEL each other; but in the nucleus of the atom God has reversed the law of nature scientists are familiar with (called Coulomb's Law). In the nucleus of the atom, in which all the protons have a positive charge, instead of repelling each other, **they are held together by some unknown force of tremendous power!** This phenomenon, to scientists, is the most mysterious thing about the atom; in fact, they call it "the basic mystery of the universe" (World Within Atoms).

In an article on "Pions," by Robert E. Marshak, in the "Scientific American," we read:

"The cement that holds the Universe together is the force of gravity. The glue holding the atom together is electromagnetic attraction. But the glue that holds the NUCLEUS of the atom together is a mystery that defies all our experience and knowledge of the physical world. It is a force so unlike any we know that we can hardly find words to describe it."

Seeking to explain this mystery, modern atomic physicists have come up with the theory that "in some way, not yet understood, pions (nuclear particles) are certainly involved in the nuclear binding force."

"The proton and neutron, once supposed to be the ultimate building blocks of matter (are now believed) to consist of a core surrounded by a fluctuating cloud of pions—an arrangement that reminds one of the atom with its nucleus and planetary electrons"—and that gives us this phenomenon: an infinitesimal 'planetary system' within an already infinitesimal planetary system! And it's all so small that this inner core (nucleus) "occupies only a thousandth of a millionth of a millionth of the space within the atom" (World Within the Atom, Columbia University Press).

This terrific energy, coiled like a spring, within the nucleus of the atom, is known the world over as NUCLEAR ENERGY. Nuclear energy is so tremendous that

"If it were possible to convert one pound of any matter entirely into electrical energy, you could run with it ALL THE ELECTRICAL APPARATUS IN THE U. S. FOR A WHOLE MONTH."

This "force," great and mysterious as it is, is spoken of in the Word of God. CHRIST is the source of this power; He has the answer.

"For in Him (Christ) were all things created . . . all things have been created through Him and unto Him; and He is before all things, and in HIM all things consist (Gr., 'hold together')" (Col. 1:16, 17; A.S.V.).

And so we see, inside each tiny atom, the reflection of both the power and wisdom of our triune God!

Each passing year brings new discoveries about the atom, "the basic miracle of the Universe." So great are the "miracles and mysteries" of the atom, one must conclude that only GOD could make an atom—a tiny power plant so minute and intricate that it can not be seen by the human eye, nor can its marvels be fully comprehended by the human mind—and yet out of these small particles GOD CREATED HIS UNIVERSE.

The fact that all matter is made up of invisible particles is intimated in the Bible, and was revealed many hundreds of years before the modern "atom age."

"Through faith we understand that the worlds were framed by the Word of God, so that things which are seen were not made of things which do appear" (Heb. 11:3).

How marvelous is our God and how wonderful are His ways —ways past finding out!

"Dost thou know . . . the wondrous works of Him which is perfect in knowledge? . . . Great things doeth He which we cannot comprehend" (Job 37:16, 5).

III. THE UNBRIDGEABLE CHASM BETWEEN THE NON-LIVING AND THE LIVING

In general, scientists call living things, or things derived from living things, "organic"; and the non-living "inorganic."

The galactic universe—the macrocosm—is certainly vast and majestic—and it speaks to us of the power and glory of our God—but it is inorganic, lifeless in itself. How can life exist in the intense heat of a "burning" star?

The atoms speak to us of the miracle of God's creation in the microcosm, the world that approaches the infinitely minute, but there is no life in either the whirling electrons or the center core, the nucleus—even though there is plenty of mystery, action and power in each atom and its particles.

But when we look around on earth we see the phenomena of LIFE on all sides: plant life, animal life, life in the sea, in the air, and on earth. It exists in a million different forms from invisible viruses and bacteria to highly complex and well organized life in the higher animals and man. Where did life on earth come from? How did it all start?

Since there is no such thing as spontaneous generation—life must **always** come from life—we conclude that life on earth, as the Bible says, was created by God.

"And God said, Let the earth bring forth grass, the herb yielding seed, and the fruit tree yielding fruit after his kind, whose seed is in itself, upon the earth: and it was so.

"And God said, Let the waters bring forth abundantly the moving creature that hath life . . .

"And God created great whales and every living creature that moveth . . ." (Gen. 1:11, 20, 21; see also vs. 24-27).

The ETERNAL GOD, who made the Universe and all things in it, is the true "secret of the Universe." All things were made BY Him and FOR Him. This is the true philosophy of life and its origin.

Pre-organic Condition on Earth and the Requirements of Life That Demand Creation

All scientists agree that there was a time when there was no life on earth. Fred Kohler says:

"According to the best estimates, living matter began to develop about one to two billion years ago from the then existing non-living material" (p. 12, "Evolution and Human Destiny").

But according to this same author—who is, by the way, an ardent evolutionist—the prerequisite for life on earth is **the presence on earth of some form of "organic compounds."**

"The non-living material which existed on this planet at the time at which the first structures that can properly be termed "living" developed, must have included some organic compounds of a high order of complexity." . . . And "as the organic compounds that gave rise to living structures could not have existed at the time the earth began to solidify, they in turn must have developed from simpler substances. It is consequently apparent that a 'pre-organic' evolution of chemical complexity must have preceded the evolution of life" (Evolution and Human Destiny, pp. 12, 13).

Now the question logically arises, What gave rise to these **"pre-organic compounds"** that had to be on earth before life could either come to pass, or be sustained after it got here?

Fred Kohler frankly admits the evolutionist faces a real problem here. He says:

"Life represents matter organized into systems of great complexity. How such orderly aggregates could develop in the first place, persist and continue to become more complex, is not so easily explainable in terms of the generally accepted laws of the physical sciences" (Ibid., p. 14).

Then he goes on to tell us WHY the presence of life on earth can not be easily explained.

The "Second Law of Thermodynamics"

Mr. Kohler says:

"One of the most fundamental maxims of the physical sciences is the trend toward greater randomness; the fact that on the average things will get into disorder rather than into order if left to themselves. This is essentially the statement that is embodied in the **Second Law of Thermodynamics**" (Ibid., pp. 14-15).

This "Second Law of Thermodynamics" is very interesting. It infers, as Mr. Kohler says, that "things will get into DISORDER rather than order, if left to themselves." This law infers and involves the fact of "the universal tendency toward decay"—and all nature demonstrates it!

Now note this well: The Second Law of Thermodynamics infers and teaches EXACTLY OPPOSITE TO WHAT EVOLUTION TEACHES! **It infers universal decay rather than universal development.** Careful, honest observers admit that the law of nature in both the inorganic and the organic world tends toward degeneracy rather than toward improvement.

This Second Law of Thermodynamics* is of universal application. The very universe itself is "running down." Instead of the sun and stars conserving their energy, they are gradually losing it.

"Astronomers tell us that the sun is gradually losing its heat and also its weight. The loss in weight is at the rate of 250 tons a minute, or 120 million tons a year."

The Second Law of Thermodynamics can also be seen at work in the radio-active elements in the atomic world. Uranium is in a constant state of decay even though its rate of disintegration is very slow.

"Lord Rutherford's group at Cambridge proved that the radio-active elements uranium and thorium decay ultimately into helium and lead" (The Age of the Solar System, April, 1957, "Scientific American").**

And so the Second Law of Thermodynamics—a law of nature—becomes a witness for the need of DIVINE INTERVENTION before life could come. ALL THINGS—ALL FACTS—drive us back to GOD, the Original Cause.

The quotations given above by Dr. Fred Kohler actually give us a perfect case for creation even though he is arguing (in his book) for evolution. Let us summarize our arguments:

---

*The First Law of Thermodynamics deals with heat transfer; the Second with Entropy, or heat loss; and the Third relates to the behavior of chemical substances at low temperatures.

** This gradual decay of the elements not only has enabled science to set an approximate date of "4.5 billion years ago when the earth and its neighbors were formed" ("Age of the Solar System")—but it also is the positive proof that our earth and our universe HAD A BEGINNING. Matter is NOT eternal; if it were, uranium and thorium long ago would have deteriorated into helium and lead, and all stars ages ago would have burned out. GOD CREATED ALL THINGS IN THE BEGINNING!

(1) There was a time on earth when there was no life; now there is abundant life.

(2) Before there can be life on earth, there must first be on earth "organic compounds of high complexity." But

(3) The Second Law of Thermodynamics sets forth the truth that things left to themselves will certainly NOT develop into "a high state of complexity" but will tend to "decay" and to degenerate into "more randomness."

(4) Therefore we must conclude that a Power greater than and apart from nature stepped in and created life. This Power, this original Cause, is of course the living God.

Complex Proteins

The "organic compounds of high complexity" that must precede life on earth are proteins. Proteins are "the basic material of life." **But proteins are always and only made by living organisms!*** So here again the evolutionist faces an impasse, a situation that stumps him. If proteins are necessary for life, and proteins come only from living organisms, where did the original proteins—without which there can be no life—come from?

Paul Doty, writing on "Proteins" in the "Scientific Monthly," says:

"Thousands of different proteins go into the make-up of a living cell. They perform thousands of different acts in the exact sequence that causes the cell to live. How the proteins manage this exquisitely subtle and enormously involved process WILL DEFY OUR UNDERSTANDING FOR A LONG TIME TO COME (caps ours). . . . Protein molecules are giant molecules of great size and complexity and diversity. . . . Proteins are polypeptides of elaborate and very specific construction . . . (with) the long chains of each protein apparently folded in a unique configuration which it seems to maintain so long as it evidences biological activity (life)."

For proteins to "just happen" or develop by natural processes is as unlikely as getting a Gettysburg Speech together

---

*A few years ago a chemist, Stanley Miller, working at the University of Chicago, put into a flask what evolutionists believe to have been the chief elements of the atmosphere two or three billion years ago: methane, ammonia, hydrogen and water. He exposed them repeatedly to an electric spark. In a week he succeeded in producing three amino acids, which are essential constituents of protein. From this experiment evolutionists presume that "lightning acting on the earth's atmosphere" may have formed the first protein molecules necessary to life. But this is wishful thinking. Proteins are highly complex substances requiring "twenty different amino acids," not three; they are far too complex and involved to "just happen."

by stirring a million macaroni "letters" in a bowl of soup! For proteins are unbelievably complex. Some protein molecules actually have "hundreds, even thousands, of atoms in formations which stagger the imagination."

"Proteins, the keystone of life, are the most complex substances known to man. . . . For more than a century chemists and biochemists have labored to try to learn their composition and solve their labyrinthine structure. . . . In 1954 a group of investigators finally succeeded in achieving the first complete description of the structure of a protein molecule. The protein (they studied) is insulin, . . . one of the smallest proteins. Yet its formula is sufficiently formidable. The molecule of beef insulin is made up of 777 atoms, in the proportion of 250 carbon, 377 hydrogen, 65 nitrogen, 75 oxygen and 6 sulphur. . . . Of the 24 amino acids 17 are present in insulin" ("The Insulin Molecule," Scientific American).

Frederick Sanger, of Cambridge University, one of the group that finally worked out the "labyrinthine" structure of the insulin protein molecule actually "spent ten years of study on this single molecule"! (Scientific American). Only a trained bio-chemist can appreciate how involved a protein molecule is. The layman is impressed, but not sufficiently, by descriptions of the insulin protein molecule. But let us quote a little:

"The insulin protein molecule consists of 51 amino acid units in two chains. One chain has 21 amino acid units; it is called the glycyl chain. The other chain has 30 amino acids; it is called the phenylalanyl chain. These chains are joined by sulphur atoms."

We will not burden the reader by more quotations as to the intricate nature of the protein molecule, except to say that anyone who takes the time to look into the structure of the protein molecule must be convinced that such a fantastically complicated structure could hardly come about by mere chance: it is far too complicated. And so the PROTEIN MOLECULE—essential to, and a prerequisite of, life on earth, becomes a most effective witness for GOD, ITS CREATOR.

Proteins are of special interest not only because of their vast complexity of structure, but also because of their great variety and versatility in nature.

"There are tens of thousands, perhaps as many as 100,000, different kinds of proteins in a single human body. They serve a multitude of purposes". (The Structure of Protein Molecules, Scientific American).

It is obvious to us, and we trust it is also to the reader, that only God could bridge the chasm between non-living atoms and life, even in its lowest forms, and that only God could and did create the intricate proteins that are necessary for life on earth.

Chapter 5

The Witness of MICROSCOPIC FORMS OF LIFE to the Fact of Divine Creation

The Ladder of Creation

There are nine steps in the Ladder of Creation: (1) the **Atom**, the basic building block of the physical universe; (2) the **Molecule**, the basic particle of any chemical compound, made up of two or more atoms; (3) the **Protein Molecule**, derived from either plant or animal life, and a prerequisite for life on earth; (4) **Viruses**, the smallest, simplest and "most primitive" of all living things; (5) **Bacteria**, single-celled, miscroscopic plants usually without chlorophyl (6) Single-celled **Algae**, plants having chlorophyl—one of the lowest forms of self-sustaining plant life; (7) **Protozoa**, most of which are single-celled, microscopic animals; (8) **Metazoa**, animals higher than protozoa, made up of more than one cell. The higher animals and man have many, many billions of cells in their complex bodies. The metazoa include the higher animals above the single-cell group; (9) The complex **body, mind and soul of man**—created in the image of God. **Each of these nine steps bears witness to the fact of Divine creation.**

The Witness of VIRUSES to the Fact of Divine Creation

Viruses (L., poison), are the smallest and simplest and "most primitive" of all living things. Viruses are essentially a protein molecule, containing protein and nucleic acid. They are ultra-microscopic in size—so small they can be seen only with an electron microscope. Viruses are halfway between the molecules of the chemist and the organisms of the biologist. Viruses are parasites on both plants and animals. Their three most common shapes are those of a rod, a sphere and a tadpole. The most minute viruses are unbelievably small, each one weighing only 1/1,000,000,000,000,-000 of a gram. Viruses, as is well known, are responsible for such diseases as smallpox, yellow fever, mumps, polio, and many other human diseases, as well as scores of mosaic diseases of plants.*

---

*In 1901 Walter Reed and his coworkers discovered for the first time that yellow fever in man was caused by a virus. Since that time "more than 300 different diseases of animals, man, plants, and even bacteria have been found to be caused by viruses" (Smithsonian Report, 1955; pp. 357-368).

"Creation of Life in a Test Tube?"

A few years ago a flurry of excitement was caused by newspaper stories about "the creation of life in a test tube," at the University of California Virus Laboratory. Actually, all the laboratory had done, or claimed to have done, was to split the tobacco mosaic virus into its two components—protein and nucleic acid—and then they re-combined these particles into what looked like and acted like the original virus! If that is "creating life" then the act of cutting a shirt in two and sewing it together again makes one a first class designer! Beware of misleading, sensational newspaper accounts. **No man has created or can create life.**

Let us now take a look at the SEVEN WAYS in which the submicroscopic virus WITNESSES FOR GOD.

1. **Many Viruses are a Deadly Poison.** What a strange start for evolution to take, in its FIRST attempt at creating life, to begin with a deadly poison! IF in its first step evolution developed a rank poison, what would the second step be? and the third? and the following?

The Biblical explanation of the presence of DISEASE and DEATH in this world is far more reliable—and it fits all the facts. God pronounced judgment on the ground when Adam sinned; because of man's Fall, sickness and death came into human experience. (See Gen. 3:17, 18; Rom. 6:23; 8:20, 21). The virus, placed in the world by the Almighty, is part of His restraining "curse" and part of the universal penalty of "death" on all mankind (Rom. 5:12).

2. **All Viruses are "Parasites"—that is, they are utterly dependent on a "Host Cell."** "No virus has yet been grown in the absence of living cells" (Smithsonian Institute Report, 1956). This simply means that the "host cell," a higher form of life than the virus, **had to be created first.** This is such a damaging fact to the evolutionary theory that they have had to invent a theory of "evolution in reverse" to account for the little virus. Marianna R. Bovarnick, writing in the "Scientific American," suggests that

"Viruses (are either) aberrant derivatives from cells or they are 'degenerate end-products of an evolution from some higher form'."

Remember, a parasite is always a lower form of life than its host; and a parasite can not live aside from its host. Obviously, the "host cell" **had** to be in existence before its parasite, the virus.

3. **The Virus has a unique method of Reproduction.** Most

protozoa and all body cells reproduce by a simple method of division called mitosis. Gunther S. Stent, writing on "The Multiplication of Bacterial Viruses" (Scientific American) says, of the method of reproduction of Viruses:

"The process of heredity—how like begets like—is one of the most fascinating mysteries in biology. . . None is more exciting than (that of) bacterial viruses. Here is an organism that reproduces its own kind in a simple and dramatic way. A virus attaches itself to a bacterium and quickly slips inside. Twenty-four minutes later the bacterium pops open like a burst baloon, and out come about 200 new viruses, EACH AN EXACT COPY OF THE ORIGINAL INVADER. What is the trick by which the virus manages to make all these living replicas of itself from the hodgepodge of materials at hand? What happens in the host cell in those critical 24 minutes?" (Caps ours).

No magician ever pulled rabbits out of a hat with a greater sense of magic and surprise than the miraculous transformation of one virus into 200, in a matter of twenty-four minutes!

This is not only a miracle of reproduction that science is at a loss to understand or explain, it is reproduction "AFTER ITS KIND," following the Biblical law laid down by the Creator in Genesis, chapter 1. And all through history the simple viruses have been reproducing in this miraculous way "after their kind," with NOT THE SLIGHTEST CHANGE. According to evolutionists viruses have been reproducing for a billion years or more—and there is NO EVOLUTION IN SIGHT YET! If the tiny little virus is ever going to evolve it better get going soon—don't you think?

4. **Viruses can not create themselves.** Scientists have been trying desperately to get a virus to emerge out of a man-concocted brew of amino acids, proteins, nucleic acids, and what have you. But many are convinced the attempt is hopeless. Dr. Fred Kohler, a leading advocate of evolution, says in his book, "Evolution and Human Destiny,"

"A virus can not create amino acids; viruses can not synthesize their structure (make themselves) from a mixture of amino acids." Then the doctor again reminds us that "Amino acids unless synthesized by plants are now only available in nature through the breakdown of living material." (P. 22).

If they can not make themselves, and evolution can not make them, who DID make them? The answer is simple: "In the beginning GOD CREATED" . . . and His work of creation includes all things, including viruses.

5. **A virus is one of the most Mysterious of all forms of life.**

Viruses can be put into crystalline form resembling salt. This crystal appears to be dead; it is dead; it can be kept almost indefinitely without apparent change. But put it into a living tissue—and something happens. The viruses start to eat, grow and multiply, in their host cells! "Even after **repeated** crystallization, a treatment no other living substance has ever been able to survive, viruses resume their activities and multiply when returned to favorable conditions (host cells)." (Animals Without Backbones).

"If we can but discover the secrets carried within the virus structures, we will have gone a long way . . . It may appear amazing that Nature selected the borderline between the living and the non-living worlds to house secrets of such great importance, yet sober reflection will reveal the wisdom of this course of action." (Smithsonian Institute Report, 1955).

"While inside a host (cell), the virus is intensely alive . . . but between invasions, say while it lies on a kitchen table top, the virus can be thought of as essentially no different from an inert grain of sugar. This double existence affords a great scientific challenge." (The Physics of Viruses, Ernest C. Pollard, Dec. 1954, Scientific American).

6. **Viruses show a most Amazing Design, a truly wonderful Architecture.** A virus is designed so that (1) It can attach itself to the surface of a bacterial cell; (2) This contact with a living cell immediately "uncorks" an enzyme in its tail, which probably has the function of opening a hole in the "skin" of the bacterium; (3) the virus pours its own DNA (desoxyribonucleic acid) into its host; (4) this DNA then induces the synthesis of new protein in the host cell; (5) finally, units of the protein combine with the DNA to form 200 or so exact copies of the parent virus!

So involved is the protein in a virus that one authority says,

"The protein of the virus can be broken down into subunits, each of which is a single peptide chain containing about 150 amino acids." (Rebuilding a Virus, Heinz Fraenkel-Conrat, June 1956, Scientific American).

This whole amazing machine, so small it has to be magnified 100,000 times before man can see it and study it, HAS ABOUT 150 AMINO ACIDS (the material from which proteins are made) IN IT; and it has the ability to pierce the tough wall of a cell, enter the cell, take possession, and transform the contents of that cell in a few moments time INTO 200 OR SO REPLICAS OF ITS OWN IMAGE! That such a miracle could "just happen" is beyond belief. The very intricacies of this "most primitive" (Fred Kohler) form of life REVEALS THAT IT IS THE HANDIWORK OF AN ALL-WISE, ALL-POWERFUL CREATOR.

7. **Viruses are capable of Mutating, but not of "Trans-mutat-**

ing." Because polio viruses stay polio viruses, and because yellow fever viruses stay yellow fever viruses, etc., doctors can successfully wage war against them—and our own Dr. Salk can develop a vaccine that successfully fights the polio virus!

But viruses do "mutate" under certain conditions. A host cell may contain "a mild or latent virus, with the possibility of a very virulent virus strain forming" (Chemical Studies on Viruses, Stanley). Viruses do mutate (change) some, and adapt themselves to new environments; BUT, it is impossible to get a virus of one disease-producing culture (like polio) to TRANSMUTATE into an entirely different type of virus that will produce another disease. And so the minute viruses act as do all other forms of life: they may and do "mutate" within certain limits—but they NEVER transmutate into an entirely different type of virus.

Take another look at God's wonderful Witnesses: the infinitesimal ATOM, with its astounding powers and mysteries that demand a Creator and the ultramicroscopic VIRUS, with its fascinating design that demands an all-wise and an all-powerful Designer! The very least of His creatures give a very powerful testimony to Infinite Creative Intelligence!

Let us consider next

The Witness of BACTERIA to the Fact of Divine Creation

Bacteria, one step in the scale of creation above viruses, are as a rule microscopic, single-celled plants, without chlorophyl. Most plants, other than bacteria, that do not have chlorophyl are called "fungi."* Without going into unnecessary repetition about the wonders of bacteria that parallel those of viruses, we would call attention to three facts of supreme importance:

(1) **Most Bacteria, like viruses, must depend on a higher form of life.** Each bacterium consists of a single cell (without a definite nucleus) much smaller than any other plant or animal. Many bacteria being fungi, live on dead matter, or as parasites in the bodies of plants or animals. This means, of course, that **the higher forms of life had to be created first.** This fact is damaging to the theory of evolution.

---

*Most plants (there are exceptions) have chlorophyl, the substance that enables them, through photosynthesis, to manufacture their own food from air and sunshine. Bacteria as a rule live largely by katabolism (destruction) instead of by anabolism (construction). Most bacteria must live on organic matter.

(2) **Bacteria, like viruses, have distinctly different shapes and forms,** though they are microscopic in size. Round bacteria are called **cocci;** these in chains are termed **"streptococci"**—source of the familiar "strep sore throat." Bacteria shaped like tiny rods are called **bacilli**. Still others are shaped like a comma (,) and are called **spirilla**. This variety in form suggests design for an intended purpose.

(3) **Bacteria, like viruses, have a predetermined, planned economy laid out for them.** In general, they were created to be scavengers—"to break down the bodies of the dead"—so that the vital organic elements in bodies of dead animals and plants might be returned to the soil, to be used by future generations. Bacteria, generally, are "saprophytes," that is they live on the dead. Probably no creature, if the choice were left to it, would choose such a humble career; but GOD, in His supreme wisdom, planned an economy in nature **that works;** and He gave bacteria a very definite place in His over-all plan.

It takes an over-all Superintendent to design and plan and put into execution such an involved plan of life as we find in this world. Each bacterium has a work to do. Were it not for bacteria, the bodies of dead animals and plants would not decompose or rot, and so return to the soil to make food for future generations—they would simply accumulate. Bacteria are constantly at work decomposing dead leaves, carcasses, manure, etc.

"The microbes of putrefaction (bacteria) resolve dead bodies and plants into sulphates, phosphates, nitrates, etc.," that return to the soil, "and so the cycle of life is complete." (The Great Cycles of Life).

That such a wonderful system in nature, with each form of life having its necessary function, all working together in a state of perfect balance, should happen by "chance" and not by design, is unthinkable.

(4) **Bacteria, like viruses, exhibit a most amazing stability,** popularly called "Fixity of Species." Bacteria, since their advent in the dawn of time, are still with us as bacteria!

The next step in the Ladder of Creation is the single-celled "algae," microscopic cells found in almost all waters of the world. These primitive plants lack roots, stems and leaves—but they do possess the magic chlorophyll, which enables them to get food directly from inorganic matter, through photosynthesis. We pass on to discuss some of the Protozoa in the animal kingdom, as

Witness for God, though algae, too, are marvelous witnesses for God.

The Witness of Protozoa* to the Fact of Divine Creation

Of the fifteen thousand and more protozoa that have been classified and described by scientists, we select one, the AMEBA, as the best known, as a WITNESS FOR GOD.

The common ameba is found in fresh water ponds, and ranges in size from an invisible microscopic animal to one that reaches a diameter of about half a millimeter, visible to the naked eye as a tiny white speck. Each ameba is a little mass of clear gelatinous protoplasm, containing many granules and droplets. The protoplasm is covered with a delicate cell membrane. In many ways this strange little creature bears witness to the fact of a Divine Creator. We select but a few.

(1) **The Ameba is gifted with many Strange Abilities for a Microscopic Animal.** It can crawl; it can breathe (though it has no lungs or gills); it can distinguish inert particles from the minute plants and animals on which it feeds; it can thrust out its jelly-like body at any point to lay hold of its food; it can digest and absorb its food; though it has no feet, it can crawl by projecting "pseudopods." Such a strange little creature could not "just happen." One cannot fail to see in these abilities the Hand of the Creator who equipped this little animal for its environment and its station in life.

(2) The Ameba moves around by means of "Ameboid movement," projecting a "pseudopod" (false foot) from any part of its body. Because of this it changes shape when it moves or engulfs food, hence its name—"ameba" (derived from a Greek word meaning "change"). The "legs" of an ameba are temporary, and soon flow back into its body, when it stops walking or completes the ingestion of food particles. This is totally different from the muscular movements of other animals. **Who designed it?**

Moreover, if the ameba is about to "swallow" an active organism, the pseudopods are thrown out widely and do not touch or

---

*PROTOZOA is the name of the first of the "phyla" into which the animal kingdom is divided. The **Phylum Protozoa** (meaning "first animals") is made up of microscopic (generally) single-celled animals. The more than 15,000 species of protozoa occur everywhere in fresh and salt waters, in damp soils and dry sand, and even as parasites inside or on the bodies of other animals.

ameba is about to ingest a quiescent object, such as a single algal irritate the prey before it has been surrounded; but when the cell, the pseudopods surround the cell very closely. Apparently the ameba can "think" even though it has no brain.

The ameba gets around by means of the strange "ameboid movement"; but another protozoan, the paramecium, has its body covered with about 2500 short "hairs" (called **cilia**) which beat in the water somewhat like the motion of one's arms in swimming with the crawl stroke, so providing locomotion. Now the question arises—why do these two protoza, living in a similar environment, have two such utterly different means of locomotion, IF they developed from the same source (possibly the flagellates?) through the processes of evolution? One is faced by an enigma of vast proportions, and an unanswerable problem. However, he who believes in Divine creation has the answer: the same God designed **both** the ameba and the paramecium, giving each a body "as it pleased Him." Evolution has no adequate answer to this problem: the problem raised by **the great diversification of life in the same phylum, and in a similar environment.** A thousand similar questions could be asked about different animals and plants that are in the same phylum and that live in a similar environment, with similar life processes, and yet are so vastly different in structure!

(3) **Through past ages the lowly ameba has been absolutely static, showing no signs whatever of evolutionary change. This is a phenomenon to the evolutionist that baffles the thinking man.** If the so-called "law of evolution" has not succeeded in changing the simple ameba into a higher form, in the last billion years or so, where and when and how will it start? This is the more remarkable when one considers that there are countless numbers of amebas in the waters of the earth. And protozoa tend to multiply rapidly.* Through countless billions of generations, involving countless trillions of individual amebas there has been NO EVOLUTION WHATEVER IN THE AMEBA; amebas we still have with us as amebas, the same as they were when God first created them. There is a graduation of all life, from lower to higher; but there is no evidence what ever that the higher forms "evolved" from the lower forms.

Without hesitation we assert that EACH OF THE MORE THAN 15,000 SPECIES OF PROTOZOA GIVES A DISTINCT WITNESS TO THE FACT OF DIVINE CREATION. Each one has some

---

*One paramecium could multiply to many billions in one month!

peculiarity that is **distinctive,** that could NOT have evolved from anything, and can not be accounted for except on the ground of special creation.

So we produce as our next witness the universal somatic CELL.

The Witness of CELLS to the Fact of Divine Creation

All life—plant and animal—has as its primary building blocks the body CELL. Cells are microscopic in size,* and this enhances their wonders. The basic material in cells is called "protoplasm," described as "the most mysterious substance in the universe." Cells are of two main types: germ cells (sex cells) and body cells (somatic cells) Cells multiply by division; when a body grows, the cells do not get larger, but they multiple in numbers.**

A cell is made of (1) outer membrane; (2) a nucleus, in which are the chromosomes and "genes"; (3) cytoplasm, the gelatinous mass of the cell. In the cytoplasm are tiny substances, having peculiar and very definite functions; they are known as centrioles, centrosomes, mitochondria, plastids, Golgi bodies, etc.

(1) **The Intricate Structure of the Cell is a Witness to its Divine Creator.** From the chemist's viewpoint, a living cell is made up of carbon, hydrogen, oxygen, nitrogen, sulphur, phosphorus, chlorine, potassium, sodium, calcium, magnesium, iron, and small amounts of fluorine, iodine and traces of a few other minerals. But from the viewpoint of the biologist the cell is ALIVE with a working mechanism that is most marvelous. This working mechanism consists of (1) The nucleus, generally round or egg-shaped. It contains one or more dark bodies known as **nucleoli** and a number of extremely fine threads called **chromosomes;** these in turn consist of a large number of **"genes,"** resembling beads on a string. We will have more to say later about the "chromosomes" and "genes."

The living **cytoplasm** that surrounds the nucleus is essentially a gelatinous substance in which are dissolved proteins, fats and salts. Imbedded in the cytoplasm are several functional elements that are the working parts of the cell:

---

*Fifty cells from the human body, laid end to end, would not be as wide as the period at the end of this sentence.

**All living matter shows four distinct phenomena: (1) Irritability—the ability to respond to stimuli; (2) Metabolism—the ability to effect chemical changes in food and absorb it into its body, and to excrete the waste products; (3) Growth; (4) Reproduction.

(A) **Each cell has several hundred mitochondria, that are constantly moving about with a sort of writhing motion.** These mitochondria "play a central role in the oxidation of the cell's foodstuffs" ... hence "they supply the cell with most of its usable energy" (Powerhouse of the Cell, by Philip Siekevitz, July, 1957, Scientific American).* All of this was PLANNED by the Master Architect. Mr. Siekevitz says:

"Many experiments demonstrate that the functional units of the mitochondrion have a DEFINITE ARCHITECTURE. We may say the same of the entire living cell. We have come a long way from the time when a cell was considered a bag of loose substances freely interacting with one another. THE CELL, LIKE THE MITOCHONDRION, HAS A RIGOROUS AND COMPARTMENTED ORGANIZATION. Perhaps this is not surprising; when we build a factory we do not park its raw materials and machines at random. We arrange matters so that the raw materials are brought in near the appropriate machines, and the product of each machine is efficiently passed along to the next. NATURE HAS SURELY DONE THE SAME THING IN THE LIVING CELL."

EVERY CELL in every body in the world presents overwhelming evidence not only of exquisitely wonderful workmanship, but also of being planned to accomplish certain purposes—and all this in AN INVISIBLE WORLD so small that it extends, in some realms, even beyond the reach of our most powerful electron microscopes (that can magnify a million diameters—and more).

How can any one fail to see the earmarks of an infinite Creator as he examines the intricacies of the inner workings of a cell? Truly a "cell is an incredibly complicated structure." Let us examine these workings further.

---

*The mitochondrion itself has a most complex anatomy, recently revealed by electron microscope studies by George E. Palade, of the Rockefeller Institute of Medical Research, and others. The mitochondrion is bounded by a double membrane, often folded, apparently to increase its area. "Inside the mitochondrion are tiny bodies whose contents and function are entirely unknown. These features of the mitochondria are similar in the mitochondria of all plants and animals examined so far, from single-celled organisms to the cells in the body of man" (Powerhouse of the Cell, by Philip Siekevitz). HOW COULD EVOLUTION EVER ACHIEVE SUCH AN INFINITESIMAL MACHINE—"Tiny bodies working WITHIN minute mitochondria, that in themselves are so small that several hundred of them are found in the cytoplasm of each cell—and the cell itself is microscopic. It is utterly unthinkable that any thing, or any one, less that a SUPREME BEING OF INFINITE INTELLIGENCE AND POWER COULD CREATE A LIVING CELL LIKE THAT.

(B) **The process of Mitosis or cell division, is most amazing.**
When a cell divides to make two cells, each chromosome in the nucleus splits lengthwise, to form two identical new chromosomes. The chromosomes are all in the nucleus. Every species of plant or animal possesses a definite number of chromosomes in its cells.

Just outside the nucleus, lodged in the cytoplasm, is a minute body called the **centrosome** (central body.) It divides into two; these then appear to act as "captains" or leaders in the intricate and fascinating process of animal cell division. First, in this process, the two centrosomes move apart; between them a sort of fibril-like strands form a spindle; radiating strands appear around each centrosome, making them look like two stars. They now are called **asters.** Then the chromosomes split longitudinally, making identical "daughters" of each chromosome, and each half gravitates, with half of the protoplasm in the cytoplasm, toward one of the two aster-like centrosomes. This completes the process, and presto, there are TWO identical cells where only one existed before! Commenting on this amazing process of mitosis, one authority says,

"The centrosome divides, its halves part, and these two halves are then the two directing and essential bodies from which proceed the cytoplasmic threads that control the splitting chromosomes, and draw the split portions to their appointed places. This nuclear division lies at the very heart of the problems of life. . . . Quite apart from its meaning and purpose, the mere series of facts is amazing. The detail is so precise and complicated, the order and program so clearly laid down, the result so exact, and the whole process so unfailing—yet all conducted in an arena where only the highest powers of the microscope can discern anything—that it beggars all attempts to explain." (The Book of Popular Science).

Again we call attention to the fact that here is a living system that WORKS, and works in an ultramicroscopic world, with unfailing precision; and no one can explain WHY it works the way it does. Only the Supreme Designer understands the secrets of LIFE that He has injected into the tiny cell.

(C) **Mysteries of Heredity in the Cell.** Each species has its own kind, number* and assortment of chromosomes, and they differ from those of all other species.** Every chromosome in the

---

*Due to the large number of species of plants and animals, as many as twenty or more unrelated species may have identical **numbers** of chromosomes, but their chromosomes do NOT have identical shapes.

**Human body cells generally have 48 chromosomes though in some individuals there are 46 or 47. The lily has 24; wheat has 42; and some crayfish have as many as 200.

different species, "differs from every other in size, shape, or in some other respects, excepting that chromosomes always divide into pairs, and the two chromosomes in each pair are identical." So the Creator has "keyed" each species by means of differing chromosomes—much like the combinations used in yale locks—so assuring this tremendous fact: **Chromosomes forbid transmutation and establish the stability of each distinct genus.**

On the other hand each chromosome has a large number of "genes"* that lend flexibility to each species,** and give individuality to each member of each species. Genes have such vast possibilities of differing combinations that the net result in life is, NO TWO INDIVIDUALS IN ANY KNOWN SPECIES ARE EXACTLY ALIKE. No two blades of grass are exactly alike; no two dogs are exactly alike; no two human beings on earth are exactly alike. These myriads of variations in individuals in differing species are due to the subtle work of the GENES, and the limitless combinations that result from mixing genes from both parents.

Here then is a fundamental law of genetics: CHROMOSOMES GUARANTEE THE STABILITY OF THE GENUS, AND GENES PROVIDE FOR INFINITE VARIETY WITHIN THE SPECIES. Evolutionists, misled by the varieties produced by gene combinations, are ever looking for NEW SPECIES TO ARISE, as the result of these endless variations due to genes. But all they get is variations of the original species.

Modern scientists have analyzed the nucleus of the cell, and have discovered that, chemically, it is composed largely of "nucleic acid", of which there are two kinds: DNA (short for desoxyribonucleic acid) and RNA (ribonucleic acid). DNA is always found in the nucleus of the cell, and RNA is found mainly in the cytoplasm outside the nucleus. The theory now is that what are popularly called "genes" are actually the "individual molecules" in this highly complex "nucleic acid" (DNA). Genetic theorists are still uncertain as to the essential nature of the so-called genes; but

---

*In 1911 Thomas Hunt Morgan, then at Columbia University, advanced the theory that "genes are arranged in a 'linear file', or row, on the chromosomes"—like beads on a string. Genetic scientists now speak of nucleic acid (DNA) "as the genetic material."

"Species" is variously defined. By "species" we mean the "members of a population that will interbreed and produce fertile offspring." If the members of different groups do NOT interbreed, **they are of different species.

this we know: there are vital parts of each chromosome, minute units called genes, whether they be individual molecules, or in some other infinitesimal form, that "give a practically limitless range of possible variations within each species."

Scientists have recently developed another amazing technique: they have "by very elegant genetic techniques (announced by Seynour Benzer, of Purdue University) mapped a single 'gene' of a bacterial virus; Benzer was able to distinguish more than 100 different functional sites arranged in a linear order along the length of the 'gene.' Assuming that genes are made of DNA we can trace a correspondence between his map and the DNA molecule." (Scientific American).

Digging down into the ultramicroscopic world, scientists are seeking to solve the mysteries of "chromosomes" and "genes." MUCH IS THEORETICAL; but this we know: whatever genes are, or are made of, THEY ARE THE MIRACLE PARTICLES THAT TRANSMIT HEREDITY FROM PARENTS TO CHILD; and they do it in a way that to this day has defied all scientific explanation. Scientists can analyze the "nucleic acid" (DNA) of genes, and they can take photographs of its very molecules, but THEY CAN NOT DISCOVER WHY OR HOW THESE MINUTE "GENES" DO AS THEY DO, except that they know that "genes are the bearers of heredity and control nearly all the hereditary differences" that appear in individuals in a species.

And here is the climaxing miracle of all: If each gene proves to be "a single complex molecule" it possesses the amazing ability "to reproduce itself"—and this is NOT inherent in an ordinary chemical molecule. GOD THE CREATOR HAS PUT "LIFE" INTO THE "CHROMOSOMES" AND INTO THE "GENES" OF EACH CELL, and that is what makes a cell alive. No amount of chemical research will EVER come up with the answer of WHY CERTAIN MOLECULES IN CELLS (the chromosomes and the genes) REPRODUCE THEMSELVES AUTOMATICALLY. It is "life," the gift of the Supreme Being. To attempt to explain it entirely on a purely physical or chemical basis is folly.

Since "the number, kind and assortment of chromosomes tends to be **specific** for each species, 'characteristic of the species,' and that in the processes of fertilization, reproduction, and growth it is 'guaranteed' and 'insured' that the new will have exactly the same as the old, or parent, cells, it follows that 'EVOLUTION IS THROWN OUT OF COURT AS A BIOLOGICAL IMPOSSIBILITY because the only way a NEW species could arise from an older and differing species would be by changing the 'number, kind and assortment' of ancestral chromosomes.* All of this is in perfect harmony with the Bible biology which teaches us that everything should bring forth 'after its kind' if it brings forth at all" (Quoted from the book, "Checking Up on the Bible," by Dr. J. W. Simmons, Simpson, Kan.).

And so we see that every cell in every body on earth bears witness to the fact of DIVINE CREATION by giving evidence of "design" and "architecture" and "organizational ability" in a realm so small that those who take pictures of the ever-living and ever-changing drama going on in a cell have to use the electron microscope to do it! Each cell is ALIVE and life can come only from antecedent life. GOD WAS THE ORIGINAL LIFE-GIVER.

Cells bear constant witness to these fundamental facts of biology: THE CHROMOSOMES IN EACH CELL INSURE ITS STABILITY—it will and must "reproduce after its kind"; and THE "GENES" IN EACH CHROMOSOME PROVIDE FOR GREAT VARIETY "WITHIN THE BOUNDS" OF EACH SPECIES. These two facts are fatal to the theory of evolution.

The Effects of Radiation on Genes and Chromosomes

For many years scientists have experimented with the effects of physical and chemical stimuli (mostly through the use of radiation) on cells, especially on genes and chromosomes. "Abundant" gene mutations can be produced by X-rays (discovered by Muller, Stadler and Goodspeed in 1927); **"but in more than 99 per cent of cases the mutation of a gene produces some kind of a harmful effect,** some disturbance of function" (H. J. Muller, in "Radiation and Human Mutation," Scientific American, Nov. 1955). The results of such experiments on cells can be seen in the report of work done on chickens at the University of Connecticut (reported by the American Cancer Society):

"Chicken monsters with such defects as a large single eye in the middle of the forehead or eyes in their palates have been produced in genetic experiments, conducted by Dr. Walter Landauer. "Other monsters also

---

*Editor's Note: "Chromosomal inversions, omissions, duplications, translocations," etc., do occur in nature—but usually with pathological results. Then too, "polyploid" cells (having a chromosome number which is a multiple of the basic number of chromosomes for the species) and "heteroploid" cells (having chromosome numbers which differ from the characteristic number of chromosomes of the species) are also exceptions to this statement of Dr. Simmons that "the number of chromosomes for each species is specific." They do NOT however invalidate his argument, for even polyploid and heteroploid cells are "characteristic" and give no comfort to or proof of the theory of evolution.

appeared—chicks with no lower jaw, chicks with dwarfed beaks or no beaks, chicks with no ears, and chicks with no heads."

Everywhere the results are the same: In 99 per cent of the cases the use of chemical stimuli, or radiation, results in either DEATH or the "production of feeble, deformed or defective offspring." In the less than one per cent where biologists claim mutations that are wholesome, or an improvement, **in no case is there transmutation from one genus to another.** GOD HAS PUT UP "CHEMICAL BARRIERS" (Luther Burbank) BETWEEN THE GENERA that forbid transmutation from one genus to another.

"Among the hundreds of scientists currently working at the Bookhaven National Laboratory, in Upton, Long Island, where atomic energy is being studied in its many phases, there are about a dozen botanists . . . studying the effects of nuclear radiation on plants." They are observing "HOW RADIATION MAIMS AND DESTROYS LIVING ORGANISMS (caps ours) . . . for radiation has power to raise havoc among genes and chromosomes." The trees, shrubs, vegetables and flowers that are deliberately exposed "to gamma rays emanating from a captive specimen of cobalt 60, confined in a stainless steel tube, 4 inches in diameter and 9 feet high" have produced hundreds of monstrosities and freaks—and a few "mutations" of which "the scientists are proud."

"Their most striking achievement in the Brookhaven garden, has been to make two different carnations—a red blossom called the William Sim, after its originator, and a white one called the White Sim—appear on the same plant." (The New Yorker, July 20, 1957).

The fact that very rarely a "mutant gene" appears that is not suffering a "harmul effect" (less than one per cent) is grasped with the earnestness born of despair by modern scientists and hailed as "evolution in action." We quote again from H. J. Muller's article on "Radiation and Human Mutation" (Scientific American, Nov. 1955):

"Very rarely a mutant gene happens to have an advantageous effect (the result of being acted on by radiation). This allows the descendants who inherit it to multiply more than other individuals in the population, until finally individuals with that mutant gene become so numerous as **to establish the new type as the normal type, replacing the old. This process continued step after step, constitutes evolution."**

But this "occasional" wholesome mutant is NOT A WHIT DIFFERENT FROM WHAT TAKES PLACE IN NATURE ALL THE TIME—except that in Nature the process is slower. Mutants do occur in nature, from which spring new "types," but this is NOT EVOLUTION but merely the development of various breeds and types within the genus. For evolution to work, and account for the gradual production of all higher forms of life from lower forms,

IT IS NECESSARY THAT A LOWER GENUS BE CHANGED INTO A HIGHER GENUS, and this has never occurred, either in nature or in experimental gardens or laboratories.

We have already called attention to the fact that the law of nature is, MANY "VARIETIES" DEVELOP WITHIN EACH SPECIES—and these varieties come from gene mutations—BUT ABSOLUTELY NO TRANSMUTATION FROM ONE GENUS TO ANOTHER IS POSSIBLE, because of the fact that the chromosomes of each genus are "keyed" and "characteristic" and WILL NOT PERMIT INTERBREEDING WITH FERTILE OFFSPRING.

The average reader is generally aware of the harmful effects of radiation on genes, chromosomes and cells, for wide publicity has been given in the last few years to "the possible harmful effects of the 'fall-out' from nuclear explosions on mankind."

Obviously, the effect of radiation on genes is almost altogether **injurious.**

To sum up: Genes make possible the great VARIETY seen in species: and chromosomes establish the **stability** of each genus. What modern scientists label "evolution" is nothing more than what is seen daily in nature—the production of great VARIETY in species. But actual "evolution," the transmutation of one distinct genus into another is ABSOLUTELY IMPOSSIBLE. God the Creator put an impassable **barrier** between the genera* that can not be crossed. IT TOOK A SPECIAL ACT OF DIVINE CREATION TO BRING INTO BEING EACH DISTINCT GENUS. The idea that the higher genera "evolved" from the lower is without foundation in fact. It is a purely hypothetical supposition that has no proof.

How about "Spontaneous Generation"?

If man could only produce "life" out of some concoction of dead chemicals the materialistic evolutionist believes he would have proof for his theory of evolution. As a matter of fact, the

---

*The Bible calls each distinct genus, "kind." God created all life "after his kind." See Genesis 1:11, 21, 24, 25. Because each "kind" of life on earth is the product of a special Divine creation, when reproduction takes place it is **always** "after his kind" (Gen. 1:11, 21, 24, 25). So, by the very laws of God, normal reproduction between "kinds" is ruled out. EVOLUTION HAS NO PLACE WHATEVER IN GOD'S SCHEME OF THINGS. "Evolution" exists **only** in the thinking and imagination of modern theorists; it has NO FOUNDATION IN FACT.

entire theory of evolution is postulated on the supposition that life was spontaneously generated from non-living matter.

Blum wrote, "That life was 'spontaneously generated' from non-living matter at some time in the very remote past, and that this process has not been repeated for a long time are two basic tenets accepted by the great majority of biologists" (P. 251, Nov. 1957, Scientific Monthly). "The idea (of spontaneous generation of life) . . . seems a necessary part of the evolutionary concept."

In answer to the question as to how spontaneous generation was possible, they glibly say, "The general answer is that the conditions no longer exist which once made the spontaneous generation of life possible. . . . Admittedly (it is not likely that) the precise chain of molecular reactions from which life first arose will ever (again) be established. In the nature of things, 'proof' will be impossible forever." (Ibid).

In "The Science of Life," by Wells and Huxley, they say: "It is much more likely that at one moment in earth's cooling down, the warm seas provide an environment **never afterwards to be repeated,** an environment differing in the temperature, in pressure, in the salts within the waters, in the gases of the atmosphere over the waters, from any earlier or later environment. The earth AT THAT MOMENT fulfilled all the conditions which the alchemists tried to repeat in the crucibles. It was a cosmic test tube whose particular brew led to the appearance of living matter."

How can an intelligent person believe that life could be produced spontaneously FROM A LIFELESS EARTH, AND SEA, but recently cooled off from the intense heat of molten rocks? And remember, that sterilized earth had NONE of the "highly complicated proteins" essential to life! Could that sterile environment do what modern man with his vast knowledge of chemistry and physics and his well-equipped laboratories has failed to do?

Scientists who are more realistic inform us that "creating life" is much more than just pulling a live bacteria out of the lifeless seas (that had been sterilized for ages by the intense early heat on earth of many thousands of degrees). One such scientist says,

"If genes are required to produce enzymes (and they are), then LIFE BEGAN ONLY WHEN THEY (genes and enzymes) BEGAN. . . . The material of life as we know it could have come into being ONLY in a complex chemical environment" (including the highly complex proteins, which come **only** from antecedent life—Editor). ("The Gene," by Norman H. Horowitz, Scientific American).

Materialistic evolutionists say that in the remote and misty ages of the past there MUST have been spontaneous generation of life. But the conditions that made spontaneous generation possible then DO NOT EXIST NOW (even though we have a liveable world now and it was a dead world then). SO IT WILL BE FOREVER IMPOSSIBLE TO PROVE THEIR THEORY. Is **that** science? Or, is it superstition? Superstition it is, as crass and crude as any religious superstition that awes the mind of the West Indies creole voodooist, or the pagan African witch doctor.

Why is it that millions of otherwise intelligent students today are deceived by a theory that admittedly CAN NOT BE PROVED, and is based **entirely** upon suppositions of what might have happened ages ago, but for which there is no proof whatsoever today.

Lacking proof, multitudes today simply "believe" that spontaneous generation DID happen in the remote past, even though there is no proof for it. All scientific investigation to date proves there can never be any form of life without there being antecedent life. LIFE MUST COME FROM LIFE. This is as factual as any theorem of geometry, as any observed, recorded fact in the whole world of nature. This FACT of course drives us back to GOD, the Source and Origin of ALL LIFE! "In the beginning GOD CREATED . . .".

Commenting on the inability of modern science to produce life in a test tube, without antecedent life, Prof. L. Victor Cleveland, in his Anti-Evolution Compendium, says:

"Today, using heat, cold, X-rays, sunlight, radiations from **nuclear** products, chemicals, experiments galore, NOTHING REVEALS THE DEAD AS GIVING BIRTH TO LIFE, the inorganic becoming organic, except as the miracle and mystery of the living thing itself turns the inorganic into the organic. So far as all the scientists of earth can prove, THERE IS NO SUCH THING AS SPONTANEOUS GENERATION or abiogenesis—life must come from antecedent life. Life produces life of the same kind, whether you look at protozoa or elephants."

Scores of qualified, noted scientists have borne testimony to the fact that "NOT A SINGLE EXAMPLE OF SPONTANEOUS GENERATION HAS BEEN WITNESSED SINCE THE DAWN OF SCIENTIFIC OBSERVATION."

Norman H. Horowitz, writing in a recent issue of "The Scientific American," says, "Bacteria, as we know, arise only from pre-existing bacteria. We can prepare a broth that contains all of the raw materials needed for the production of bacteria, and we can provide all the necessary environmental conditions—acidity, temperature, oxygen supply, and so on—but if we fail to inoculate the broth with at least one bacteria cell, THEN NO BACTERIA WILL EVER BE PRODUCED IN IT." (The same is true of every other form of life, from the minutest virus to the most complex animal.)

John Crompton, writing in the recent book "The Living Sea," (p. 16) says,

"There is an explanation of life and there is an explanation of how it came to earth, but nobody has found it. There have been guesses and indeed assertions, but we are still no wiser than we were before these knowledgeable folk made their guesses and their assertions."

Mr. Crompton is right as far as modern "science" is concerned —NOBODY HAS FOUND THE SECRET OF THE ORIGIN OF LIFE. But what science does not know, every true Christian knows. The "secret" of life is in God, and in God's Son, the Eternal Word. Moreover, it is an open secret, revealed in the Bible so that all may read and understand:

"In the beginning was the Word (Gr., **Logos**), and the Word was with God, and the Word was God. . . . All things were made by Him. . . . In Him was LIFE" (John 1:1-3).

And speaking of microscopic life (bacteria) as Mr. Horowitz does in the quotation given above, we are reminded that

"Bacteria are actually **just as complex** as any cell of our own bodies, and their spontaneous origin from non-living material is not much more likely than the spontaneous generation of scorpions." (Ibid.)

According to this authority, it is as likely that a scorpion (or any other complex animal) could be produced by "spontaneous generation" as that a bacterium (which is also highly complex) should be produced by spontaneous generation! In view of the facts, how unreasonable for a man to believe that "spontaneous generation" happened at one time, even though it can not be done now. Sensible men are still of the opinion expressed by Lord Kelvin many years ago:

"Is there anything so absurd as to believe that a number of atoms, by falling together of their own account, could make a microbe, or a living animal?"

Chapter 6

THE WITNESS OF "DESIGN" AND "ADAPTATIONS" TO THE FACT OF DIVINE CREATION

"Adaptation" and "Design" are phenomena of nature that prove the presence of a Superintending Mind. "Design demands a Designer." The millions of queer shapes and differing habits of the plants and animals making up the teeming populations of creation were all cleverly designed to enable each species to live and survive and to reproduce its kind in a very complicated world. C. H. Waddington, writing in the Scientific American, describes "Adaptations" in these words:

"Every kind of creature is endowed with or develops qualities—we call them 'adaptations'—which are neatly tailored to the requirements of its special mode of life." He mentions the mystery and miracle inherent in such "adaptations": "How these adaptations come (or came) into being is one of the oldest and still one of the thorniest problems of biology. . . . Darwin argued that the whole of evolution depends on RANDOM CHANGES in the hereditary constitution and the selection of helpful changes by the environment. If a change, which we nowadays call a gene mutation, happens to make an animal better adapted and thus more efficient, that animal will leave more offspring than its fellows and the new type of gene will increase in frequency until it finally supplants the old."

Recognizing that many will question the adequacy of Darwin's theory of "RANDOM CHANGES" to account for the marvels of "design" and "adaptations" Theodosius Dobzhansky, another writer in the Scientific American, says:

"Perhaps the most troublesome problem in the theory of evolution today is the question of HOW THE HAPHAZARD PROCESS OF CHANCE MUTATION and natural selection could have produced some of the wonderfully complicated adaptations in nature. Consider, for instance, the structure of the human eye—a most intricate system composed of a great number of exquisitely adjusted and coordinated parts. COULD SUCH A SYSTEM HAVE ARISEN MERELY BY THE GRADUAL ACCCUMULATION OF HUNDREDS OR THOUSANDS OF LUCKY, INDEPENDENT MUTATIONS?" (Caps ours). "Some people believe this is asking too much for 'natural selection' to accomplish, and they have offered other explanations." (Dr. Dobzhansky himself believes in the theory of evolution; but at least he raises the question and shows that many do NOT consider Darwinism or any other theory of evolution as giving an adequate explanation of the fact of adaptations in nature).

Personally, we think it is absured to believe that "RAN-

DOM CHANGES" can account for the marvelous "adaptations" and "design" found in nature. The only explanation that meets the demands of reason is the fact of Divine Creation. The miracles of "Design" as found in nature and the millions of marvelous "Adaptations" **demand** a Superintending Mind. Perfect "adaptations" are the rule of life. "Always, wherever an animal appears it comes perfectly equipped for the sphere in which its life is to be lived."

Evolution claims that the myriads of forms of life, each with its perfect adaptations, came about GRADUALLY. This confronts the thinker with an **impossible** situation. Here is the argument:

Evolution teaches that fish evolved from lower animals. How did they get fins? No evolutionist claims that fins came in one generation, but rather in many generations. In other words, the development of fins was a GRADUAL development. And from where did birds get their wings. No evolutionist claims that wings came in one generation, but rather in many generations. They claim that the development of wings was through a very slow, gradual process. Now when the process, let us say, was halfway complete, the "fin" or the "wing" as the case may be, would be useless—good for nothing—for a fish cannot swim with a part-fin nor can a bird fly with a part-wing! A part-fin, or a part-wing, would be a monstrosity, not a perfect adaptation such as we see everywhere in nature. NO WHERE IN THE WORLD TODAY CAN ONE FIND PARTLY DEVELOPED APPENDAGES OR ORGANS, but everywhere there is perfect adaptation, perfect development for its intended purpose. That fact proves that EACH CREATURE, IN ALL ESSENTIAL FEATURES, HAS BEEN EXACTLY AS IT IS NOW, **and was so created in the beginning.**

Think for a moment of the trunk of an elephant; it is perfectly adapted to the use the elephant puts it to. Having 20,000 muscles, the trunk of an elephant has great versatility. With it he can lift a peanut to enjoy, or lift and crush to the ground a 600 pound tiger. He can twist his trunk in every direction; and its sensitive end is endowed with such a delicate sense of touch that he can pick up even a small pin from the ground at his feet! But what good would a "partially developed" trunk be?

What good to an eagle would a "partly developed" wing be? Everything in nature is perfect for the purpose and environment for which each creature was created. Anything **less** than perfection would be useless. On the very surface, the theory of evolution proves itself to be a nightmare, an impossible theory that does NOT fit the facts of a world that functions as this one does. For a workable world, it is absolutely necessary **that all living things be created perfect!**

"Adaptations" Everywhere!

The teeth of carnivores (flesh eaters—dogs, wolves, tigers, lions, hyenas, etc.) are especially adapted to seizing and rending prey. The South American anteater, on the other hand, has NO teeth, but a long snout, at the end of which is a small, soothless mouth with a tiny slit as the opening. The tongue of the anteater is a tubular affair about eighteen inches long that can be used either to pick ants off the surface of the ground or in the passageways of ant colonies. Think for a moment: what good on earth would an anteater's snout be **if it were only partly developed?** Imagine a monstrous creature, with a few deformed canine teeth in a half-developed snout, with a tongue not fully adapted to picking up ants! Such sorry messes would wreck the world in one generation, if such idiocies prevailed. It is a good thing the evolutionist is NOT right. Life would never have gotten beyond the single-celled algae stage, even if life had begun by "spontaneous generation"! If the evolutionist's theory be true, it would ruin every chance for life on earth.

Think for a moment of the quills of a porcupine. They are a perfect "adaptation" given to that animal as a means of protection. While evolving, if quills were NOT quills, why would an animal such as the porcupine (if it had anything to say about the matter) seek to develop quills, IF it would take a million years for the things to work as quills? To serve a useful purpose, quills either have to be quills or something else that works—but a "partially developed" quill is about as useless as a gizzard that is only a "part-way gizzard," or feathers that are half feather and half scale. In nature any imperfection is like an incomplete bridge.

Hugo de Vries aptly said, "Natural selection may explain the **survival** of the fittest, but it cannot explain the **arrival** of the fittest."

Other writers have sensed this preposterous absurdity in the theory of evolution. Dr. Criswell says,

"Take a spider. In the posterior region of the spider are highly specialized organs for the spinning of a web. He spins the web in order to gain food to eat—in his peculiar way. Now, in the millions, and millions, and millions of years it took for those modifications in the posterior regions of the spider to develop into those highly specialized organs, so he could spin a web, so he could catch his food, **why did he not starve to death while those organs were developing** so he could spin his web?" (Did Man Just Happen?)

Obviously, all specialized organs, such as the trunk of an elephant, the spinning apparatus of a spider, the eye of an eagle, the retractable claws of the tiger, the beak of a woodpecker, the tongue of an anteater, the quills of a porcupine, HAD to be created perfect and suddenly—otherwise they would never serve their intended purpose. The evolutionary idea of gradual development through "randon changes" through long ages of time can in no wise account for the facts of a practical, workable world.

"ADAPTATIONS" AND "DESIGN" SEEN IN PLANTS

In the whole panorama of plant life there are literally millions of special "adaptations" and evidences of special "design." We select but a few that illustrate this point — "Design" demands a Designer of Intelligence; and "Adaptations" have to come SUDDENLY and be PERFECT to accomplish their purpose, and that eliminates the preposterous idea of "random changes" through long periods of time as an explanation for adaptations.

Who gave cacti and other succulent plants of arid regions their unbelievable ability to store enough water during the rainy season to carry them through the many dry months? Most plants lose gallons of water daily through their leaves—but not the cactus; it has no leaves. The swollen stems function to carry on the process of food-making, and water storage. Who put the spines on cacti to protect them from being eaten by foraging animals? And who left the spines off other plants, so foraging animals could have food? Do cacti have the intelligence or the foresight to protect themselves from raiding animals? To ask the question is to answer it: the wisdom in evidence is NOT in the plant, but in the God who created the plants as well as the animals, and Who designed all things to fit into a perfect, workable economy.

Who designed the arrangement of leaves on such native trees as the beech, elm, oak and chestnut, to secure a maximum amount of sunlight for all leaves? The leaves are arranged on the vertical shoots **in spirals** so that any given leaf does not shade the leaf next below it on the shoot. That is the result of INTELLIGENCE —not "random change" or "chance mutations". In a case like this, where the individual members of the plant seem concerned with the welfare of the entire tree, one must presuppose "actual foresight", which is NOT an asset of mindless plants.

Who shaped the leaves of the teasel and compass plants so that both rainwater and dew are retained at the bottom of the leaves in a little cup, long after evaporation would have dried the leaves were they not so shaped?

Who anticipated the need of the morning glory, when the bee makes a sudden crash landing in its open mouth, for strength for its delicate flowers to handle the impact? The five corrugated blades of the flower that radiate upward from the stem, held together by tissue—thin curved sheets—take the blow easily without injury to the flower and enable the flower to deliver nectar and pollen, according to plan.

Who designed the walnut shell, making it hard to crack, so preserving at least some of the walnuts to serve as SEED for other trees, and guarantee the cycle of life?

An Engineer of great ability had to figure out the design, which no man as yet has been able to improve on "to get strength without weight." To its naturally rigid dome shape is added a compression ring around the middle. Then the surface of the shell is heavily corrugated so it can not be dented. Inside, two tension plates at right angels—the whole being of very light weight material—give still more rigidity!

Who gave countless plants their hairy stems and branches to keep off pilfering ants and beetles? This is another case of obvious **foresight,** which plants do not possess. Who thought of the idea of giving milkweeds, wild lettuce and dandelions a white, sticky milky sap, and made the surface tissue tender, so that when ants and beetles seek to climb up the plants, their pick-like claws pierce the tender tissue, letting a tiny droplet of sticky "milk" gush out? Soon the legs of the unwanted insects become covered with the sticky adhesive and further progress for them becomes difficult—and the trespassers retire in disgust.

"ADAPTATIONS" AND "DESIGN" SEEN IN INSECTS

Insect structures show vast variations—hundreds of thousands. The grasshopper's leg was designed for jumping; the leg of the diving beetle for swimming; the leg of the bumblebee for carrying pollen: and in each case the leg is perfectly adapted to its intended use.

The tongues of some moths and butterflies are as long as their bodies. The nectar, which is their food, is produced in the deep, hidden pockets (nectaries) of flowers. By unrolling the tongue and thrusting it into the far recesses of the flower, the insect is able to reach the nectar and suck it up. Commenting on this amazing fact, one evolutionist says,

"This long tube has been developed in the course of ages from the jaws of the insect." (Article on "Butterflies and Moths").

A partly-evolved sucking tongue would be useless; such a tongue is of no use whatever unless and until it is PERFECTED FOR THE JOB IT MUST DO. Surely no thinking person can be so foolish as to believe that such an intricate instrument as the long sucking tube of the butterfly, which it neatly curls up when not in use, was the product of "development through the course of ages."

Who gave locusts their "automatic stabilizer" (an aerodynamic sense organ, surrounded by hairs on the front and top of the insect's head)? and who devised an automatic "gyroscope" (club-like structures called "halters") for flies. During flight the

"tips of the halters swing to and fro in the arc of a circle. When the fly is turned off its course, the halters continue to swing in the same plane as before the turn, on the principle of a gyroscope" (Smithsonian Institute Report, 1954).

What mechanical genius devised the wings of **Glossina palpalis** (tsetse-fly) that beat 120 times **per second**—and arranged the timing of the beat so that the wing **actually rests three-fourths of that time!** (rest periods between each beat.) If that is startling, think of Who is able to create wings for the tiny midge (an insect less than one-tenth of an inch long) that beat **2,000 times per second!** (Nature Parade).

Miracles galore mark the "Nature Parade"! Dr. G. I. Watson examined a flea under a lens.

"He noticed that its body was streamlined—but for travelling **rear-end first.** Watson then watched how fleas alighted after a jump. He found they invariably landed facing the direction from which they had come. The first parts of the insect's body to touch ground were the two back legs. If the landing-ground proved unsuitable, the flea was thus ready for an immediate departure in the opposite direction!" (Nature Parade).

How amazing that the Infinite Creator should so carefully look after the safety and interests of a tiny flea!

Who first instructed the **water spider** in the art of making and using a "diving bell"?

"The tiny, inverted nest of silk is anchored firmly under water — a watertight air chamber. The spider then fills it with air by trapping a bubble between its hind legs; the air in this bubble is released into the nest. Fresh air is brought down to this 'diving bell' as often as required until the family is raised."

Who gave the lowly cricket a "built-in violin"?

"The lower part of its upper wing has a thick ridge while the upper part of the lower wing has a rough, thick vein like a file. The animal rubs the thick part of the upper wing over the file on the lower wing and this makes the chirping sound that we know as 'the song of the cricket'."

Who designed the amazing architecture of the grasshopper's hind legs? Graham Hoyle, writing in a recent issue of the Scientific American, describes the astonishing mechanism.

"The grasshopper's jump is one of the most remarkable performances in the biological world. The little animal can leap about 10 times its body length in a vertical jump or 20 times its length (almost one meter) horizontally. . . . The grasshopper weights only two grams, and its leg muscle is only 1/25th of a gram. . . . But the tiny muscle exerts the astonishing power of some 20,000 grams per gram of its own weight (10 times that developed by the muscles of man). Two features of MECHANICAL DESIGN account for the efficiency and enormous power of the grasshopper's jumping muscle. First, the muscle fibers are very short—about 1.5 millimeters, or one-twentieth of an inch. Secondly, they are arranged like the fibrils of a feather along the whole length of the femur, attached to the external skeleton of the insect's leg and to a long, broad tendon inside the leg. Thus the load is distributed evenly over the whole limb. Such an even distribution is impossible in a vertebrate structure."

Who gave to the **bombardier beetle** the formula for its **poison gas?** In its body it secretes a foul-smelling liquid which turns into a vapor as it is discharged from two glands near the anus. There is a sound like the explosion of a tiny pop-gun as the gas attack is launched against its enemy!

Who gave the strange Ichneumon fly **(Thalessa)** a drill, about 4½ inches long? The **Thalessa** "bores into WOOD and lays its eggs." Moreover, the female **Thalessa** lays her egg near the larva of the **Tremex.** When the Ichneumon larva hatches it attaches itself to the **Tremex** larva and feeds greedily on it.

Who first suggested to **balloon spiders** that they spin parachutes of silk which they use to transport themselves across fields, or as far as a hundred miles away? And who gave the **water spiders** the ingenuity to build tiny rafts, held together by the silk they spin, and on which they sail over the surfaces of ponds or calm streams?

Who equipped the water-scorpion with a snorkel-type tube so that it can breathe fresh air while submerged?

Who taught the **tent caterpillar** to use guide lines which it spins and lays down as it travels from branch to branch on an apple tree? By following these lines in the evening it finds its way back to its nest!

Who gave the female **mosquito** such an elaborate "surgical kit" that she can drill through skin and get her fill of blood?

"Nature has fitted the mosquito with a perfect midget tool kit. It is carried in the beak, which is a long, slender sort of nose. The tools are sheathed in a well fitting pocket of soft skin which is really the mosquito's lower lip. Inside the cover are six long neat tools, a pair of saws, a pair of lancets, a syringe and a syphon. . . . And in the mosquito's head, placed where they can see and supervise the drilling operation, is a pair of compound eyes. . . . We are not 'bitten' by a mosquito. The thirsty little monsters have no teeth. The damage is done with a tool kit." It literally cuts a disc out of the victim's skin.

And all this was achieved by "RANDOM CHANGES"? Tell me, thinking reader, HOW MANY BILLIONS OF YEARS WOULD BE REQUIRED "ACCIDENTLY" TO develop such a minute, intricate kit of surgical instruments—plus the 'know-how' given to the mosquito to use the kit? And, remember, if such a marvel were developed step by step—**it would be useless until perfected!** Who is so naive as to believe such puerile nonsense?

Who first suggested to the soft-bodied **leaf hoppers** that they enjoy the comfort of an air-cooled bubble bath? This they provide for themselves by sucking juices from plants, then blowing bubbles with the liquid. An investigator can readily discover this **"spittlebug,"** buried in froth, but enemies pass by and do not notice it. The **spittlebug** is vulnerable to heat—so its bubble-bath is a life saver! The merciful Creator looks after the welfare of even the least of His creatures!

For unparalleled ingenuity, though, one must study the **hunting wasp.** Let us examine the procedure used by the **Odynerus** hunting wasp. We quote from the book, "The Hunting Wasp."

"**Odynerus** (hunting wasp) makes and cements a horizontal cell, broad at one end and narrow at the other. From the ceiling of the broad end she hangs a silken thread and on the end of this thread, in mid-air, she suspends her egg. Under the egg she places two or three caterpillars, and in the narrow part of the cell she places a lot more—about twenty— jammed together so that they cannot move or wriggle away. What happens? The egg splits and a tiny yellow larva emerges. It does not fall to the ground but hangs from the end of the shell of the egg, and its weight lengthens the thread. The larva stretches its little head down and takes a small bite out of one of the recumbent forms below.

"These (stored) caterpillars are not dead; they have not been treated too severely and are not inert. Observers have found that, at a touch, up will go their tails. They have in fact only been treated in the head end. So, on the bite (from the hunting wasp larva) being taken, the caterpillar rears and the scared larva streaks up its thread out of harm's way. By and by, when all is quiet again, it steals down and takes another bite and the caterpillar rears as before. As the caterpillar weakens with this treatment the grub grows stronger. In twenty-four hours it has eaten the first caterpillar, and starts on the next. With the body of a whole caterpillar inside it, it is, of course, larger and stronger" and soon it is able fearlessly to tackle and devour, one by one, its remaining stock of caterpillars—all kept as fresh meat until eaten, by the neat process of the mother wasp's stinging the victim in one segment only—so paralyzing it but not killing it! (P. 176).

Generation after generation—without change—the **Odynerus** hunting wasp goes through this procedure. Such miracles can not be explained by evolution, or "chance mutation," or "random change," or any other theory that leaves God out. The very intricacy of the scheme used by this wasp demands that some superior Intelligence devised the entire thing, and created the wasp with those strange, yet practical, habits that provide so uniquely for its offspring.

Some will ask, Is God then the Author of such schemes that involve the killing and eating of one form of life by another? True, this is an effective way of keeping over-population of animals in check. We must also remember that we are living in a world judged by reason of the Fall of man (see Genesis 3:14-19). **Death** is the wage of sin—and the animal creation suffers with mankind in the judgments of God on sin. We read of this "sympathetic" suffering in Romans 8:

"For creation was made subject to vanity, not willingly, but by reason of him who hath subjected the same in hope.

Because the creation itself also shall be delivered from the bondage of corruption into the glorious liberty of the children of God.

For we know that the whole creation groaneth and travaileth in pain together until now." (Vs. 20-22).

All will be changed at the second advent of Christ, when the present "curse" on creation will be removed. Then Edenic conditions will be restored and the nature of wild animals will be subdued (see Isaiah 11:6-9; Rom. 8:19, 23). Wickedness among men will be put down and summarily judged; and wars will cease (Isa. 2:4; 11:1-5). Until that glorious time comes, "creation"— all life on earth—reflects the sad condition of fallen man, and shares with mankind the results of God's judgment on man's sin. Even in the state of a "cursed" creation, in which depredations, killings, suffering and death are the common lot in the animal domain, we see the perfections of the creative wisdom of God in all His works. Even a snake, associated in the Bible with the Fall of man, and the special object of the "curse" at the time of the Fall, is a beautiful creature that functions perfectly as an animal machine, adapted to its realm and role in life.*

*ALL NATURE speaks of God's Handiwork; many volumes could be written on the millions of marvels in nature that preclude "Chance mutations" as the cause for such wonders, and witness to the fact of Divine Creation and Divine Wisdom in Creation.

Chapter 7

The Perfect "BALANCE" and the Universal "INTERDEPENDENCE" of All Life on Earth Witness to the Superintendence of a MASTER MIND

All life on earth forms a wonderful unit. In nature are found many "checks and counterchecks" which keep the so-called "Balance in Nature." Bats and birds keep insects in check; large fish eat the more prolific small fish; hawks keep down the mouse population—and in a thousand other ways the "Balance in Nature" is maintained.

Here is one link in the so-called "food chain": the worm is eaten by the frog; the frog in turn is eaten by the snake; the snake is eaten by the buzzard.

Another "food chain" is in the sea. One authority says,

"Ten thousand pounds of diatoms are eaten to make 1,000 pounds of copepods; 1,000 pounds of copepods when eaten produce 100 pounds of smelts; 100 pounds of smelts, when consumed produce 10 pounds of mackerel; the 10 pounds of mackerel when eaten by tuna make one pound of tuna. Caught, canned and eaten by man, one pound of tuna increases man's body weight by one-tenth of a pound." Such "food chains" not only illustrate the way "balance" is maintained in nature, but also the interdependence of all life.

"Nature's world," says Robert S. Lemmon, writing in **Nature's Wonders,** "is peopled by no random assemblage of isolated, unrelated forms of life jumbled together without rhyme or reason. Rather, it is an inconceivably vast and integrated organization, a network composed by myriads of . . . vital connecting strands."

"The living things of a community," says another author, "form a natural balance, which is often upset but just as often restored. . . . If we consider the living world we see a vast number of species, animal and vegetable, high and low, some numerous, some scarce, some spread everywhere, others confined to limited parts of the earth. On the whole, these proportions, numbers and particular distribution of species remain constant; there is a balance maintained between them which we are wont to call the 'BALANCE OF NATURE'." (Book of Popular Science).

The major controls in nature that are used to maintain this "balance" are (1) predators; (2) starvation; (3) disease; (4) weather hazards (such as extreme cold or heat). All four of course result in the death of the members of a population that get too numerous.

Animal and Plant Characteristics are given by the Creator

Though Dr. Irston Barnes, president of the Audubon Society of Washington, D. C., does not give the Creator the credit, he speaks of the "fixity of character" in animals and plants that assures the maintance of the "balance in nature." He says,

"Each animal is chained . . . to an instinctive pattern of behaviour. . . . Thus a hawk is powerless to alter its tastes or its manners. This dictate of nature asserts that each form of life shall fulfill its destiny, that no chaos of individual choices shall destroy nature's balance. . . . Each form of life has its essential role in a community."

Such an intricate system in nature could NOT have been evolved by "blind chance." It demands a thinking, planning Architect who is both all-wise and all-powerful to put it into effect.

"Handicaps" and "Safeguards"

The Creator's Hand can easily be seen also in the many "handicaps" and "safeguards" found in nature. "Balance" in nature is maintained by the "handicaps" placed on certain creatures that otherwise would kill off all weaker species. The poisonous snake might become a great threat to all other forms of life, were it not handicapped by having to grovel along the ground without feet. Poisonous snakes in India also have a natural enemy—the mongoose. The stronger predators like tigers and lions breed more slowly than the prolific smaller animals, such as rabbits, that are eaten in large numbers by predators such as foxes and wolves.

Certain "safeguards" also are given to forms of life that otherwise would be at a great disadvantage in the struggle for existence. The cactus is provided with spines; some plants have a strong poison in their leaves; other plants are protected by disagreeable odors; the slow turtle is placed in a heavy armored plate; the small fish are very prolific; the dumb porcupine is given quills. All such ingenious devices—handicaps and safeguards— are evidence of a Creator who put each form of life here to perform a predetermined function in the scheme of things. To do this successfully, someone had to have an over-all view of things, with sufficient knowledge to plan an integrated whole, with all parts functioning together, and supplied with the right checks and balances so that no one section of life would wipe out the rest. Only GOD could do that!

"Balance" Maintained in the Insect World

Insects multiply at an unbelievable rate. For example, a female house-fly can lay 500 eggs in one season. Each egg develops into an adult fly in one week. Each of these adult female flies can then lay 500 eggs of its own. If all eggs hatch and if all the newly hatched survive, the original fly would have some 200,000,000,000,000,000 descendants at the end of the season! That means little to the average person—the number is simply too big to be comprehended; so let us state it this way:

"If all the offspring of a single pair of common houseflies lived to mature and reproduce, the earth would be blanketed beneath a layer of flies nearly fifty feet deep in less than six months!" (U.S.D.A.)

One great enemy of insects is the spider.

"The insects—with nearly a million different kinds—might dominate the land if the arachnids (spiders and their relatives) were not pitted against them. From time immemorial spiders and their clan have killed and eaten insects of all kinds and sizes. One female spider is reported to have destroyed 250 house flies, 33 fruit flies . . . during its lifetime. . . . More than 60,000 different species of arachnids are known, and they live about us in vast numbers. An Englishman calculated that there 'are probably more than 50,000 spiders per acre in England'." ("Spiders and Their Relatives").

In a thousand other ways, "insects are pitted against insects" to keep the populations of insects in control. For example, "the horse guard (a wasp) kills horse flies; the Microgaster (ichneumon) helps preserve our cabbage gardens from destruction by the larvae of the cabbage butterfly; the "lady bug" beetles (called the **bete a Dieux**—God's creature—by the French) destroy destructive garden aphids by the millions; a tiny wasp **(Apanteles medicaginis)** each year saves thousands of acres of alfalfa by preying on the caterpillar population that destroys alfalfa; and so the story goes.

Birds and small animals too are pitted against insects.

One toad will eat as many as 30 flies an hour; and a giant toad was observed snapping up mosquitoes at the rate of 50 a minute. A swallow will devour as many as 2,000 mosquitoes per day, in addition to large numbers of flies and other insects. Some birds, like the wren, will destroy their own weight of pestiferous insects and larvae in one day. If God had not provided such "hungry" insect eaters, disaster would soon overtake the world. Birds as a whole are God's agents, His "police," to keep insects and weeds in check. (They keep weeds in check by eating the weed seeds).

"We have birds for every place where insects and worms might be found.

Up in the air we see Swallows, Swifts and the Martins catching flying insects as they sail along. These birds have wide mouths so they can catch flies, bugs and beetles as they fly. At a little lower level we have Kingbirds, Flycatchers, Woodpeckers and others darting out of trees to chase passing insects. In the treetops Warblers and Flycatchers take care of the insects. Birds like the Sparrows and the Maryland Yellowthroats take care of the insects in the bushes. The Quail, Robin and Meadowlark, and many others, make their meals of insects and worms they find on the ground. Snipes, Sandpipers and Herons are the "waders" on stilt-like legs walking around, catching insects along the shores of lakes and streams. Ducks, Geese and others guard the surface of the waters and dig into the mud to keep insects and worms from over multiplying. . . . It has been said that if birds were taken from this world, IN LESS THAN ONE YEAR NEARLY ALL PLANTS WOULD BE DESTROYED BY INSECTS. Birds spend most of their life hunting for food." (School Textbook).*

What a thorough job the Creator did in providing birds for EVERY ENVIRONMENT—air, earth, water and under the water— to keep down the insect population! EVOLUTION COULD NOT BE RIGHT—for if insects had arrived as little as three years before birds, their destroyers, all the earth would have been denuded of vegetation and so all higher forms of life would have been impossible. In fact, insects would eat themselves out of food, and literally destroy themselves! One can easily see that it was absolutely necessary, as the Bible teaches (Genesis 1) for ALL LIFE ON EARTH, WITH NATURE'S MARVELOUS SYSTEM OF CHECKS AND BALANCES, TO HAVE BEEN CREATED AT ABOUT THE SAME TIME. It has been truly said, "Many minds have overlooked the fact that if life had evolved on the earth without a Master Mind it would have evolved its own destroyers" (Dr. B. H. Shadduck).

And if insect would stop eating insect, it would take only ONE year to upset the balance, and "in a single season insects would denude all living plant life from our planet" (Science Digest). One can not help seeing the Hand of the Creator and the Mind of the Supreme Being behind His creation.

What Keeps the Seas From Overflowing with Life?

Near the surface, the ocean waters will produce "400 million diatoms per cubic yard." What keeps the oceans from becoming

---

*There is a rather large group of small animals (the insectivores) that specializes in eating insects. It includes the hedgehogs, shrews, moles, tenrecs and solenodons. They are active mostly at night, and consume enormous quantities of insects.

clogged with diatoms? A small copepod will have about 120,000 diatoms in its stomach; then the herring comes along and eats 6,000 copepods at a feeding; that takes care of getting rid of the surplus diatoms—the "hay" of the sea.

A codfish will release 4,000,000 eggs in a season, an oyster 100 million, and a sunfish 300 million! How is it that the ocean is not soon filled to overflowing with oysters? or sunfish? For the simple reason that many fish eat these eggs by the hundreds—and so the "balance" of life in the oceans is maintained. This shows the handiwork of an allwise creator.

How Rats, Mice and other Small Animals are kept in Check

One of the most fascinating of African birds is the **secretary bird**. It stands about three feet tall. Stalking through the bush, it captures and eats snakes, scorpions and lizards. Our American roadrunner bird does the same in our Southwest.

"One pair of meadow mice could be responsible for one million offspring within a year, if their fecundity were not disturbed. Nature has wisely provided controls for the mouse population. Not only birds of prey, but a wide variety of mammals eat mice as staple food."

Rats also are exceedingly prolific; they generally bear five or six litters a year, with an average of ten young in each litter.

"Fortunately for us, rats and mice have a great many enemies. These enemies include mammals such as foxes, coyotes, badgers, skunks, weasels and wild cats; birds such as hawks, owls, crows and ravens; and reptiles, such as snakes and certain lizards."

The barn owl has aptly been termed "a living mouse trap"— it eats so many mice. Now here is a revealing fact: "Owl breeding closely follows the mouse population. In seasons when mice are abundant, tawny owls may attempt to rear TWO broods instead of one."

And so we see that God created a self-regulating system in nature that tends to balance itself. Anteaters eat ants*; owls eat mice; birds eat insects; spiders devour insects; ladybugs eat aphids; big fish eat little fish, etc. The entire self-regulating system shows a Master Mind behind the whole scheme of things.

---

*Loren C. Eiseley, Professor of anthropology at the University of Pennsylvania, calls attention to this fact: "Consider the disaster that would overtake an animal like the tubular-mouthed toothless anteater if extinction overtook the social insects (ants and termites). The anteater could never re-adjust. He would starve in the midst of food everywhere available to the less specialized. He will last only as long as his strange environmental niche remains undisturbed." (Nov. 1950 **Scientific American**).

This is an indirect admission that evolution won't work. Dr. Eiseley

Nature's "Undertakers"

Plants and animals live and die. If there were not some means of clearing the earth of lifeless tissues, life on earth would soon become impossible, for undecayed leaves, branches, and corpses of animals would pile up and choke out all possibility of new life. God has, in His economy for nature, created a number of "undertakers" who take care of the dead. In His employ, at this lowly yet necessary work, are bacteria fungi (molds), the **Necrophorus** beetle, vultures, hyenas—and others.

Bacteria and fungi are great agents of decay. They reduce anything that dies to simple chemical substances that can then be used by green plants.

"Many molds are equipped with a powerful arsenal of enzymes which bring about the rapid decomposition of woody plant materials. Fungi rot leaves, dead branches and tree trunks. In so doing they build up the humus layer and enrich the soil for the growth of future generations of trees."

Since bacteria and fungi depend for their food on dead tissues of plants and animals, we see another illustration of "interdependence" as well as "maintaining balance."

The **Necrophorus** beetle, popularly called the sexton, is invaluable as one of God's undertakers. Working mostly at night, it will bury a small dead animal, such as a rabbit—then use it for food for itself and for its offspring. So it has food for itself and the next generation, and it also clears the surface of the ground of what would soon become a foul-smelling, unsightly cadaver.

The **bluebottle** flies are also on "Nature's payroll as qualified undertakers." They and their grubs help dissolve the meat of

---

knows that any radical readjustment necessary to save the anteaters if their present food supply failed, would have to be immediate—and evolution does not work that fast.

Anteaters not only help keep the "balance in nature" by keeping down an excess ant population, but ANTS AND TERMITES ARE THEIR SOLE FOOD SUPPLY. So "balance" and "interdependence" are woven in a practical way—an evidence of Divine oversight.

We already have called attention to the fact that the long, toothless tubular snout of the anteater could NOT have been evolved through long ages—for a 10% or a 20% snout would certainly NOT serve its specialized purpose. Anything less than perfection in a specialized organ, such as this, will not work. Evolution is a meaningless fake; it cannot explain the facts of "design" and "adaptation," "balance in nature" and "interdependence" of all life.

corpses, such as dead horses and cows or other animals that would soon fill the earth unless disposed of quickly. Bluebottle grubs, by the thousands, live on putrefying flesh and help get it out of the way. These grubs "exude a liquid of great potency" that dissolves the meat; for the grubs can eat only "liquid" foods.

If insects had anything to say about it, what one would want to be a "sexton beetle" or a "bluebottle fly," with their revolting task of consuming dead, decaying, putrefying flesh? God, who had to plan all things to keep nature solvent and liveable, created these humble "undertakers" for a distinct purpose in the scheme of things. And, He created them without the possibility of "evolving" into something else—their station in life is static. In fact, ALL life is limited by the laws of the Creator to reproduce "after its kind," and **only** after its kind. In that way the Creator, who planned every form and phase of life, **keeps** the "balance of nature."

Vultures and other birds of prey such as Marabout storks that live on carrion are also God's "undertakers." They have a keen sense of smell and extraordinary vision, for "the smallest dead snake or mouse does not escape detection by these birds several hundred feet up in the air." After the vultures* have cleaned off the flesh from a dead animal, hyenas (in some areas of the world) come along, crack open the bones and eat the marrow. This hastens the decay of the bones, and adds greatly to the accomplishing of the gruesome task assigned to God's "undertakers" of clearing the face of Nature of the corpses of dead animals.

Nothing was forgotten, nothing was neglected, in the "scheme of things" to keep nature solvent and keep one phase of life (or death) from destroying the rest. Such an over-all Plan had to be the work of a Master Mind.

The "Interdependence" of All Life

Not only is there designed "balance in nature" that keeps nature solvent and functioning century after century, but also all

---

*The turkey vulture of North America, has rusty black plumage with a bare head and neck. Vultures have weak bills and feet, hence usually they are incapable of killing live animals. Virtually all their food is carrion—dead livestock or wildlife. Obviously, vultures were **"designed"** the way they are, to fill a particular niche in Nature—that of undertakers.

life is "mutually dependent." As an illustration, let us take Darwin's classic example of the effect of cats on red clover in England.

"If field mice are not kept in check by cats, the nests of bumble bees would be destroyed by too many mice. With no bumble bees the red clover could not be fertilized, and would soon die out."

Now here is the miracle Mr. Darwin did not mention at the time: "The closely packed tubular flowers have nectar concealed at the bottom of each tube. To get his head inside, the bumblebee must push the petals apart. As he does so, the pistil springs up, its stigma brushing his face and picking up the dose of pollen he acquired from his last clover. Then the shorter stamens pop up and dust his face with pollen to be carried to the next flower. The size of the bumble bee's head, his weight and the pressure he exerts are all PRECISELY BALANCED FOR THIS FLOWER," and no other insect can do the job! They are **designed** for each other as much as a threaded bolt and the nut that fits it.

Red clover is DEPENDENT on bumble bees—and bumble bees get their nectar from the red clover: a beautiful example of "interdependence" in nature. Here is forceful evidence of both "design" and "adaptation" in the perfect balance that exists between the size and weight of the bumble bee and the flower of the clover.

All species serve one another, as is the case of the birds "who eat stone-fruit, thus serving themselves to food, but also effecting thereby a scattering of the seeds—so serving to scatter the very species it attacks!" Small creatures like squirrels are the means of planting many of our large forests; they eat some of the nuts, and scatter others in their travels from tree to tree. They often bury nuts in the Fall and neglect to dig for them later. Many thousands of other illustrations can be given of the "interdependence" of life. Here is another illustration.

"Vast numbers of one-celled plants and animals (protozoa) live in the stomachs of cattle and obtain nourishment from the contents of the digestive organs; at the same time they break down the cellulose in the plants on which the cows feed, and as a result the cows are able to make use of the various nutritive elements contained in the cellulose."

Speaking of this fact of "interdependence" in nature, one authority says,

"The variety of interplay between the great branches of the living world is endless and inexhaustible. We need only recall two salient illustrations: (1) the formation by plants of chemical compounds which animals can consume; and (2) the formation by animals of carbon dioxide which they pour into the air, and which is a source of food for all the higher forms of plants. Thus plants serve animals; and animals, though absolutely dependent on plants, serve plants."

See here the wonderful "balance": animals breathe oxygen and exhale carbon dioxide; green plants (in sunlight) take in carbon dioxide and give off oxygen! Without plants in the world, the time would soon come when ALL ANIMALS AND ALL MEN WOULD USE UP THE AVAILABLE OXYGEN IN THE AIR AND WOULD PERISH.*

Miracles of "Interdependence" seen in Cross Pollination

Flowers supply bees with nectar; bees in turn transfer pollen from one flower to another, thus preserving the life of the species.

Dr. Arthur I. Brown said, "God devised a curious and altogether marvelous plan whereby pollen dust would be carried to the proper destination and reach its appointed place safely. He called on the insect world to undertake the job. The bees are His chief agents, and naturally the Creator, wishing to bring them to the flowers, gave the flowers color and fragrance as definite attractions, along with nectar for food for the bees, to attract them."

Bees, with their long, slender tongues, can reach the nectar, but most other insects can not. Nearly all species of moths and butterflies also have long tongues for sucking nectar.

Writing on "The Fertilization of Flowers" in the June, 1951 Scientific American, Verne Grant tells an intriguing story. We quote in part:

"As the bee takes the nectar, its body hairs inevitably pick up pollen from the flower's stamens. In some bee flowers the stamens have special lever, trigger or piston devices for dusting pollen on some particular spot of the bee's body. When the bee has finished working on one flower, it flies rapidly on to another. BEES HAVE AN INSTINCT TO CONFINE THEIR ATTENTION TO FLOWERS OF ONE SPECIES AT A TIME. . . . This assures that the bee will deliver its load of pollen to another flower of the same species which the pollen can fertilize." (Caps ours).

"Adaptations" and "design" in the realm of the cross-pollination of flowers are so evident and so wonderful that they

---

*There are, as we have said, a thousand and more illustrations of the fact of interdependence and interplay in nature. Here is one more of these interesting facts:

"Give the plants a free hand and the water would in time become so alkaline as to destroy them. Give the animals a free hand and they, in the end, would be killed by the acidity they themselves produced; but the two working against one another insure the maintenance of conditions vital to both. All life is like that: a thousand interacting and balanced forces, like the flying buttresses of a towering Gothic Cathedral; destroy one and the whole graceful fabric comes down in irreparable ruin." (Creation's Amazing Architect).

seem almost unbelievable. The Creator has devised scores of odd means to insure the pollination of certain flowers. We mention some of the more interesting:

Some flowers are pollinated by beetles. "Several of them hold the beetles in a trap while the stigmas receive pollen and the stamens sprinkle a fresh supply onto the bodies of the prisoners. Then they open an exit by which the beetle escapes.

"The flowers pollinated by sunbirds, which settle on the plant, usually stand erect and provide a landing platform. . . . The petals of flowers pollinated by birds are fused into a tube which holds copious quantities of thin nectar. THE PROPORTIONS OF THE TUBE OFTEN CORRESPOND TO THE LENGTH AND CURVATURE OF THE BIRD'S BILL.

"The tree-borne bat flowers of the tropics are large, frequently a dirty white in color and open only at night. (Bats fly about only at night). They attract the bats by a fermenting or fruit-like odor, which is GIVEN OFF AT NIGHT.

"Some flowers are pollinated by flies. Since flies derive their food usually from carrion, dung, humus, etc., THE FLOWERS THAT ATTRACT FLIES CARRY SIMILAR ODORS. . . . Rafflesia, a large-blossomed fly-pollinated flower of Malaysia, smells like putrefying flesh; another flower, pollinated by flies, black arum, has the odor of dung; . . . there is a species of Dutchman's-pipe that smells like decaying tobacco." (Ibid). (Caps ours.)

Let us give a few more examples, in detail, of the various ways the miracle of cross-pollination is carried out.

"When the beetle **Catonia** lights on a magnolia flower, its weight springs a triggerlike trap that releases a sudden shower of petals that frightens the insect and causes it to take off to another flower, with pollen from the first flower stuck to its back. When it alights on the next flower its back rubs against the stigma and leaves the pollen where it should be!"

And what can match the resourcefulness displayed in the lovely Iris? Attracted by the beauty of the flower, the bee lights on a flag and follows a distinct line that leads to the nectar well in the center of the flower. As she does so she must move under the drooping stigma which rolls the pollen off her back by a little pedal curved downward like a bent finger. When the bee then goes on in to suck the nectar, her back picks up a fresh load of pollen from the anther under which she is forced to stand while sucking nectar. Meanwhile the 'finger' has straightened out and the stigma has moved up out of the way so that the new cargo of pollen cannot be scraped off as the bee backs out—and so self-pollination of the flower is prevented. Neither bee nor flower designed this.

In the **Speedway** flower the system is different, but just as effective. This flower has a mechanism so delicately balanced that it tickles the stomach of the hovering fly as he dives into the flower. In that fraction of a second the stigma of the flower receives the pollen that the fly brought from another flower; and a fraction of a second later the flower's anthers brush the fly's stomach and deposit a fresh load of pollen before the insect can back out and go to another flower!

"An example of the perfectly exquisite way in which insects and some flowers are adapted to help each other is seen in cases where the scent of the flower BECOMES OBVIOUS EXACTLY AT THE TIME WHEN THE FLIGHT OF CERTAIN INSECTS BEGINS. Some of the honeysuckles and petunias, which have a very faint smell, or none at all, during the day, are powerfully scented in the hours of the evening, at which time the particular insects which visit them are on the wing.

"Most important 'conveniences' are offered by flowers to their pollen carriers, such as landing platforms or a 'door step' for short-legged insects like bees. . . . Another convenience is guide-lines of different colors which converge and point like so many fingers to the opening that leads to the nectar well. . . ." ("Plants and Their Partners").

"Such stories (as quoted above) are limited in number only by the flowers studied, since each one has its own story to tell to eyes keen enough to see. The bleeding heart, the sweet pea, the columbine, the sage, the hollyhock, the laurel, all have very interesting devices for securing cross-pollination and they and their composites have an equally wonderful mutual arrangement." (Ibid).

"And as you study them," writes Rutherford Platt in **This Green World**, "you cannot help feeling that sense of incredulous awe which prompted Jean Henri Fabre, the Homer of insects, to say of cross-pollination: 'Before these mysteries of life, reason bows and abandons itself to adoration of the Author of these miracles'."

For such "adaptations" and such miracles of "design" to develop through the "natural processes of nature" — "random changes"—would require billions of years and **then** could not be developed by mere chance. To believe that "evolution" developed such marvels is credulity stretched to the point of the preposterous.

The Case of the Yucca and the Pronuba Moth

In all nature there are few cases of such obvious "interdependence" as exists between the Yucca plant and the Pronuba moth. It is most amazing.

"The yucca is a bright and popular desert flower which seems tough and independent, sending up flowers of white lilies from a cluster of sharp leaves like wicked swords pointing out in all directions. But this beautiful, boastful lily's life **hangs on one little white moth** that hides underground in the daytime and comes out and flaps around, without ever eating, in the desert night. Yucca buds open at nightfall and pour out their white flowers which, on certain nights, give forth a strong fragrance.

"AT THIS EXACT MOMENT the pronuba moths break out of their cocoons beneath the sand. They struggle up into the air and are led by the odor straight to the flowers. The moth goes to the top of the stamens of the first flower it reaches and scrapes together a wad of pollen three times as big as her head. Carrying this big load in her jaws and tentacles, which are specially enlarged for this purpose, she flies to another yucca plant. Still holding the pollen, she backs down into the bottom of a flower, pierces a hole with her egg-laying needle and lays eggs among the seed cells in the green pod at the base of the pistil.

"Then she climbs to the top of the same pistil where there is a cavity just the right size to receive the wad of pollen. She stuffs this full, pushing down the pollen and padding it to make sure that plenty of pollen tubes will grow quickly and spark the seeds where she has laid her eggs.

"The mother moth plans far ahead. . . . She has deliberately bred the plants so that her babies will have a supply of food when they are born. While the pronuba eggs are getting ready to hatch, the yucca's seed are ripening. When the moth's larvae (caterpillars) finally emerge from their eggs, they find themselves surrounded by delicious food. They eat their fill of seeds, grow and finally cut a hole through the pod and lower themselves to the ground by spinning a silk thread.

"The mother moth never eats. She just lays eggs, pollinates the yucca to make the seeds ripen, and dies. As the moth babies eat only about a fifth of the seeds in the pod, the rest of the seeds mature successfully and go on raising more yuccas which in turn will raise more pronubas.

"No one can say how and why this vital partnership of the lily and the moth was planned. WHY DOES THE MOTH COME OUT ON THE NIGHT WHEN THE FLOWERS BLOOM? Why does she do things in the right order? WHAT TELLS HER TO CARRY POLLEN FROM ONE FLOWER TO ANOTHER INSTEAD OF POLLINATING THE SAME FLOWER? What prompts her to work so hard to drive home the pollen in just the right spot? Why don't the caterpillars eat all the seeds? These are just a few of the questions that must go unanswered when we look for reasons in nature's order of things." (Rutherford Platt, in an article on "POLLINATION.") (Caps in quotation are ours).*

Mr. Platt may not know the answers to his questions—but the believer in God does. The only possible explanation for such an intricate series of actions, in perfect coordination between a plant incapable of thinking and a moth incapable of reasoning power or foresight, is that GOD MADE IT SO. And so the little

---

*There are other amazing facts not mentioned by Mr. Platt. One is, there are several species of the Yucca plant and each species has its own species of moth. The flower is so constructed that it can only be pollinated by this particular moth—and that moth is as dependent on the yucca plant as the yucca plant is on the moth!

The Pronuba moth "is provided with special tentacles covered with stiff bristles and obviously designed for the purpose of collecting pollen from the anthers of the yucca flowers. . . . There are years in which the yucca plants in a given locality do not flower. . . . It has been observed that in those years when the yucca does not bloom the moths remain dormant in the pupa stage; but when the flowers appear again on the yucca plants THE MOTHS EMERGE AT THE RIGHT TIME TO CARRY ON THEIR PART" in this amazing scheme! This is altogether wonderful, and utterly inexplicable, except on the grounds of DIVINE CREATION. God alone can devise and put into effective operation such miracles!

Pronuba moth, and the lovely Yucca plant of the desert both are mighty witnesses to the fact that GOD MADE THEM SO. Throughout all creation there are a million voices—voices that rise from the throats of songbirds, insects, animals, flowers and fishes of the sea—that bear testimony to the power and wisdom of God in His magnificent creation.

We would like some evolutionist to answer this question: HOW COULD THE YUCCA PLANT AND THE PRONUBA MOTH **BOTH** EVOLVE, by "chance variation," "random changes," "mutations," or by any other chance method, IN SUCH A WAY THAT **BOTH** ORGANISMS WERE PERFECTED SO AS TO BE DEPENDENT ON EACH OTHER SO COMPLETELY AS ARE THE YUCCA PLANT AND THE PRONUBA MOTH? Such a mutual relationship of inter-dependence and helpfulness could not possibly come about, when carried out to an extreme as is seen in the "Yucca-moth" relationship, except by outside intervention; and that outside intervention was and could only be One of supreme power and intelligence. EVOLUTION HAS NO ADEQUATE ANSWER FOR THE "YUCCA MOTH" MYSTERY.

Chapter 8

THE ENDLESS VARIETIES IN NATURE AND THE "PERSISTENCE OF SPECIES," AND THE MANY STRANGE AND ODD SPECIMENS OF LIFE ARE WITNESSES TO THE FACT OF DIVINE CREATION

THE WORLD OF NATURE is most fascinating and intriguing, because of the great versatility in creation. The Supreme Architect apparently delighted in creating a well-nigh endless variety of life, including forms of life that are unbelievably odd and so strange as to defy explanation. These unusual creatures, and the vast numbers of varieties of life, are in the air, on the land, in the earth, and in the sea. And here is the miracle of this intricate and involved creation: every one of the myriads of forms of life on earth is "whole, complete and perfectly fitted to the environment in which it was made to live and function." Moreover, each distinct genus is essentially **static,** breeding "after its kind" generation after generation, with absolutely no evidence of transmution from one genus into another. Nowhere do we see "partly developed" creatures or any "half-made" or "10% or 20% developed" specialized organs (such as the beak of a woodpecker, the trunk of an elephant, the spinerettes of a spider, the fangs of a snake, etc.) ALL specialized organs, and ALL species are perfectly designed and perfectly adapted to the niche in life in which the Great Designer placed them. Even life in its embryonic forms is "perfect and complete" and perfectly adapted to its environment as far as it has progressed in its development from its original parental cells to the time of its birth.

The record of prehistoric life in the rocks is the same:

"When a family appears it appears whole and complete and fitted for the environment for which it was made to live" (**Did Man Just Happen?** p. 68).

Endless Varieties of Life on Earth

More than a million different animal species have already been described, classified and named — "and it is probable that many, thousands more are still to be discovered." In the world of insects alone there are at least a million different kinds.*

The beetles alone include some 250,000 species! There are more species of beetles than there are species of all other animals known, outside the insects. Butterflies and moths total over 110,000 species! Bees, wasps and ants number over 10,000 species! Beside all the insects, there are multiplied thousands of different kinds of "insect-like animals," often confused with insects, such as spiders, the sowbugs (land crustaceans), centipedes, millipedes, etc. Here are the questions that arise: Since evolution demands such long periods of time for the development of species, by the slow processes of "fortuitous changes" and "natural mutations," how can it possibly account for such a vast number of species, and why did such an incredible number of species evolve in the same environment? Why, if evolution did it, did not all beetles evolve into a few primary varieties? If it took millions of years to develop one type of beetle, and millions more to evolve another, how many years did it take to evolve 250,000 species? Then think of the other thousands of species of life on earth. And remember, the 250,000 species of beetles are **distinct species,** each an interbreeding population, and NOT just "varieties." Since science has set the age of our earth at from four to five billion years, ALL EVOLUTION MUST HAVE TAKEN PLACE IN THE LAST TWO TO THREE BILLION YEARS AT THE MOST. So the whole theory collapses in view of the vast variety of life on earth, and the tremendous time needed by evolutionists to account for even minor changes.

Wherever one looks in nature, he is confronted by innumerable varieties of life—especially in the lower echelons. There are over 100,000 known species of fungi; 5,000 species of green algae; 3,000 species of sponges; 5,000 species of corals and their kin; 25,000 species of crustacea (barnacles, crabs, lobsters, shrimp, etc.); 80,000 species of mollusks or shellfish; and there are over 180,000 species of plants!

Many of these multiplied forms of life exist in great profusion.

"Two hundred million insects may inhabit a single acre of pasture"—and the same acre will harbor trillions upon trillions of bacteria.

---

*Insects are as remarkable for their variety as for their numbers. There are tiny wasps less than one-one hundredth of an inch long. There are thin insects, fat insects, meek insects, fierce insects, flat insects, cylindrical insects, insects that seldom move and others as fleet as the wind. WHICH CAME FROM WHICH—or, did the Creator design and make them all?

Many insects multiply with unbelievable rapidity. We already have spoken of how prolific the common housefly is. Consider the aphids.

"If all the progeny of a pair of aphids survived for ONE SEASON they would number 1,560,000,000,000,000,000,000,000." Thanks to the Creator's marvelous provisions of maintaining "balance in nature" their natural enemies do not permit a run-away development of aphids, fleas, flies or any other prolific insect.

Most amazing of all—EACH OF THESE MYRIADS OF SPECIES IN NATURE HAS ITS OWN DISTINCTIVE CHARACTERISTCS. For instance:

"Each of the giant silkworm moths of North America makes its own distinctive cocoon, and many of them have a distinctive leaf for food. The cocoons of the **Luna** and **Io** moths are merely spun in a leaf, with which they fall to the ground. On the other hand, those of the **Promethea** moth (which feeds on **spicebush, sassafras** and other trees), are spun in a leaf which is securely bound to its twig."

The pupa of the Imperial Moth buries itself in the ground to await its transformation into a gorgeous moth. But the caterpillar of the pretty Royal Walnut Moth is "a fearful looking creature," that stays above ground. It is colored gray with black and white bars on its sides, while its ugly head is decorated with spines that resemble horns, making it even more hideous. Why does this caterpillar have such a fearsome look? There are a thousand and one different types and "styles" of caterpillars, as well as moths, for which fact evolution has no adequate explanation.

Speaking of only ONE group of Hunting Wasps (the **Crabros**), an authority says,

"It is impossible to lump the **Crabros** (hunting wasps) together—the species are too dissimilar. Some are big, others are hardly larger than midgets, many are square-headed and thick-waisted, while a few possess long, thin waists. They are no more constant in the colors they wear: some are yellow and black, others are black and scarlet, still others are pure black. As for their habits, some make their nests in the ground, others bore into wood—and even here they differ, for one species must have rotten wood, and another sound wood, and yet others bore into stems of briars, brambles, reeds and similar plants." (The Hunting Wasp, pp. 113, 114). And miracle supreme: each of these distinctive species keeps breeding, generation after generation, "after its kind," with only minor variations, and so each distinctive species preserves intact and unchanging its own peculiar characteristics.

Variety Among Beetles

Nowhere in nature can one see such vast variety as exists in the 250,000 species of beetles. Some beetles are as small as a pinhead, and others, like the elephant beetles (the Goliath and the Titan) are a full six inches in length! Some, like the weevils, destroy our foods and crops, and others eat vast quantities of destructive insects.

The so-called "death-watch beetle" (Anobium) will bang his horny head upon the wood where he has made his home, making a noise to attract his lady-love. Because of this strange ticking sound, the superstitious call it the "death-watch beetle."

The bombardier beetle, when apprehensive of attack, "will audibly eject with explosive force a fluid which volatizes, upon emission, into smoke."* From what did this strange creature evolve? What made it get this strange ability? And why did it lose it (the ability to eject a volatile fluid that vaporizes into smoke, when ejected) when it "evolved into a higher form"? And why did some beetles evolve into bombardier beetles and some did not? The very asking of these questions reveals the utter fallacy of evolution. EVERY ONE OF THESE QUESTIONS COULD BE ASKED ABOUT EACH OF THE 250,000 SPECIES OF BEETLES, and in every instance one draws the same inference: Blind evolution is helpless to produce such endless intricacies of design and adaptation. Why—how—could chance evolution, by "random changes," produce an intricate, living, workable mechanism like a bombardier beetle—then suddenly and radically change the pattern and produce, let us say, a "water beetle" (adapted to aquatic life), and then repeat this process 250,000 times!

Other distinctive beetles are the "violin" beetle (so named because of its peculiar shape), the "dung" bettle that rolls up dung into a ball several times bigger than itself, "water" beetles, the so-called "drug-room" beetle — so named because "it waxes fat on chemicals strong enough to poison an army." Another strange beetle (the **Sitodrepa panicla**) thrives on cigars.

And now, more questions for evolutionists to answer: How many generations or ages did it take unguided evolution gradually to make the change in eating habits from cotton or corn to cigars and poisonous chemicals? And why do these beetles persistently refuse, generation after generation, to change their eating habits, whether their diet is cotton, corn, cigars or poisonous

---

*The **Guiana termites** have an equally strange ability. Members of a "soldier caste" among them have a sort of squirt gun on their heads, through which they squirt a sticky liquid over raiding ants that have invaded their colonies. How long would it take "evolution" to develop that "squirt gun"? Who gave them the chemical formula for this "sticky liquid" so well suited to their defense needs?

chemicals? The truly logical and satisfying answer is, GOD MADE THEM SO IN THE BEGINNING.

Why is it that the larvae of the "stag beetle" passes years in timber oak trees before it attains its adult form? whereas the larvae of the so-called "oil beetles" have the "incomparable instinct" of gaining a living in bee hives! And note well the complicated procedure these larvae go through to get their daily rations:

"When hatched they climb without difficulty the stems of flowers where they calmly await the arrival of a bee. As soon as a bee comes along the oil beetle larva takes a firm hold of the bee's hair and rides along to the hive where it first diets on the **eggs** of the queen bee, and later (it undergoes two distinct stages of development) it subsists on the **honey** of the hive." This phenomenon is so utterly unaccountable that evolution is at a complete loss to explain this mystery!

Who taught this helpless beetle these maneuvers so necessary to its survival? Why will it in its first stage eat only bee's eggs, and later only honey? Who gave this humble beetle the wisdom to climb flower stems, then attach itself to bees—not moths or worms? Which came first, or which came from which—the chemical-eating beetle or the bee-egg-eating beetle? Such facts as these demand an adequate explanation, and the only adequate explanation is DIVINE CREATION.

The life cycle of the **Stylopidae** (parasitic beetles) is equally mysterious. These weird beetles actually live in the bodies of their hosts—bees, wasps and bugs.

"Here the female lays her eggs from which emerge active little six-legged larvae. These move to a NEW host and complete their growth in another body. The male develops into a smart little winged beetle, living a short but active life of only a few hours; but the female never advances beyond a grub-like condition, and is eyeless, wingless and limbless."

There is Great Variety Everywhere

Wherever one looks in nature there is great and pleasing variety. The cheerful songs of birds are as varied as their colored plumages and their nests. The intriguing subjects of sound and color in nature disclose variety without end.

We have the chirp of the cricket, the babbling of the brook, the swish of the wind, the crackling of the fire, the melody of the songbird, the bark of the dog, the mooing of cows, the neighing of horses, the bleating of sheep, the gentle purring of the cat, the roar of the lion, and a thousand and one other sounds we all are familiar with and that help make life pleasant, interesting and adventurous.

In the realm of color we have the same phenomenon: endless, pleasing variety. Who has not been charmed by the subtle colors of the rainbow, the ever-changing color schemes of the sky at the time of the setting of the sun? What is more fascinating than the gorgeous display of colors in a flower garden or the richly colored plumage to be seen in an aviary?

Who gave sea shells their delicate colorings and symmetrical designs? Who designed the "spiral staircase" shells, so delicate, symmetrical and attractively colored? Who created the "Glory of the Sea," an exquisite sea shell from the East Indies? The Fingers of Omnipotence can also be seen in the pinkish shells of the "Angel's Wings" bivalves, that live about a foot below the surface of the mud. Why such ornate beauty covered by mud—unless the Great Designer loves the beautiful and made this shell so that man too might enjoy its loveliness with Him? Who first put on the drawing board the intricate designs of such shells as the limpet, whelk, moon shell, the fascinating helix, and the charming periwinkle? Who tinted the "Queen conch" with delicate pastel shades of pinks and yellows and light browns? It would take volumes even to begin adequately to describe all the marvelous, symmetrical and beautiful sea shells that are subjects of study and admiration to students of conchology. But Whoever made them left evidence in His Handiwork that He is an Engineer **par excellence** and an Artist without a peer!

Why is it that all things in nature are not a dull slate color, or a listless gray? Such infinite variety and pleasantness as exists in the color schemes of the world—in sky, earth and seas—witness to the fact of purposeful design, by One who loves the beautiful, and shows His desire that His creatures too enjoy the lovely and the beautiful things He has made.

Let us now discuss the fact of

The "Fixity" and "Constancy" of Species

If ever the evolutionist had an opportunity to demonstrate his theories, it would be in the realm of some lower forms of life, such as bacteria or aphids, or flies that multiply so rapidly and with but brief periods of time from one generation to the next. And yet, THERE NEVER HAS BEEN ANY DEMONSTRABLE CHANGE OF ANY GENUS INTO ANOTHER, NO MATTER HOW BRIEF THE TIME FROM ONE GENERATION TO ANOTHER, IN ANY OF THE INNUMERABLE GENERA ON EARTH.* On the

---

*Some bacteria and protozoa, under favorable conditions, can mature and reproduce a new generation in 30 minutes or less. Thus over 17,500 generations could appear in one year! Yet we have definite proof that many kinds of these minute creatures have persisted without change for many thousands of years. Ancient Greeks and Egyptians suffered from the same bacterial diseases we do. Moreover, paleontologists have found numerous examples of diseased conditions among fossils of ancient rocks, proving that certain pathological conditions of the bones, caused by the same disease-producing microbes that are active today, have persisted without change for countless generations!

other hand, each genus demonstrates a most stubborn "fixity" and absolute refusal to change, even through long ages. And this well-known fact can be proven abundantly from the writings of evolutionists themselves! Consider well the force of the facts presented in these quotations. (The time elements are the estimates of the authors of the statements, and do not necessarily reflect our beliefs as to time involved).

"Mollusks are one of the oldest and largest groups of animals. For over a half a billion years species of mollusks have been common in the seas."

"Some of the oldest known fossils are corals that lived about 500,000,000 years ago."

"At the base of the family tree of animal life are single-celled animals (Protozoa) WHICH STILL LIVE in warm shallow seas as they did many millions of years ago." (Caps ours).

"As long as 200 million years ago, roaches and other insects were common. . . . Most of the 12,000 kinds of fossil insects identified are similar to living species." (This reminds us of the statement of Huxley, "The only difference between the fossil and the animal of today is that one is older than the other."

"Gastropoda (shell fish, limpets, winkles, whelks, etc.), are very old inhabitants of the sea and have lived there without undergoing much change for from three to four hundred million years." (The Living Sea, p. 202).

"One hundred twenty million years ago, oysters lay in quantities in the shallow seas." (Ibid., p. 211).

"Chitons (amphineurans) first appeared nearly 500 million years ago, yet they have remained unchanged to this very day." (National Audubon Society Nature Program).

"Scorpions can boast of the longest family line of any land animals. They have changed hardly at all during a span of 400,000,000 years" (W. J. Gertsch).

"As far as the earliest geological records of the existence of algae (sea weeds) allow us to see, THEY HAVE NOT CHANGED MUCH IN EITHER FORM OR LIFE ACTIVITIES; some species of today may well be identical with ancestors that lived in the Archeozoic sea about 1.2 billion years ago." (Francis Joseph Weiss, in "The Useful Algae," Scientific American, 12-'52).

"Sharks appeared on earth 300,000,000 to 350,000,000 years ago" (T. H. Eaton, Jr.)

"In the rocks of the earliest period for which we have good fossils (The Cambrian Period), **all of the important invertebrate phyla are already represented.** So that . . . the fossil records have nothing to say about the order in which the phlya arose" ("Animals Without Backbones"). (What a confession, coming from an evolutionist!)

"Turtles . . . have come down almost unchanged in form and habits since the great Age of Reptiles, more than 160,000,000 years ago." (Book of Popular Science, p. 2075).*

"Modern species of **Lingula** (one species of Brachiopods) are almost identical with species which we estimate, from the fossil record, to have lived almost 500,000,000 years ago. This is a record for conservatism among animals, and **Lingula** has the 'honor' of being the oldest-known animal genus." (Animals Without Backbones, p. 178).

"The Ginkgo, or maiden-hair tree, flourished during the Jurassic period, and was the first broad-leaved tree. It still exists today in China and Japan, **having undergone no change** for more than a hundred million years" (Book of Knowledge, Vol. 5, p. 1545).

Speaking of the "King Crab" (Genus **Limulus**), one authority says, "These animals are often referred to as 'living fossils' because they have changed so little from the earliest fossil representatives of the group." (Animals Without Backbones, p. 271).

Grasshoppers in Glaciers and Ants in Amber

Two most remarkable witnesses for the "persistence of Species" are GRASSHOPPERS IN GLACIERS AND ANTS IN AMBER.

There is a so-called "Grasshopper Glacier" of the "Pleistocene age" (one to two million years ago), in Montana. In the Glacial Period, these grasshoppers fell by the millions into a lake; they froze there and the lake became a part of the glacier. "One can see those grasshoppers in the glacier today, and **they are the same kind of grasshoppers we have today**". (Did Man Just Happen, p. 74).

Many scientists have written on "Insects in Amber." (See "Insects in Amber," by Charles T. Brues, Scientific American; "Evolution of Insects," by Carpenter, in the 1953 Annual Report of The Smithsonian Institution; The Living Sea (chapter on Time Periods and Fossils). Here is the unbiased witness of Chas. T. Brues:

"There is a deposit vault where we can find ancient insects, more beautifully preserved than any fossil ever disinterred from the rocks. This reservoir is amber: an ancient tree-sap which trapped insects like fly-paper and then hardened to preserve the insect intact for millions of

---

*Many authors speak of this fact. Doris M. Cochran, writing in The National Geographic, says, "Turtles are reptiles of ancient and honorable lineage. Their fossil ancestors are found in rocks at least 175,000,000 years old."

years.* . . . It is now possible to compare the insect life of 70 million years ago with that of today. . . . Among the earliest insects were some hardy types, such as the cockroaches, THAT STILL EXIST IN MUCH THE SAME FORM. . . . Some 70 million years ago (insects) WERE PRESENT IN NUMBERS AND VARIETY COMPARABLE TO THE PICTURE THEY PRESENT TODAY. The insects of that period, as preserved in the Baltic amber, were very similar to those that now inhabit the temperate regions of Europe and North America" (Insects in Amber).

". . . More remarkable still is the occurence in the amber (Baltic amber) of certain species of insects, mostly ants, which are apparently identical with some species now living. The Baltic amber has also furnished **proof of the existence of social habits among the insects of that time,** for the ants that occur there include, in addition to males and females, major and minor workers. The extent to which the complex habits of living ants had already been acquired in the early Tertiary is shown by the presence of plant lice attended by ants in search for honey dew, and by the presence of mites attached to the ants in the same manner as is characteristic today." (Annual Report, 1953, Smithsonian Institution).

And so there is proof that through the ages ants have not changed either their form or their social habits—not even their custom, still practised today by many ants, of using aphids as "cows" as a source of honey dew for their own use! IN ALL THE WORLD AND IN ALL THE HISTORY OF THE WORLD THERE IS NO PROOF ANYWHERE OF EVOLUTION, i.e., the change, or transmutation, of one genus into another. On the contrary, there is overwhelming proof, from scores, even hundreds, of scientists, that GENERA ARE STATIC—they tend, for ages on end, to reproduce "after their kind," showing no change in either form or habits for periods of time running into the millions of years— even hundreds of millions of years, according to evolutionists.

We feel constrained to quote the words of our friend, Prof. Leroy Victor Cleveland.

"Not a bacterium, nor alga, nor salp worm nor anything else ever "evolved higher." Check the facts and see. The Rhodesian algae are supposedly 'three billion years old.' WHEN, pray tell, are they going to evolve higher? Or when will ameba, stentor, volvox, ascidian larva, ant, moss, gnat, clam or bedbug 'evolve higher'?"

---

*Again we want to remind our readers that though we quote these authorities, giving such ancient dates and vast periods of time, we do not necessarily agree with their "time element" deductions or statements. It is our conviction that NO MAN KNOWS FROM FOSSILS HOW OLD THEY ARE. But the ancient dates they quote become a most powerful argument for they themselves admit that they believe many species HAVE NOT CHANGED MATERIALLY FOR HUNDREDS OF MILLIONS OF YEARS!

Yes, we ask again, WHEN will turtles, oysters, bacteria, amebas, sharks, ants, chitons, cockroaches and the Ginkgo tree, begin to evolve? One might ask the same of EVERY OTHER GENUS IN THE WORLD TODAY. All true genera are static—reproducing only "after their kind"

Another Question for Evolutionists to Answer

Now, we believe, is the time for us to ask this question:

"How could it occur that one individual, or a few individuals, of a given genus, or population, should advance toward a higher type, WHILE ALL THE REST OF THE SAID SPECIES SHOULD REMAIN in status quo?" For example: Amebas we have still with us today, and they still multiply true to form. Yet the amebas are supposed by some evolutionists to be one of the earliest forms of organic life. IF ONE AMEBA "EVOLVED" WHY DIDN'T ALL OF THEM EVOLVE? HOW IS IT THAT THERE ARE ANY AMEBAS ON EARTH TODAY, IF THEIR TENDENCY IS TO EVOLVE TO HIGHER FORMS?

THE MANY STRANGE AND ODD SPECIMENS OF LIFE ARE WITNESSES TO THE FACT OF SPECIAL, DIVINE CREATION

One of the strangest creatures God ever made is the Australian Platypus. We believe He purposely made it to confuse and confound the evolutionists. It is a squat, heavy-bodied animal about eighteen inches long. It weighs three to four pounds. It has a deep rich brown velvety fur (gray or white underneath) like the fur of a seal or a mole. It has a flat bill, like a duck, with no teeth after it reaches maturity. It has five toes on each foot, which is webbed—a cross between the feet of a duck and an animal adapted to scratch and dig. It is one of the only two mammals in the world that lays eggs.* Unlike other hatched animals, their young nurse. But instead of nursing from "conventional" nipples or breasts, the young simply lick the mother's belly fur, and the milk follows the hair ends.

The male platypus has a hollow spur on the inside of its heel, which connects with a gland as poisonous as most poisonous snakes. So **it is the world's ONLY venomous furred creature.**

---

*The other mammals that lay eggs are the **Echidnas,** toothless, spiny anteaters, also of Australia. **Echidnas** do not in other respects resemble the Platypus. Instead of having a covering of fur, they are covered with sharp, hard spines. Their snouts are long and slender. They live on ants and termites.

Unlike most mammals, its limbs are short and parallel to the ground—like the limbs of a lizard. Their eyes are small, while their external ear is only a hole, and not the customary ear-lobe such as mammals usually have. In habits it is nocturnal.

To help hold its food, which it catches under water (worms, snails, larvae, insects, etc.) it has large cheek pouches like those of a monkey or a squirrel.

It lives in burrows, which start from a point below water level, in rivers or ponds. The platypus can dig well despite the fact that the web on its front feet extends out beyond the claws. The webb folds back, like a small umbrella, into the palm, leaving the sharp claws exposed, ready for aggressive digging. The unique foot of the platypus is "an amazing contraption" and gives clear evidence of design and adaptation for an intended purpose— to dig and to swim.

What did the platypus evolve from? Let us imagine an Evolutionists' Round Table Discussion of this problem.

"He must have got his bill from the duck," suggested one. "That is obvious."

"Think so?" asked the second. "But a duck has feathers, not fur. It seems to me his fur indicates direct descent from some animal like the beaver—but then a beaver doesn't lay eggs."

"Wait a minute," interposed a third. "He's toothless and has spurs: that could suggest an ancestry from the chicken—and remember a chicken lays eggs, too." He caught his breath, thought for a moment, then changed his course. "But then, a chicken doesn't have fur either. That pesky fur eliminates descent from either a duck or a chicken. Quite confusing," he mumbled. But he started in again. "The female lays eggs, but she isn't a bird. Then too, those poison spurs present a problem—no other furred animal is venomous."

"Yes, and she suckles her young, but has no breasts," interrupted the first speaker. "Whales suckle their young—but then they don't lay eggs. Confound this problem, anyhow! Everywhere we turn we meet a roadblock. Let's try another line. The male has poison-dealing spurs, something like a snake, but they are spurs and not fangs. And everyone knows it couldn't come from a snake anyhow, for a snake doesn't have webbed feet."

Speaker number two had been engaged in deep thought. He was now ready to theorize again. "How in the name of common sense did it get its nipples—I mean its milk hairs, or what do I mean." He was clearly confused. Presently he reassembled his wits and continued. "And from what did its webbed, clawed feet develop—from ducks or muskrats? We have already eliminated ducks, and I guess we'll have to throw out muskrats, because they don't lay eggs." He started to scratch his head.

It may have been that unconscious gesture that caused another cogitating disciple of Darwin to suggest that "there might possibly be some distant relationship between the platypus and the monkey—for both have pockets in their jaws to carry food in." But on second thought he opined that "that isn't possible because the monkey is higher up the ladder of evolution than the platypus."

"Then where will we place this evasive critter" asked one of the Discussion Group who up to this time had felt that silence was the better part of rushing in where angels fear to tread. Being a neo-Darwinian, he had mentally recoiled from their naive and hasty suppositions. "We'll have to look into this matter from the viewpoint of heredity and genes," he reminded them, with an evident air of superior knowledge.

"But his genes must be as mixed up as he is," countered the first theorist. How could sensible genes and chromosomes come up with a conglomeration like this thing? He isn't a duck, or a beaver, or a snake, or a bird, or a monkey, or a lizard, much less a whale—but he seems to have been assembled from parts from all of them!"

At this point a "theistic evolutionist," who usually keeps his opinions to himself, suggested somewhat shyly, "Well, maybe this is where God stepped in and helped evolution along."

At this the rest of the group chuckled and the first speaker said in a superior tone—"certainly you don't believe that an all-powerful and an all-wise God is going to waste time guiding evolution! If a 'Supreme Being' had any thing to do with it, it is much easier to believe in 'special creation' than what you suggest."

The second speaker chimed in: "I agree; it seems to me that if a 'Supreme Being' had anything to do with it He wouldn't follow such a devious route that requires such a waste of time; but my main objection is that no one yet has told us WHAT the platypus came from—or how it got along before its organs were fully evolved."

"I'll tell you what" said the neo-Darwinian, with a twinkle in his eye, "He probably was dropped down from Mars!"

They all laughed and were about to give up the discussion, when an even tempered egghead, who had been listening up to this point, spoke up: "Don't give up; remember evolution does its work slowly—through millions of years. Just give us more time, perhaps a few million years more—and we'll come up with the right answer. After all, the platypus is HERE, and it HAD to evolve from something"—and then he suddenly recalled the animated discussion they had had—"or shall I say, 'from some **things,'** didn't it?" More laughter, and then they quit the discussion, this time for good.

Clearly, the conglomerate platypus defies all explanation, from the viewpoint of the theory of evolution. It is one of God's road-blocks, warning the theorists of the blind alley ahead that they persist in going down. It is therefore a living Witness for God and Creation, shouting to all who will listen to facts and common sense: "GOD designed my perfectly adapted feet for the niche in life He created me to fill; my webs are to swim with and my claws are to dig with, and they work, even if it is a novel arrangement. GOD gave me my bill so I could secure my food from the mud from under the water; and GOD put those pockets in my jaws so I could hold more food at each diving, and then come to the surface and enjoy my meal at leisure. GOD gave me my fur to keep me warm after my repeated immersions. GOD put claws and poison fangs on me so I could protect myself against my natural enemies. GOD gave me the knowledge to build my well-designed home underground, with an underwater entrance that helps keep my family from many dangers. And I rather suspect that God made me as He did, equipping me with "impossible" combinations, in a most unusual departure from normal routine, to confuse and confound those who ignore HIM. I tell you GOD MADE ME AS I AM—and I want the world to know it!"

Plant Oddities

Every plant in the world is a miracle and a mystery, with a thousand and one functions, characteristics and abilities that defy all explanation: all life is like that. Life itself is the most mysterious thing on this planet, for it is the gift of GOD, the infinite Author of life. Some forms of life deviate so from conventional types that they seem to defy the very laws of life.

Some bacteria can live in hot springs at a temperature of 175° F., while spores of other bacteria have survived after being exposed to the temperature of liquid air (-310°). Some flowers push their way up through snow and ice, while others lie dormant in desert sands for years, then carpet the desert valleys after a rain that may come but once in several years.

Many deadly poisons (some of which are useful drugs) are extracted from delicate plants with beautiful flowers, such as aconite from monkshood. Strychnine, opium, cocaine, digitalis and belladonna are but a few of the many others.

Some plants die as soon as they have flowered,* while some trees (the Joshua trees and the giant sequoias) live 3,000 years and more.

The great water lily of the Amazon and Indonesia has leaf blades five feet in diameter, while some palms have leaves twenty

---

*There is a bamboo plant, native to the mountains of Jamaica, that takes 32 years to mature. It then flowers ONCE—and dies. No one knows why.

feet long. There are seaweeds that grow in the dim light of the ocean 450 feet below the surface. This is quite an achievement, for the light is so dim at that great depth, that the normal process of photosynthesis is greatly retarded.

There are several kinds of "epiphytes," or "air plants," that get their nourishment from the air rather than from the soil. The staghorn fern is an example.

"It grows on other trees, with its leaves pressed against the trunk of the tree. The leaves cover large masses of roots that get their nourishment direct from the air."

One could try vainly for a thousand generations to "educate" the roots of plants or trees adapted to get their food from the soil, to get their sustenance from the air only—and not succeed. How is it then that SOME plants, the "epiphytes," HAVE mastered the secret? The answer is, God in the beginning made them so.

Who designed the 500 kinds of so-called "killer plants" that trap, kill and eat insects? We already have mentioned some of these, such as the famous "pitcher plant."

What could be more ingenious, complicated, designed for a purpose, and with apparent "intelligence," than the machinations of the sundew?

The sundew plant has about 200 tiny red filaments on the upper surface of each leaf. Each filament is club-shaped at its free end and carries a refractile globlet of fluid, that is a sticky substance from which it is impossible for an insect to free itself.

Movement of wind, rain or dust, or falling bits of mud, sand or leaves, or even small bits of sugar, placed on them by human hands, on the leaves of the sundew, cause the leaves and the filaments to re-act. The filaments will secret an acid fluid, but there is no attempt whatever at "capturing" the non-living object, nor is there any attempt to digest it. "BUT LET A SMALL INSECT LIGHT ON A SUNDEW LEAF, AND—wonderful to relate — THE CHEMICAL COMPOSITION OF THE SECRETION OF THE FILAMENTS IS AT ONCE CHANGED INTO A DIGESTIVE FERMENT, and the process of appropriating the unfortunate insect as food begins." Where did a lowly plant get such "intelligence"?

The bladderwort, that grows in water, is equally amazing. It is equipped with traps that look like small bladders floating in the water. These traps are cleverly designed to catch small aquatic animal life.

An opening exists at one side of a bladder. Around this opening is a set of radiating hairs set diagonally outward. These serve to guide the unsuspecting victim into the mouth of the trap. The opening is provided with a hinged, transparent door, **which opens inward, but not outward.**

Once a creature has entered this door, his doom is sealed, for the door closes and he cannot get out. He becomes food for his captor—a PLANT showing more intelligence than an animal! Or, is it that the Creator has made the plant function in such a way as to seem to have intelligence. In any event the whole scheme shows DESIGN that could only have been achieved by a Designer. It is utterly impossible for such a well-functioning trap to come to pass "by mere chance," or by unguided "random changes."

Nature Teaches Man Many Moral Lessons

Did the Creator make the bladderwort so that it could ensnare and kill and eat small animals only? Or, is there a moral lesson in it for mankind? In a world judged by reason of the Fall of Man, there is much "evil" in nature—the reflection of the evil in mankind. We personally, however, have no doubt that God deliberately created many animals and plants **for the express purpose of teaching mankind some important lessons.** Both the spider's web and the bladderwort's trap are graphic illustrations of temptation and the resulting ruin. The fox is the age-long illustration of cunning and rapaciousness. The lamb is the picture of non-resistance to evil. The lion speaks to us of powerful leadership. The poisonous snake reminds us of deadly cunning. **As a matter of fact, there is an important lesson inherent in practically every creature God ever made.** The Son of God while on earth spoke often of lessons from nature, as the branch "abiding" in the vine, the adornment God gave to the lillies of the field, the shortness of the life of the grass of field, etc. Then too, He was spoken of as the "Lamb of God" that takes away the sin of the world.

More About Queer Insects

We already have called attention to a few of the many strange "beetles." Let us list some other strange facts about insects.

Though the great majority of insects come from eggs, through a larval stage, the aphid, a tiny plant louse, sometimes gives birth to living young!

In Java there are strange earthworms that sing—and even whistle!

Consider the "misplaced" ears of the grasshopper.

"The ears of the grasshopper are either at the base of her abdomen or in her forearms, according to her species. . . . What surprises me is that Nature . . . has had the imagination to put ears anywhere else than on the side of the head" (The Hunting Wasp, p. 53).

What quirk of Evolution could move ears from the side of

the head to the rear of the abdomen, or to the forearms? And how many million years did it take to make such a change? And why, in the next genus higher than the grasshopper, were the ears moved back again to the conventional position? For the sovereign, all-powerful God to change the style occasionally creates no special problem—but for unguided "evolution" to be given the credit for such a radical change creates an unanswerable problem.

Who instructed the **Difflugia,** a free-living ameba, to gather sand grains, cement them together with a sticky secretion and build them into a kind of house having a definite design (it looks like a ball) which it carries about with itself and into which it withdraws when disturbed? (See "Animals Without Backbones"; chapter on "A Variety of Protozoa," p. 49). And who imparted to this small ameba the secret formula for this cement?

There are, moreover, various species of **Difflugia** that can be recognized by the specific shapes of the protective covering which they construct! Such miracles in the miniature world of the protozoa demand an adequate explanation. Evolution, that always demands "vast eras of time", and works entirely by "random changes," and "chance mutations," fails utterly to account for such phenomena.

A female water bug, "not trusting her husband's voracious appetite," cements her eggs directly to his back, where he can't reach them! How long would it take this lowly water bug to think up this scheme, and put it into practice by training her husband to cooperate and stand still while she did the cementing? And how much longer would it take her to design and install in her tiny body a chemical plant capable of manufacturing the proper kind of cement to make the eggs stick there? If the female water bug had to rely on the uncertainty of "random changes" to produce such a wise scheme as she has to protect her young, she would NEVER attain her end; as a matter of fact, she would NEVER have the foresight to think up or desire such a scheme in the first place. ALL OF THESE TENS OF THOUSANDS OF CLEVER ARRANGEMENTS IN NATURE DISPLAY FORETHOUGHT AND "DESIGN" WITH A PURPOSE IN MIND, and as such they must be the work of a Great Intelligence.

The female spittlebug has another scheme. She makes a froth on stems and grasses to cover her eggs. The young nymphs make a froth also to cover themselves while feeding. Open the small mass of bubbles and you will see the small, squat insects inside.

Where, how, when did this spittlebug get the necessary chemical equipment to "blow bubbles" that STAY as bubbles? Who or what gave her such an idea in the first place; why should she **want** to make an elaborate bubble bath for her offspring? Such strange, elaborate schemes did NOT originate with nature or with the insects of nature, but with the Creator of those insects!

The Strange Cicadas

These peculiar insects, the cicadas, are sometimes called "Seventeen-year locusts," though the 75 species of cicadas differ widely in the time they take to mature. Their life cycle is very strange—and utterly unaccountable, aside from the miracle of Divine creation.

The females cut slits in young twigs and deposit eggs in them. As the wingless, scaly young hatch, they drop to the ground, burrow in, AND STAY THERE FOUR TO TWENTY YEARS, according to their species—and no one can ever guess why. Those that stay in the ground four years breed a new generation that also stays in the ground four years; and those that stay in 17 years breed a new generation that stays in the ground 17 years!

As nymphs underground, they live on juices sucked from roots. When its predetermined time cycle elapses, the full-grown nymph emerges and climbs a tree trunk. Its skin splits down the back, and the adult emerges. These adults live about a week—long enough to mate and start another brood.

Why do they stay underground for several years? Who designed them and gave them the necessary adaptations for such a long underground existence? If they enjoy underground living so well, why do they **ever** come out—why not just live and die underground? Evolution has no answer.

The Extremely Odd "Praying Mantis"

The so-called praying mantis is an "insect nightmare" if ever there was one. It is commonly about 2 inches long. Their spiny, ferocious forelegs, their protruding eyes that pop out from their head that appears to be a caricature of a snake's head, their long bodies and ambling gait, and their bony "armor" suggest "a prehistoric reptile in miniature." They have no voice, and lack real ears. Their closest "relative" in nature is the grasshopper—but they are so unlike the grasshopper, there is a "gulf" between them impossible to bridge by any evolutionary theory. They are cannibals; and in the natural state their prey must be "alive and moving." In the fall the female lays hundreds of eggs in a frothy mass that dries like hardened brown foam. "After mating, the female dispatches her mate with a well-placed bite and devours him at her leisure." (National Geographic Magazine article on "Praying Mantis").

Both in appearance and in habits (characteristics) the Praying Mantis can not be explained by evolutionary theories. The variations in the over 1000 species (15 of which may be found in the United States) may be accounted for on the basis of "gene mutations," but WHERE DID THE ODD CREATURES COME FROM TO START WITH? Who gave the mantis the uncanny ability "to thrust forth her spiny forelegs" with lightning speed and grab her victim (a fly or other insect) as in "a toothed steel trap"? From what ancestor did the female learn the revolting art of beheading her husband, who is smaller and of slighter build? The mantis, in looks and in habits, is like a lone island in the midst of a vast ocean of creation, with NO CLOSE CONNECTION WITH OTHER INSECTS. "Divine creation" answers the problem of their origin—but evolution has no adequate answer.

STRANGE FISH AND OTHER ODD INHABITANTS OF THE SEAS: WITNESSES TO THE FACT OF DIVINE CREATION

It has been said that "the body of the tuna fish represents one of the most perfect streamlined contours known to Nature." But then, other fish also draw forth enthusiastic comments about their perfect streamlining:

"An adult swordfish may measure 15 feet from tip of sword to end of tail. It is shaped on the lines of a mackerel and is the epitome of streamlining. The pointed head . . . the sharp, backward rake of the dorsal fin, the long, lithe, powerful body, sloping gradually to the great crescent-shaped tail, fit it for the most rapid and forceful movement through the water."

The air bladder of the bony fish is obviously **designed** for an intended purpose.

"Most of the bony fishes possess air bladders, containing oxygen—sometimes undiluted—to enable the fish to float at certain depths. By regulating the gas pressure the fish can readily move about on a horizontal plane at any reasonable depth." (The Living Sea).

As a fish rises toward the surface, the pressure of the surrounding water decreases, and consequently the gas in the bladder expands and the body of the fish tends to rise too rapidly. But then gas is absorbed by the appropriate parts of the bladder wall, so that equilibrium is restored. When the fish descends the system works in reverse—and it is all automatic.

There are features about the anatomy of the fish not yet fully understood by modern science. For example:

"In a cavity on each side of the fish's skull are two chambers, each containing a small stone. These are the ear stones, or 'otoliths,' and these chambers and stones constitute the ears of the bony fish. These ears are very different from the ears of land dwellers, **and quite how they operate is not known."** (The Living Sea, p. 141).

To believe that an intricate and working mechanism for hearing and "balance," designed for use under water, should have just "happened" or came about by "chance mutations" is absurd. The thinking man can see the Hand and the Intelligence of the great Creator behind this especially designed ear mechanism.

Whales and some fish do not have the "air bladders" that the bony fishes have; and how they (whales) endure the tremendous pressure changes involved in dives of several hundreds of fathoms is a mystery.

Every type of marine life is especially "adapted" to its own environment and to the place in the scheme of things that the Creator assigned to it. It is not necessary to illustrate this fact by many cases, but let us give one.

The weevers, arrow-like fishes with hatched flanks as though streaked with rain beaten down at an angle by the wind, are **Trachinidae** (from the Greek **trachus,** or stinging) and they spend most of the time more **or less** buried in sand. This way of life determines their three fundamental characteristics (or, do the 'characteristics' the Creator endowed them with determine their habitat and manner of living? We believe the latter). All of these characteristics are excellently "adapted" to their life: eyes directed upwards to spot their prey from their hiding place in the sand: a mouth with a vertical gape made to snap at any prey coming within reach; and dorsal fins with long and venomous spines to protect them against **their** enemies. (The Underwater Naturalist, p. 209).

Instead of trying to delude ourselves into believing that these fish (weevers) lived in the sand, at first totally unprepared for such an existence, and that these special "adaptations" developed through the ages, it is much more reasonable to believe that the Great Designer made the weevers to suit the habitat He put them in. A fish does not plan ahead; how then could the fish itself embark on a program of "evolving" that would end up with those three features: eyes directed upwards, a mouth with a vertical gape, and dorsal fins with venomous spines, for protection?

If a fish tried to live in an environment and was totally unprepared for its hazards, it (the species) would soon become extinct, and never arrive at a state of adaptation. This fact is one of the principal arguments against the fallacy of evolution: Adaptation, to be workable, must be **perfect;** an imperfect or partial adaptation is unworkable and ruinous. And since "adaptations" according to evolution do not appear instantly, but are the result of gradual changes, through many ages. ALL "EVOLVING" CREATURES WOULD BE IN A STATE OF TRANSITIONAL IMPERFECTION—hence chaos. But such is not the case; ALL life throughout the entire realm of nature is perfectly adapted to its environment and gives indisputable evidence of being "designed" and hence created for its place in the world of nature. Moreover, each genus is **static,** persistently so, and gives no evidence whatever of change from its "kind" except in minor "variations" within the confines of the genus. The happy state of "workability" and "dependability" that exists in nature could not exist if evolution were true.

There is no proof in the whole realm of nature that any animal anywhere ever acquired "special organs"—such as the venomous spines of the weever, the suction discs of the starfish, the razor-sharp lancet of the surgeon fish, or the marvelous drilling mechanism of the oyster drills—by mere "chance mutations" or through a process of gradual change.

Great Variety of Life in the Plankton

Consider now the "miracle" of the profuse and fascinating **variety of life in the plankton.** The drifting animal and plant life of the oceans near the surface, that is food for the larger fish and marine animals, is called "plankton"* (from a Greek word meaning wandering.) One authority says,

"Any one may find in the surface waters of the sea animals (mostly microscopic) that hold their own with those in Fairy Tales." (The Strange World of Nature, p. 19). Some of these strange creatures "are wholly un-

---

*Plants and animals that live in the water are divided into three main groups: the plankton, the nekton and the benthos. **Plankton** is the name for those forms of life that float at the surface. They include an enormous variety of tiny animals and plants as well as larval forms of many other animals. The **Nekton,** made up of creatures that swim actively, include most fish and also squids, whales, porpoises, and shrimp of many kinds. The **Benthos** includes those countless animals that creep on the sand and bury themselves in the mud, hide in crevices or fasten themselves to the rocks.

like any known animals from land or even fresh water" ("Strange Babies of the Sea," by Hilary B. Moore, in the July, 1952 "National Geographic").

Included among these strange creatures are weird specimens as the transparent Salp; arrowworms (named from their shape); the trumpet-like Stentor; the unbelievable **Siphonophores** that lay eggs in one generation and develop bud-like plants in the next; tiny creatures with ghost-like and nightmarish shapes, as the thin, transparent, baby lobster, needle-nosed babies of Porcelain Crabs and a thousand and one other oddities that defy description.

Myriads of Marvels of the Deep

A visit to an **"undersea garden"** as seen through the bottom of a glass-bottomed boat "is like a scene from fairyland, with strange-patterned fish darting about sea flowers of every description." (H. J. Shepstone).

"The coloring of the corals, sea flowers and other varied marine life is almost beyond description. The coral polyps themselves are of every conceivable color—brown, violet, pink, white, yellow, purple, bright blue and vivid scarlet. The anemones, sea cucumbers and sea urchins are also of many varied tints. There are sponges of black and purple, covered with a thin sheen of emerald green. And darting hither and thither are troops of fishes having color patterns which are exquisitely beautiful—tube worms with brilliantly-colored crowns of tentacles, and innumerable starfish, crabs, and crustaceans, many of them also highly colored. Truly, a reef of living coral with its gorgeously-colored inhabitants is a sea garden, more interesting then any garden of flowering plants."

Another author speaks of "the strange and fascinating world that exists down in the ocean, in which beautifully colored sea anemones and brilliant corals make gardens of fairylike loveliness. Here are to be seen the SEA CUCUMBER, which, when it moves, looks something like a giant caterpillar covered with warts or spines . . . the pugnacious SWORDFISH, with its long 'sword' always ready for battle, fastened to the end of its 'nose.' This 'sword' is a prolongation of the bone of the upper jaw, and is so sharply pointed and so strong that, when driven with terrific speed, it can penetrate the oaken planks of an ocean-going vessel." "Steel experts agree," says Ben Berky, that "the sword of the swordfish is a natural phenomenon. Its power of penetrating several inches of solid oak is not surpassed by any similar weapon of steel." Where did the swordfish get its formula for the fabrication of this extremely hard sword—stronger than steel? And where, by the way, did it get the idea in the first place of "developing" a sword, while the rest of its neighbors did not? And what did it have for a "sword" —how hard was it, what did it look like—while the "sword" was going through the long process, of many ages, of evolving into a sword? Obviously, a cartilaginous, half-soft "sword" would be no sword at all! It had to be created as a sword, complete and perfect, for a partly-developed "sword" would be no "sword" at all!

Nowhere in nature will you find "partly developed" "swords."

Let us present some more of these

"Strange Sea Creatures"

(1) The Sea Horse

"Mother nature outdid herself when she assembled the sea horse. This bizarre creature has the arching neck and head of a stallion, the swelling bosom of a pouter pigeon, the grasping tail of a monkey and the color-changing power of a chameleon. It has eyes that pivot independently, so that when one eye scans the surface, the other can be directed underwater. To top this fantastic make-up the male is equipped with a kangaroo-style pouch from which the little ones are born."

This four-inch-long sea horse is **the only fish that swims upright!** He can look forward and backward at the same time (having eyes that pivot independently), or he can look up and down at the same time. He has a special "gas bladder" that enables him to keep his upright position. If this bladder is damaged and he loses even a tiny bit of the gas, he sinks to the bottom, there to lie helpless until death overtakes him or his bladder heals.

But the most amazing feature of all is that it is the male sea-horse that "goes into labor and gives birth to its young. This strange division of the sea-horse's reproductive functions, UNPARALLELED IN NATURE, is the peak of this tiny fish's paradoxical make-up."

"The female sea-horse provides the eggs. During courtship, the female actively pursues the male, deposits her eggs in a pouch on her mate's belly, and then swims away. In the pouch the eggs are nourished on the father's blood for 45 days. . . . after a series of parental convulsions (with apparently every muscle brought into play) the pouch is emptied and the baby sea-horses (from 300 to 600) are born!"

Evolution is utterly at a loss to account for such unorthodox procedures, and such a strange creature. The "Sea horse" is in a similar category with the platypus, as far as evolution is concerned: it presents an enigma that baffles and frustrates all theories that seek to account for it! Admit the Divine Designer, and all is accounted for.

(2) The Improbable Sting Ray

For perfection of movement, look at the ray.

"Among the movements of all living things in the sea the most perfectly harmonious is undoubtedly the swimming of the ray. When this 'bird of the depths' beats its wings, the fleshy wings themselves undulate. A sinuous movement takes place from back to front, being most supple at the edges, creating a movement as of frills and scallops reminiscent of the waving of a silk handkerchief, or an Egyptian dancer. . . . This improbable-looking bat-like creature, looks like a monster from another world, a demoniacal phantom in violet or dark grey, but with a pure white patch on its belly, flying silently and mysteriously through the water. . . . But when the ray comes to rest it appears to be an almost deformed-looking beast, flopped down." (The Underwater Naturalist, pp. 228, 236).

What Architect designed the supple movements and the perfect rythm of this dread beast of the sea? The coordination of muscles and the rhythmic movement through the water could never be achieved by the trial and error method; the grace and suppleness and perfect rhythm of the ray demand an Architect of supernatural ability, a Worker with Infinite perfections.

"Its long tail stretches out behind it like a whip. It does not possess the slightest vestige of a dorsal fin. It has only one or two 'spines' which have been enlarged, hardened—tempered one might say—into formidable spikes or stilettos. At the same time they are jagged in order to aggravate the wounds they deal and, in addition, they carry one of the most venomous poisons of any fish in the sea."* (Ibid. Pp. 235, 236).

It is impossible for us to believe that the "sting ray" developed by "chance mutations" through many ages from some other creature. Let us ask the evolutionist, from what did this strange creature evolve?

(3) The Humble Oyster: the Brainless Wonder

In the November, 1953, "Scientific American" is an intriguing article by Pieter Korringa, on "OYSTERS." We quote:

"The existence of the oyster is so different from a vertebrate's experience that even with the most unprejudiced study we find it hard to understand. Although thousands of investigations have been made of the bivalve, its life is still mysterious. The creature defies many elementary rules of animal biology. . . . Even anatomically we cannot make head or tail of the oyster, for it possesses neither of those organs. Yet in spite of its lack of a brain and its seemingly poor equipment for survival the oyster deserves our boundless admiration. It has senses (chemical and tactile) which are extremely acute, a feeding system which is extraordinarily delicate and effective, a metabolism which ministers to its needs in a highly versatile way and a bagful of other resources which enable it to survive even though it seems one of the most defenseless of creatures, a passive thing altogether at the mercy of its environment." . . . .

The oyster has an intricate pumping system far more involved ("more delicate and complex") than was previously supposed. "With its pumping system the oyster couples a filtering system, for which it uses mucus. Very thin sheets of mucus pass continuously over the oyster's gills. This mucus traps food particles, and conveys them to the oyster's mouth. Both the pumping and the filtering mechanisms are sensitive to environmental conditions; the oyster does NOT feed continuously: it tests the water from time to time, and it sets its intricate feeding mechanism into operation ONLY

---

*We do not know the chemical formula of this poison.

WHEN THE QUALITY OF THE WATER MEETS CERTAIN REQUIREMENTS. . . . Its chemical receptors apparently warn it not to feed when certain organic excretions or other poisons are in the water. And its filtering mechanism enables it to segregate from its intake and throw out organisms or particles which it presumably recognizes as inimical.

"The manufacture of the oyster's shell is an intricate, fascinating operation. The mollusk has herds of small glands which secrete calcite. . . . It deposits the calcite on a thin network of protein, steadily enlarging and thickening the shell as it grows. The oyster does NOT use dissolved calcium carbonate, which is rather sparse in sea water, but it captures calcium ions. JUST HOW THE OYSTER CATCHES THOSE IONS AND POURS THEM OUT AGAIN THROUGHT ITS SHELL-SECRETING GLANDS TO FORM THE CALCITE LAYER OF ITS SHELL IS UNKNOWN.

"The oyster has to create a home of a very definite shape . . . and its construction must be right the first time, for the shell cannot be broken down or remodeled. Investigators have been amazed to find that the oyster pads out the thick places in the shell with 'cheaper' construction—a chalky, porous deposit which requires only about one-fifth as much building material. JUST HOW IT CONTROLS THE MAKING OF THE DIFFERENT TYPES OF SHELL IS HARD TO UNDERSTAND.

"The two valves of the shell are hinged by a rubberlike elastic ligament which pushes the valves apart when the oyster does not hold them closed; the closing of the valves is controlled by a powerful central adductor muscle. This muscle has a 'quick' part which can open or snap the valves shut very rapidly, and a 'catch' part which can keep the shells closed for a long time, apparently without getting tired . . ." (Caps ours). (The rest of the article by Pieter Korringa gives many more fascinating facts).

Not the least of the achievements of the humble oyster—the Brainless Wonder—is the creation by the oyster (starting with an irritant: a grain of sand) of a pearl—"the queen of gems." By what legerdemain can the oyster after many months transform an irritant into a "perfect, fully formed jewel, the iridescent pearl, that never requires polishing, cutting or other artificial methods to improve its beauty."

Where did this "brainless Wonder" get such wisdom? Who taught it how to make its shell, and make it right the first time? Who gave the oyster the secret of capturing calcium ions—and also a high concentrate of copper, zinc, iron, manganese and rare metals (in concentrations thousands of times higher than in the surrounding sea water)—and put them into an easily digestible form for man, making the oyster a rich and succulent food for man? Who taught the oyster how to create the matchless pearl? Surely One infinitely Higher than any intelligence on earth designed this marvel and MADE IT FOR A PURPOSE. So the humble oyster—the brainless wonder—becomes a most effective WITNESS FOR DIVINE CREATION. The oyster just **cannot** be accounted for on any other basis than Divine creation.

As a matter of fact, we have not presented one-tenth of the "marvels" to be observed in the life history of the common oyster: the "Brainless Wonder."

(4) **The Incredible Dance of the Grunion**

The Bay of Fundy, Nova Scotia, lies at the end of a huge ocean basin, and the tidal movement is greater than anywhere else on earth: the water rises and falls through 50 feet or more. In Hawaii, on the other hand, the tides rise less than a foot. These Hawaiian tides are controlled almost entirely by the sun. The many influences that control the movement of tidal waters—the pull of the sun, the pull of the moon, the pull of both together or one against the other, storms at sea—have made theoretical predictions of tides a very intricate mathematical feat. But trained oceanographers can predict with accuracy how tides will run in different parts of the world.

How then can grunion (small silvery fish), without study or training, forecast the ever-changing tides? Yet they do, and with amazing accuracy!

"Grunion runs" are found only off the coast of lower and southern California, beginning in March and continuing through July. Thousands of grunion appear on the beaches to lay their eggs in the sand three or four nights after the new or full moon. "The forecasting of the hour and minute when grunion will run is reached by adding fifteen minutes to the time the tide reaches its nightly peak. In other words, there is a margin of safety: they come ashore AFTER the turn of the tide, and on nights when the tide reaches a little less high than on the preceding night. . . . Thus the eggs are laid in sand which will NOT be reached by the tide for about two weeks.

"The female, heavy with eggs, permits herself to be washed in by the tide and strands herself. She energetically burrows into the sand tail first to a depth of two or three inches. The males then, in a horizontal position, curl their bodies around the partially buried females and discharge milt which runs down along the female body and fertilizes the eggs which are being laid in the sand." The whole process lasts about 30 seconds. Both grunion then flop back into the sea, but the eggs, deposited on a night when the tide has begun to recede will NOT be washed out until the next high tide two weeks later. During the two weeks between the laying of the eggs and the next high tide, the eggs are incubated in the warm, damp sand. When the next high tide erodes the beach and uncovers the eggs, the eggs hatch explosively and the new-born fry swim out into the ocean!

Who teaches each NEW GENERATION of grunion how to time the tides, and know when it is 15 minutes AFTER high tide, the night after the fortnightly high tide? Who designed the grunion eggs to hatch in two weeks? and in a nest of damp sand? Who taught the female grunion to place the nest in the exact locale where the tide will expose the eggs two weeks later? One bows in awe before such a miracle and concludes that the Creator of all so equipped the grunion with the necessary abilities that all people might have a constantly recurring demonstration of DIVINE CREATION. There just isn't any other plausible explanation for this phenomenon.

(5) The Spectacular Swarming of the Palolo Worm

The grunion is not alone in its uncanny time sense. The palolo worm puts on a similar demonstration. We refer to the **Eunice viridis** (palolo worm) of the south Pacific. This worm lives in deep, cavernous hollows at the base of sunken coral reefs in the ocean waters around Samoa, Fiji and some other Pacific islands south of the equator.

"Once each year, at a definite time, the palolo appears in myriads at the surface of the sea to perpetuate its species in a spectacular swarming. This takes place in the early spring, **exactly one week after the full moon in November** (Springtime, south of the equator) and occurs with such regularity each year that 'palolo time' is the outstanding date of the native calendar. . . .

"The worms grow to a maximum length of eighteen inches. As November approaches the hind part of each worm, which is about three times as long as the fore part, becomes filled and distended with minute eggs in the female and sperm in the male.

"When the moment arrives each worm crawls backwards out of its hole deep in the coral and the hind part breaks away and wriggles up to the surface. The fore part of each worm remains in the coral and grows a new hind end which, the following November, again supplies the eggs or sperm for the perpetuation of this strange species.

"Almost immdiately when the hind end of the palolo reaches the surface, it bursts and the eggs or sperm are fired into the water 'like an explosion'. The empty, shrunken remains of the worm then sink down to die on the sea-bed. The great majority of the countless millions of the palolo worms inhabiting the coral reefs in the South Pacific behave in this way once a year, in the **early morning of the seventh day after the November full moon.** Burrows says, 'the palolo makes its annual rising AT AN ACTUAL DATE BY THE MOON AND THE TIDE" year after year, without change or failure. (Animal Wonder World, Pp. 153, 154).

Who taught the lowly palolo worm how to discern "times and seasons"? How can it tell when it is exactly one week after the November full moon? And why does the new generation of palolo worms, the next year, and the next, and the next, never miss the date by even one day? Did the Creator have the natives in mind too, when He created such a huge stock of the edible Paolo worms, that they might know when to catch them? Such miracles as the "spectacular swarming of the palolo worms" cannot possibly be the result of "blind chance," or the final outcome of many ages of "fortuitous changes" or chance mutations, for whoever or whatever caused this phenomenon DESIGNED it so—and then when the pattern was set, He made it static, so that generation after generation, century after century THERE IS ABSOLUTELY NO CHANGE WHATEVER IN THE PROCEDURE. The palolo worm is a powerful witness for the miracle of Divine Creation.

(6) Fish with Built-in Dynamos

The electric eel **(Electrophorus electricus)** is a native of the backwaters of the Amazon. Four-fifths of the length of his stubby body contains electricity-generating tissue, which enables him to send out discharges up to 500 volts many times each minute!

"When a piranha or other foe of the eel comes too close, **Electrophorus** builds an electric fence around himself by switching on his generators and charging the water in the vicinity with electricity. With his enemies stunned by the shock, it is an easy matter for the eel to escape. This eel's truly shocking weapon is a good example of the clever devices which nature furnishes to many animals to help them fend off trouble."

"The current from the electric eel may be released from any part of the fish with equal intensity; it is directional, having one polarity at the head and another at the tail; the fish can regulate the amount it discharges."

The electric catfish has a novel way of getting its meals—even though it is somewhat revolting. Swimming boldly up to a large fish, it slyly touches it on the stomach with a fin and gives it an electric shock—a shock that does not kill it, but causes it to disgorge any half-digested food before the stunned fish seeks to make a hasty get-away. The catfish eats the free lunch and then looks for another victim of its practical joke. (See p. 17, "Nature Parade").

But the champion electrician of all is the Electric Ray. Its electric equipment is so astounding, we must quote a detailed description.

"The electric organs of the electric rays are exceedingly complicated and only a genius in the field of electricity could fully understand them. ... These electric organs are a complicated wet battery. There are about 450 special tubes in each of the organs supplying the positive and negative currents, all separated from each other by special insulating tissue. There are many electric plates, and the wet medium (corresponding to the acid solution in a man-made wet battery) is a clear jelly. There are special nerves going to every plate which comes from a main nerve which itself is connected to a separate section of the brain that deals solely with electricity. ... The electric ray's 'batteries' consist of two nodes, one positive and the other negative, which have to be connected before a discharge takes place. A powerful shock is then given of a frequency as high as one hundred fifty per second. This kills small creatures and is quite enough to knock a man flat on the ground. ...

"Nature, in fact rather surprisingly, has shown herself here (and in other electric fishes) to be a skilled and inventive electrician. I say surprisingly because with most other animals she has given no hint that she knew anything much about the subject." (The Living Sea, Pp. 130, 131).

Can any honest, thinking person read that description and not come to the conclusion that the God who created all things, who knows all the secrets of electricity, as well as gravity and all other natural laws, is the One who made the electric ray? For such an involved mechanism to evolve by "chance" is utterly impossible: and any mathematician who is conversant with the laws of "chance" will tell you that it is as possible for thirty pounds of the raw, unformed materials that make up a battery (lead, acid, etc.) to be agitated by a windstorm and produce a wet battery as for unguided chance to produce the involved mechanism of the electric eel or the electric ray.

(7) **The Strange Case of the Fish Hatched in Father's Mouth**

In the fish world parenthood at times is more trouble to the father than the mother. We already have spoken of the male of the sea horse that carries the eggs of the female in a pouch on its belly. The male **Tilapia macrocephala** is also an exceptionally devoted father—or shall we say he is a hen-pecked husband. The **Tilapia,** about three inches long, lives in the rivers of Africa.

"After the female has laid the eggs and the male has fertilized them, the male picks up the eggs and carries them around in his mouth like a bunch of marbles. He keeps them there until they hatch and the young **Tilapia** are large enough to fend for themselves. During this two-week period the father cannot eat a bite, and he has to exist off his own tissue. The family life of the **Tilapia** has been studied for 15 years at the Museum of Natural History in New York City by Dr. Lester R. Aronson, who also has been to Nigeria observing them in their natural habitat.

The female **Tilapia** scoops a hole in the gravel at the bottom of the river with her mouth. She then lays eggs, about 80 of them, in this nest. The male drops sperm on the eggs, then darts headfirst toward the nest, scooping up a few more eggs with each plunge, until he finally has gotten them all into his mouth. If he overlooks a few, the female slaps him with her tail to remind him he has left a few in the nest—but this happens only rarely. Crammed with eggs, the male's mouth bulges. The eggs hatch in about five days, but he usually keeps his youngsters in his mouth for about six days more."

What strange power keeps this fish from swallowing those eggs as other hungry fish would do?

Anyone who knows the tendency of the typical male to shun household duties can see in this nothing short of a miracle! And all who know the male appetite can see in this a **double** miracle—for by what natural power was the male Tilapia ever persuaded to keep from eating for eleven days? This is the more wonderful when one remembers that many species of fish eat not only their own eggs but also their own fry as well! Going counter to ALL NATURAL TENDENCIES, the male Tilapia actually performs, without remonstrance, a specialized function in the propagation of the species that MUST have been "born into it" by a Superintending Providence, and could NOT have been evolved by natural processes. And so the male Tilapia shouts out to all: "I am the way I am because **God made me so,** and not as the result of mere natural processes—for had it been up to me, I would from the very beginning have gulped those eggs as soon as I got them into my mouth; you know how hungry we male fish get! And after all, to us hungry fish, an egg's an egg, whether it is from our species, or from some other fish."

(8) **The Mystifying, Clever Crabs**

Crabs, lobsters, oysters—all are relatively low in the scale of life—and yet the crab gives evidence of cleverness bordering on apparent intelligence. The whole life story of the crab is unbelievable.

"When the egg of a crab hatches, a speck emerges that moults within an hour and turns into a tiny creature THAT BEARS NO RESEMBLANCE WHATEVER TO A CRAB. It is only about one-twentieth of an inch long, translucent, and carries two long spears, one on the middle of its back and the other projecting in front, like a beak. It has large eyes, set flat and NOT on the tips of stalks like those of its parents. (This incredible fact is an insoluble riddle to all naturalists). It swims actively (most adult crabs do not swim at all). Other moults (stages of growth) take place during which the baby crab presents an astonishing variety of COMPLETELY DIFFERENT SHAPES. (This periodic metamorphosis is absolutely inscrutable to zoologists). Finally, however, it loses the gift of swimming and sinks to the sea bed, still exceedingly minute, but now a replica of its parents in every way—stalked eyes, pincers, and the rest.

"The last creature from which one would expect intelligence is a crab. Yet if one judges by behavior, certain crabs possess considerable intelligence and cunning. It has been said that men are the only animals to have learned to carry weapons. The monkey may hurl a coconut from the top of a tree but it never carries a stick. Man however was not the first creature to carry weapons; the crab had been doing this long before. . . . To what extent the crab knows what it is doing does not concern us: it **does** it. It will be said that only those crabs survived that carried these weapons and so the process became automatic and instinctive. But the many species that did NOT carry weapons also survived. (Note the argument he gives against evolution—editor). . . . Such actions are instinctive now; it is the ORIGIN of these schemes that give rise to thought.

"Consider the crab named Dromia. . . . In nature a sponge is not soft and sweet-scented, but is covered with fine needles of lime or silica. It also has an offensive smell. Consequently it is given a wide berth by all forms of life except worms and other small creatures who live securely in its tubes. The crab, Dromia, takes advantage of the sponge's unpleasantness and converts it to its own use. It takes a living sponge, cuts and trims it to the size of its own back, places it on its back, and holding it down with its last two pairs of legs proceeds about its business both camouflaged and protected at the same time. . . . It has to keep two pairs of legs permanently employed clamping the sponge down, but the other legs suffice for its other affairs and the protection the sponge gives amply compensates for their absence from normal duties.

"Take the case of the **spider crab.** . . . It cuts off seaweed and other marine growths and attaches it to its back. 'Nonsense,' you say; 'seaweed and other marine growths fasten themselves to anything at the slightest opportunity. That crab simply happened to have got overgrown with the stuff without knowing it.' You are wrong. The seaweed was **deliberately planted by the crab.** To prove it, catch one of the crabs, remove all growth from it, and place it in an aquarium among seaweeds. Here you may observe it pluck pieces of weed and place them on its back where they are held by the crooked hairs until they take root.

"But that is not the end of the spider crab's repertoire. Take this same crab with its trailing garden attached to it and put it in an aquarium that is full of vegetation or other growth of another kind, small sponges, maybe, anyway, things different. Here in this different environment, the little forest on its back is more an advertisement than a camouflage, so the crab picks off every bit of weed it carries and implants instead bits of the growths amongst which it finds itself. Such cleverness is rather bewildering, and it takes place with one of the commonest of our crabs."

". . . . Some tropical crabs carry over their backs (held there also by the last two pairs of legs) the shell, or rather half-shell, of a bi-valve mollusc such as an oyster or a scallop.

". . . The sea anemone . . . is avoided like a plague by practically every creature that swims the sea. . . . The unpopularity of the sea anemone has not been overlooked by the crab, who, as usual, has turned it to his own advantage. So we find in warm seas certain crabs that pluck off (for defence) a small sea anemone which they hold and carry about wherever they go. Nothing could be better from the viewpoint of protection.

The crab's cleverness goes even further. "The **hermit crab, Lupagurus prideauxi,** living in a borrowed shell, invariably has a sea anemone attached to the shell in such a way that the mouth protects the crab and also being in a position to take in any food the crab may drop. Both, in fact, get bits of food from each other. When the crab grows too large for the shell and moves into another, it detaches the anemone from the old shell and puts it on the new."

Most mysterious of all is the uncanny "light sensitivity" possessed by the HORSESHOE CRAB. Prof. Talbot H. Waterman, of Yale's Osborn Zoological Laboratory, has shown that these lowly animals "**can detect the position of the sun, even if they cannot see the sun,** by the degree of polarity or angle of vibration of its light in the sky. . . . The horseshoe crab has been using the polarity of sunlight for some hundreds of millions of years, whereas we (men) became aware of this basic principle of light only during the last half century."

Who placed in the lowly horseshoe crab the necessary "scientific know-how" to be guided by a principle of light that man has discovered only recently? Who taught the hermit crab to use the sea anemone, sea weeds and the sponge for its own protection and advantage? All thinkers must admit that the "intelligence" and "cleverness" reflect the work of the Creator, and are not acquired or inherent, for a crab is a relatively low form of life.

(9) **Strange Fish of the Deep Sea**

The vast world of the deep seas "is in many ways as strange and remote as another planet. It is a world of total darkness, eternal cold and enormous pressure—up to 1,000 atmospheres and more. As Charles Wyville Thomson remarked when he first sounded the great ocean depths in the famous **Challenger** expedition, it is almost as hard to imagine life existing in these conditions as in fire or in a vacuum. But we know today that there are forms of life—strange forms, to be sure—which thrive in the very deepest trenches of the ocean bottom." ("Animals of the Abyss," by Anton F. Brunn, in The Scientific American).

"The deep-sea angler is a bizarre food trap. It has a cavernous mouth filled with long, sharp teeth. It looks vicious. . . . Like some of its deep-sea associates the angler has a pair of large, well-developed eyes. Extending out from its upper jaw is a process (looks like an antenna) that bears a luminous bulb at its end. This bulb may serve to lure unwary animals within easy range of the angler's teeth. Hanging from the angler's lower jaw is a beardlike mass of luminous tissue, adding to its bizarre appearance." Another "angler" has a more slender body, with a "fish line" that is actually FOUR TIMES the length of the fish's body, with a luminous tip at the end of the line! This "line" grows out from the front end of the fish.

One of the most remarkable of all fish anglers is **Lasiognathus**. "This fish carries a fishing-rod armed with hooks at the end, and a light. Another fish, seeing the light makes for it and is hooked. But a fish hooked at the end of a long rod in front would be of no more use to **Lasiognathus** than a carrot at the end of a stick is to a donkey—so the rod is provided with a hinge in the middle." (P. 219, The Living Sea).

The various and sundry types of "angler" fish are obviously "designed" and so created.

The DEEP-SEA SQUID and the DEEP-SEA HERMIT CRAB. A squid that lives near the surface can eject an ink cloud to escape from its enemies. "But an ink cloud in total darkness would be wasted effort, so certain species of the squid that live in the depths eject a LUMINOUS cloud, while the HERMIT CRAB of the kind that carries two sea anemones goes one better by carrying, in the deep, TWO ILLUMINATED ANEMONES, which not only give it protection but serve as torches too." (P. 219, The Living Sea).

Some of these weird deep-sea fish have telescopic eyes, set on long stalks; certain others are "equipped with headlights like a car." These lights are placed just in front of curved, glistening reflectors near the eyes and are projected as two beams of light. Others have huge mouths, with fearsome, fang-like teeth, and some have the added horrible aspect of having illuminated teeth. In the deep sea is a peculiar "Scarlet Shrimp" that shoots forth a cloud of luminous fluid to blind its assailant. The Five-lined Constellation Fish has five rows of illuminated spots, that resemble a "pulsating aurora boreallis," on each side. The Great Gulper Eel **(Saccopharynx harrisoni)**, 55 inches long, has a flaming-red light organ near the tip of its tail." (National Geographic Magazine).

We quote a most interesting description of luminous deep-sea creatures, found on p. 218, in "The Living Sea." "The deep-sea creatures, strange enough in appearance, most of them, as they are, are also able to light themselves up. How this lighting-up is accomplished we do not really know. So to talk about the bottom of the sea being a region of everlasting darkness is not quite correct; illuminated fishes, etc., are always moving here and there. In fact, if a number of these were to gather together that place would resemble Broadway or Piccadilly Circus at night, a sort of fairyland, for many of the lights carried by the bottom dwellers are COLORED. The pattern of lighting varies. Some species have a row of lights along their bodies, others whole tiers of lights along their sides, making them look like ocean liners at night, and which they can switch on or off as they desire; some have illuminated circles around their eyes and mouths, some illuminated heads and faces, and some are illuminated all over, some glow from inside. Particularly strange are the fishes carrying a light at the end of a long rod in front of them."

There are thousands of living witnesses in the fearsome depths of the sea for God and Divine creation!

Because of the great changes in water pressure these weird deep sea fish CAN NOT LIVE near the surface; neither can surface fish live that far down. In other words evolution is ruled out, for if a surface creature descends to a great depth (excepting whales and squids, and a few others) it is crushed to death, and if a deep-sea fish is brought to the surface waters, it explodes. Obviously, it took an act of special creation to adapt these deep-sea fish to their environment. And their peculiar "adaptations" to life in the deep and their special "design" for their niche in life, give full and abundant proof of Divine Creation.

We say with the Psalmist, "Thou art the God that doest wonders" (Ps. 77:14)—not the least of which are His "wonders in the deep" (Ps. 77:14). Again we say, "Bless the Lord, all His works in all places of His dominion: bless the Lord, O my soul" (Ps. 103:22).

(10) Some of the Thousand and One other Strange Species of Life in the Seas

Two kinds of fish—photoblepharon and anomalops—carry "lanterns" which are made of luminous plants in the form of a tiny species of bacteria. Just below the eyes of the fish are receptacles especially designed for carrying the lanterns and there is a mechanism for turning the light on and off. "Divine Design" is the only answer to this phenomenon.

In the Mediterranean is found a peculiar, slender creature called the **Venus-girdle.** It looks like a ribbon of light as it glows in the water. Evolution has no adequate explanation for this phenomenon.

There is another little sea creature called the **Sea Gooseberry.** It is about the size of a sparrow's egg. At night it shines brightly, but in the daytime it is a lovely mass of beautiful colors like the colors in the rainbow. Who designed the unique **Sea Gooseberry?** What Genius made it look like a rainbow at day and a glowing variegated gem at night?

Who gave the arrow-shaped SQUID the secret of jet propulsion, ages before man discovered this principle? "The squid is a rocket. Jet-propelled by the muscular ejection of water, shaped like a rocket, it has 'vanes' on each side like a rocket, and it moves fast." (The Living Sea). Water is taken in near the front end; contraction of the body suddenly compresses this water and forces it out of a tube-like funnel, pushing the animal in the opposite direction. It determines its direction by bending the tip of its funnel.

For "sound" in fish, consider these strange facts: The trumpet fish toots like a horn; the booming whale has a love song that can be heard for miles; the taps of the drum fish can be heard at a depth of sixty feet; the singing catfish emits sounds that are deep and penetrating; the Croaking Gourami sometimes makes a purring sound. What is the ORIGIN of these abilities? The versatile God has made such a great variety of life, in the sea, as well as on the land, to give tangible expression to a wee bit of His greatness.

For "beauty" in fish, think of the exquisite "Gold Butterfly" fish from near Ceylon, with its orange-gold, dotted with black spots. Consider the Rock Beauty (**Holocanthus tricolor**) from the West Indies: it is robed in three major colors—yellow, black and scarlet. Think of the Queen Angelfish with its orange-yellow pectoral and tail fins and black ocellus on the nape. Few fish are more colorful than the Moonfish (**Platypoecilus maculatus**); "it comes in all colors of the rainbow, and in a few others nature never got around to including in the rainbow." The male of the Iridescent Barb (**Barbus oligolepis**) is a brilliant light red-brown, dark above and silvery below, with orange-red fins. The fierce little Siamese Fighting Fish, in gorgeous red, green, blue, lavender and orchid varieties, is said to be the "most beautiful" fish in the sea! WHO DESIGNED THESE LOVELY CREATURES? Who gave them their exquisite colors? The God of Glory has seen fit to put in the seas a little reflection of His infinite glory.

From whence came the unique Paddle Fish? It has an over-sized snout, a broad thin plate of bone, one-third its length, with which it scoops up mud and gravel in search for food. "Evolution" would have kept this fish eating smaller fry, the same as most other "sensible" fish do.

Where did the fish that "looks like a swimming pine cone" come from? It has recently been identified off the coast of Chile, and sent to the Smithsonian Institution. Belonging to a distinctive genus this PINE CONE fish (**Monocentris**) has been described as having "an isolated niche in ocean life." Dead or alive it looks like a pine cone. Evolutionists do not know what to do with it or where to place it.

The "Leaping Spawner" (**Coeina arnoldi**) has been described as "a real show stopper" because of its strange breeding habits. When the time for breeding comes "the male and female clasp each other firmly and leap out of the water. The eggs are laid and fertilized on a leaf, an overhanging branch or some other spot ABOVE the water level. The eggs would die if they fell back into the water, but they must be kept moist, so the male spends the two or three days it takes the eggs to hatch splashing them with water." Who taught this fish that it must keep splashing the eggs? What ever possessed it in the first place—if Evolution is responsible—to jump out of water to lay its eggs? And why didn't these eggs, when it first happened (say a half-billion years ago?) die, as they would have, if it had been a "chance" performance. WHY WOULD THAT SPECIES OF FISH GO TO ALL THAT TROUBLE OF LAYING ITS EGGS OUT OF WATER WHEN OTHER FISH HAVE GOOD SUCCESS LAYING THEIR EGGS IN WATER? It is successful and SO MUCH EASIER to lay the eggs IN the water.

Who gave the "ARCHER FISH" of the East Indies its uncanny accuracy in hitting insects with a stream of water that it ejects? "This little yellow-and-black-barred fish taxis to a position below an overhanging twig and parks there until a tempting insect settles above on the twig. Then pushing his mouth out of water, he takes careful aim and spits a stream of water at his quarry, knocking it into the water where he gobbles it up." "In the mouth of the archer fish there is a deep groove, and when the fish's tongue is placed against the roof of its mouth this groove is converted into a "blow-pipe" about 1/16th of an inch in diameter. When shooting, the fish compresses its gill covers, and water is forced under pressure into the blow-pipe. The thin rounded tip of the fish's tongue acts as a valve, and the fish can thus expel the water in a single drop, a succession of drops, or, if the valve is left open, in a continuous jet. . . . Normally, one or two of the discharged pellets of water are sufficient to bring down its prey. . . . The archer-fish nearly always scores a direct hit when its prey is within four feet." (P. 19, Nature Parade). How could mere evolution groove the mouth of the first archer fish. How did it first learn to make a "blow pipe" by arching its tongue over the groove in its mouth? Who taught it accuracy in shooting? Who gave it the wisdom and skill to use its tongue as a "valve" and so enable it to eject one drop, many drops in succession, or a stream? SUCH DELICATE ADJUSTMENT IN DESIGN, FOR AN INTENDED END, IS ENTIRELY OUTSIDE THE REALM OF CHANCE. Clearly, the archer fish was CREATED.

Nature does not lack the ludicrous nor the miraculous in its well-nigh infinite repertory of oddities. "The file-fish, who feeds among clumps of eel-grass, stands on its nose in times of danger, with fins gently waving to imitate a clump of grass. Its mottled green color matches the flora perfectly." But the wrasse, however, is perhaps "the master 'quick change' artist. It can change its brilliant colors to that of any fish with which it comes in contact. Likewise, in a twinkle of an eye, it can completely vanish from sight by taking on the color of any underwater object." (American Mercury). THERE ISN'T A SCIENTIST ON EARTH THAT CAN EXPLAIN THE ORIGIN OF THAT ASTONISHING PHENOMENON! No magician that ever strutted a stage can duplicate such astonishing legerdemain! How could senseless evolution ever produce such a miracle as the wrasse, nature's astonishing quick-change artist?

If the wrasse is gifted one way to a point of mystifying perfection, the equally odd GLOBEFISH is gifted in another—and it is just as inexplicable. "In tropical seas there is a globefish which, when inflated, resembles a miniature balloon. This strange creature of the deep dilates its gullet with air, and its body swells up in the shape of a globe or balloon. In this inflated condition it rises to the surface of the water and each passing breeze blows it along." (American Mercury). Even more grotesque are the PORCUPINE FISH. Normally, they inhabit the bottom of warm seas, but when a potential enemy approaches, things begin to happen. The porcupine fish is covered with rather fearsome-looking spines. When it is threatened by danger the fish swallows water. Or if it is near the surface at times of danger it swallows air and floats belly upward. Now it is ball-like in form, with the spines pointing outward, and it presents a difficult object for other types of fish to swallow." The mechanism to make a "globefish" or a "porcupine fish" work had to WORK THE FIRST TIME: such oddities couldn't possibly have been evolved "gradually." Creation explains these strange creatures; evolution can not account for them. Let us ask the evolutionist—From what did the globefish and the porcupine fish evolve? And how many millions of years did it take? And what sort of a creature was each while in the process of "evolving"?

The CLIMBING PERCH of Burma often leaves the water, travels inland and actually climbs trees. At each side of its head there is a built-in storage tank where it can hoard supplies of water to keep its gills moist, giving it a chance to breathe until it reaches the safety of another pool. If it is a long time in locating another pool THE WISE LITTLE FISH KNOWS THERE IS OFTEN WATER TO BE FOUND IN THE HOLLOW OF A TREE. "If its store of water is nearly exhausted, it begins to climb a tree, hoping to find water. . . . It clings to the bark with its gill-covers and uses its spiney fins to help it to climb. . . . At last the perch reaches its goal and is rewarded by finding the precious water which means new life to it." It is unreasonable to believe that an ordinary perch of the sea EVOLVED into the tree-climbing perch. Everyone knows that an ordinary perch could never develop into a tree-climbing perch, for each perch that tried it would DIE, no matter if a trillion perch made the attempt! Who first taught the climbing perch—before it became the climbing perch—that there is WATER in the hollow of some trees in Burma? Who first trained this little creature to climb trees, when, naturally, a fish is anything but a tree-climber? Who first put those water-storage tanks in each jowl? This unbelievable fish, the CLIMBING PERCH of Burma, is a perfect witness for God and Divine creation. No amount of evolutionary argument can persuade the man who thinks that such a miracle in nature developed merely through the "natural processes of evolution."

Let us consider also the BUBBLE-NEST BUILDERS (**Osphronemidae family**). "These fish are equipped with accessory breathing gear that enables them to draw oxygen from the air. They, much like birds, hatch their eggs in nests. These nests are built out of bubbles that the males supply. The male gulps a mouthful of air from the surface, darts down below to coat it with a sticky secretion from his mouth, and then releases it as a bubble. These small bubbles, loosely joined together, form a raft or floating nest to which the eggs are later attached. The eggs are laid and fertilized under the nest, and the male quickly darts down after each batch of falling eggs, catching them in his mouth. He then blows them up into the nest where they adhere and hatch in two or three days." This unusual procedure is a phenomenon that was so DESIGNED, and the actors (the Bubble-nest builders) were so created that they follow this routine by instinct. Evolution cannot account for the BUBBLE-NEST BUILDERS. It would be as easy for a tornado to produce a Michelangelo's "Moses" or "David" as for "chance mutations" to produce such a well-designed fish. Such intricate "design" demands a DESIGNER and a BUILDER, Who is God.

Chapter 9

BIRDS: "WINGED WONDERS"—
Witnesses Par Excellence for God and Creation

Perhaps in all the realm of nature there is no more forceful witness for Divine creation than birds. And here is the reason: there is a bigger gulf between reptiles* and birds than between most any two other groups adjacent in the "evolutionary ladder" —and practically all evolutionists are agreed that "birds developed from a reptilian type of animal, and the feathers probably developed from scales." Here is a summary of the argument against evolution, when considering the miracle of bird construction. Remember, evolution teaches the slow and GRADUAL change of one genus into another; how then can evolution account for all the **radical** differences that exist between reptiles and birds? How can evolution account for the complete change of the covering, from scales to feathers? And explain the many vast simultaneous changes made in body structure? For example, many of the bones of a bird are hollow, and some have air sacs, in addition to lungs. In birds heavy jaws and teeth (that would put too much weight too high and too far forward) have been removed and there is provided a gizzard that grinds the food. The gizzard is lower and farther back in the body than jaws and teeth. As we proceed with this discussion, many other radical differences between birds and reptiles and other animals will be mentioned. IF such great differences were brought to pass by gradual changes, there MUST of necessity be some evidence

---

*Evolutionists are hard-pressed to find any connections whatever between birds and their supposed ancestors, reptiles. One author says, "It is difficult to think of birds as being even remotely related to reptiles." (The Strange World of Nature, p. 52).

The **Archaeopteryx** often has been referred to as being intermediate between reptiles and birds. A careful examination, however, indicates it had very typical bird feathers, feet, and wings adapted for flight. Practically all of its aberrant features (such as clawed digits of the wings) are to be found in some form in some living birds.

somewhere of the intermediate changes—but there are none.**
A reptile is a reptile—designed and adapted for its particular
mode of life; and a bird is a bird—designed and adapted for its
particular mode of life. And there is ABSOLUTELY NO EVI-
DENCE WHATEVER OF THE SUPPOSED "GRADUAL CHANGE"
OF SCALES INTO FEATHERS, or of the development of wings,
the loss of teeth, the development of exceptional sight and the
hundred and more other colossal differences between birds and
reptiles. We repeat what we have stressed before: only a completed organism works; a partly developed organism (such as a
bird's wing, claw, bill, feather, etc.) IS OF NO VALUE WHAT-
EVER TO A LIVING ANIMAL, and such "partly developed" organisms are nowhere found in nature. Evolution exists ONLY
in the minds of its devotees. So every bird in the world, with its
symmetry, its beauty and its song, is a living witness for God and
creation, for it is a COMPLETED and PRACTICAL UNIT OF
LIFE that fits and works in its environment perfectly. And it is
so far removed from its supposed reptilian ancestors that evolutionists themselves are hard pressed to find even remote resem-

---

**J. Augusta, in "Prehistoric Animals" (p. 42) seeking to trace the
ancestry of birds, says, "the Saurian (reptilian) arche-ancestors of the birds,
which we do not yet know well . . . seem to have gone over to walking and
running on their hind legs only. **Their bodies were still covered with scales.**
At a further stage of evolution, FOR WHICH WE STILL HAVE NO PROOFS
BUT WHICH WE MUST ASSUME TO HAVE EXISTED, the 'pseudosuchian
saurian' changed into a kind of 'pre-bird'—proavis—with its scales changed
into feathers and already able to climb about on the trunks and branches
of trees. By the transformation of its scales into feathers (only in the
imagination of the evolutionist) . . . there arose in the course of the further
evolution of the proavis a kind of parachute, which allowed it to glide
smoothly down from branch to branch and from a tree to the ground.
That was the first beginning of flight, . . . with the gradual transformation
of the proavis into the archebird, and then of the archebird into the bird."

Read again the above amazing statement by Dr. Augusta, noted Professor of Paleontology. He admits there is absolutely NO PROOF of any
gradual change from "scales to feathers"—and so HE INVENTS A "PRE-
BIRD" that he calls the "pro-avis" to fill that gap! WHEN THE ARDENT
EVOLUTIONIST LACKS SCIENTIFIC FACT AND PROOF he imagines the
"missing links" and writes them in his books as though they were fact—
and our impressionable children and youth take it as gospel truth!

AS LONG AS THERE IS NO SCIENTIFIC PROOF OF GRADUAL
CHANGE FROM "SCALES TO FEATHERS" THE THEORY OF EVOLU-
TION MUST FOREVER REMAIN A THEORY, SUPPORTED ONLY BY
THE VIVID IMAGINATION OF ITS OVER-ENTHUSIASTIC ADHERENTS.

blances, much less can they trace a step-by-step development of "scales to feathers" or of solid bones to hollow bones, etc.*

Carl Welty, writing on **Birds as Flying Machines,** ("Scientific American"), sums up the "specialties" that birds have that reptiles do not have, that make birds birds:

> "Birds were able to become flying machines largely (because of) gifts of feathers, wings, hollow bones, warm-bloodedness, a remarkable system of respiration, a strong, large heart and powerful breast muscles. These adaptations all boil down to the two prime requirements for any flying machine: high power and low weight."

It has been observed many times by others that "every major transformation of an organ is, in general, correlated with a greater or lesser change OF THE ENTIRE ORGANISM. The acquisition of flight in birds, to mention a drastic case, involved A REBUILDING OF THE ENTIRE SKELETON, eventual loss of teeth, change of metabolism, change of the sense organs, of the brain, of most of the behavior patterns, etc. **The organism seems to change as a harmonious entity,** and NOT by random mutation of its parts." (Systematics and the Origin of Species, by Ernst Mayr; published by The American Museum of Natural History). In other words, slight, gradual, random mutations do NOT account for such drastic changes involved in "the acquisition of flight in birds," for, to be successful, **the entire body had to be rebuilt at the same time in order to make flight possible!** The phenomenon of radical changes, such as the development of flight in birds, precludes the idea of gradual change by random mutations. The only way a bird could possibly come to being is by a SUDDEN CREATION; so the gradual change from reptiles to birds is completely ruled out as an impossibility.

Let us further consider in detail how birds are witnesses for God and Creation.

---

*Because of their traditional belief that evolutionary changes are a "SLOW PROCESS" evolutionists themselves find it hard to believe their theory. Ernst Mayr, writing in "Systematics and the Origin of Species, from the Viewpoint of a Zoologist" (American Museum of Natural History), says, "It must be admitted that it is a considerable strain on one's credulity to assume that finely balanced systems, such as certain organs (the eye of vertebrates, or bird's feathers) could be improved by random mutations. . . . However, the objectors to random mutations have so far been unable to advance any alternative explanation that was supported by substantial evidence." WHY DO THEY INSIST ON REJECTING THE TRUE EXPLANATION OF DIVINE CREATION!

(1) **Birds are "Miracle" Creatures that give most forceful evidence of Special Design; they are the work of a Master workman.** Note:

(A) **The bones and skeleton of birds**

A mammal bone is heavy, dense; but the bones of a bird are hollow, filled with spongy network and engineered for air capacity and strength. As a bird breathes, it is inflooded with air to its very marrow! The air cavities in the bones are directly connected with its lungs. Yet, strength has not been sacrificed, but the light, hollow bones are stiffened with ridges, where needed, according to advanced engineering principles. We have before us a drawing of a longitudinal section, showing the internal structure of the metacarpal bone of a vulture's wing. "The braces within the bone are almost identical in geometry with those of the Warren truss commonly used in steel structure."

"Combining both lightness and strength, surely the bones of a bird could not have been more wonderfully engineered." (Eugene Burns, Ranger-Naturalist).

"Although a bird's skeleton is extremely light, it is also very strong and elastic—necessary characteristics in an air frame subjected to the great and sudden stresses of aerial acrobatics." (Carl Welty, in "Birds as Flying Machines," in "The Scientific American"). Mr. Welty in his article shows a picture of a cross-section of the frontal bone of the skull of a crow, revealing the hollow bone, with a marvelously intricate and obviously designed braced interior. The result is, says Mr. Welti, "The skull of a crow achieves the desirable aerodynamic result of making the bird light in the head. Heavy jaws are sacrificed: their work is largely taken over by the gizzard. The skull of the crow accounts for less than 1% of its total weight."

Evolutionists recognize the difficulty of accounting for the phenomenon of the bird's light bone structure. C. H. Waddington, writing in The Scientific American, says,

"There are adaptations of such a kind that it is difficult to see how they could ever be responses to external circumstances. For instance, birds tend to have hollow bones, by which they gain in lightness without losing strength. It is impossible to see how external conditions could directly produce hollowness of bones."

If they would only acknowledge the fact of a Divine Designer and Creator, their many problems would be solved!

(B) **The feathers, wings and flight of birds**

A bird is actually a "living airplane." "It flies by the same aerodynamical principles as a plane," says **John H. Storer** ("Bird

Aerodynamics," in The Scientific American); "and uses much of the same mechanical equipment—wings, propellers, steering gear, even slots and flaps for help in taking off and landing."

"Where is a bird's propeller?" continues Dr. Storer. "Astonishing as it may seem, every bird has a pair of them. . . . They can be seen in action best in a slow motion picture of a bird in flight. During the downward beat of the wings the primary feathers at the wing tips STAND OUT ALMOST AT RIGHT ANGLES TO THE REST OF THE WING AND TO THE LINE OF FLIGHT. These feathers are the propellers. They take on this twisted form for only a split second during each wing beat. But this ability to change their shape and position is the key to bird flight. Throughout the entire wing beat they are constantly changing their shape, ADJUSTING AUTOMATICALLY TO AIR PRESSURE AND THE CHANGING REQUIREMENTS OF THE WING AS IT MOVES UP AND DOWN. This automatic adjustment is made possible by special features of the feather design. The front vane of a wing-tip feather is much narrower than the rear vane. Out of this difference comes the force that twists the feather into the shape of a propeller. As the wing beats downward against the air, the greater pressure against the wide rear vane of each of these feathers twists that vane upward until the feather takes on the proper shape and angle to function as a propeller. . . . (So) with their specialized design the primary feathers are beautifully adapted to meet the varied demands of bird flight."

That is a rather long quotation, but we thought it important—because it shows MARVELOUS DESIGN for an intended purpose. A bird's wing is self-adjusting, as though it were controlled by a highly complicated, automatic electronic machine that reacts in a thousandth part of a second! Honestly now, could such an intricate, complicated, self-adjusting arrangement in the wings and feathers of a bird, that make flight possible, come to pass by "random mutations"? or, must it have been DESIGNED and CREATED by an infinitely wise and all-powerful Creator?

The feathers are miracles of ingenuity. Allen Devoe, writing on **The Miracle of Birds** (American Mercury, Oct. '53), says,

"A feather may seem to be only a central shaft with projections on either side. It is much more. Each projection (called a vane) from the feather stem is composed of numbers of parallel rods, the barbs. A barb is itself virtually a complete miniature feather, with extremely fine side-projections called barbules. Look still closer with a lense and it is revealed that on these barbules are tinier barbicels, and on these are almost infinitesimal hooklets. The hooklets mesh the barbs; the whole vane is one light, perfect interweave. Barbules and barbicels **on a single feather** MAY NUMBER OVER A MILLION!"

No wonder Elliott Coues, the famous ornithologist, said, "A bird to me is as wonderful as the stars!"

Every feather is a mechanical wonder. The quill is strong,

light, hollow, tough, elastic, and tapers to a fine point with geometrical precision—exactly what is needed.

The miracle of a bird's feathers is further seen in this quotation:

"Feathers, the bird's most distinctive and remarkable acquisition, are magnificently adapted for fanning the air, for insulation against the weather and for reduction of weight. It has been claimed that for their weight they are stronger than any wing structure devised by man. . . . When a bird is landing or taking off, its strong wingbeats separate the large primary wing feathers at their tips, THUS FORMING WING-SLOTS* which help prevent stalling. It seems remarkable that man took so long to learn some of the fundamentals of airplane design **which even the lowliest English sparrow** demonstrates to perfection." (See "Bird Aerodynamics", by John H. Storer, Scientific American, April, 1952).

"Beside all this, feathers cloak birds with an extraordinarily effective insulation—so effective that they can live in parts of the Antarctic too cold for any other animal.

"The streamlining of birds of course is the envy of all aircraft designers. . . . The feathers shape it to the utmost in sleekness."

Dr. Gray, writing on "The Flight of Animals," in the 1954 Annual Report of the Smithsonian Institute, p. 290, says,

"A bird's front limbs have been COMPLETELY specialized for flight." (Caps ours; note, the transformation—according to evolution—from legs to wings is COMPLETE, not partial). "Each wing forms a structure of peculiar beauty and complexity. . . . UNLIKE THAT OF ANY OTHER FLYING ANIMAL, the wing surface in a bird is made up of feathers, all fitting together to form an efficient lifting surface and yet capable of being neatly furled when not in use." Again we see, from expert authority, that birds' wings are specialized organs that are COMPLETE and perfectly designed for their intended use, with no evidence whatever of being in the process of gradual change from one form of life to a higher form. We ask the evolutionist, WHEN and HOW did this transformation take place from legs to wings, and **where is the evidence of transitional forms?** There is NO evidence of transitional forms from saurian (lizard-like) limbs to birds' wings, save in the imagination of evolutionists!

Birds are by far the fastest creatures on our planet. The streamlined peregrine falcon can dive on its prey at speeds up to

---

*Many birds have on their wings a little group of feathers known as the alula or 'bastard wings.' These come into operation . . . when the bird is in danger of losing lift or stalling. The 'alula' then acts as a safety device. Actually, it was not until Sir Frederick Handley Page invented the now famous 'slotted wing' anti-stalling device, which has done so much to make airplanes safer, THAT IT WAS REALIZED THAT BIRDS HAVE HAD IN THE ALULA THE SAME SAFETY GEAR FROM TIME IMMEMORIAL." (P. 193, "Nature Parade.") (Caps ours.)

180 miles an hour (some authorities say 250 miles an hour). And yet their great speed is under perfect control! The African eagle, swooping down at its prey at a speed of over 100 miles an hour can brake with "such stunning skill, by spreading wings and tail in an aerial skid-stop, that it comes to a dead halt in the space of 20 feet!"

Authors wax eloquent indeed as they describe the wonders of bird's wings and bird flight.

Actually, hundreds of "special adaptations" in as many different birds, have been observed by naturalists. For example, the wing and tail feathers of most owls are covered with a soft pile—an effective "silencer" equipment: quite necessary when one remembers that a large part of an owl's diet consists of mice, whose ears are very sensitive, hence the owl's need for silent flight becomes apparent. Incidentally, the Indian fishing owl **(Ketupa)**, which lives primarily on fish, does NOT have this "silencer."

Note again, the shape of a bird's wing is clearly related to its habitat and manner of life. Thus the 11-foot-long wing-span of the wanderer albatross, makes it one of the most efficient soaring birds in the world; the albatross lives in regions where there is always a strong wind to enable it to rise. But such wings would be useless where there is not a prevailing wind. In fact, the albatross is so poor at taking-off that it can with difficulty get off the ground without the aid of a wind. On the other hand, birds which live among trees or underbrush have, of necessity, short rounded wings. What they lose in flight ability is compensated for in safety: long wings would get caught in branches more readily, and lead to their destruction. ALL THINGS IN NATURE REVEAL THE PROVIDENTIAL CARE OF THE HEAVENLY FATHER: He made even the birds so that they could live and thrive in their peculiar habitat.

(C) **The sight and hearing of birds**

An owl scans the dark woods with eyes ten times as sensitive to faint light as ours. Most birds have prodigious eyesight. In some birds the eyes are so big in relation to the head that there is scarcely room for them in the skull! Nature has also endowed them with a third eyelid that can be drawn back and forth across their eyes as a "windshield wiper" as they rush through the high sky, constantly encountering bits of dust and other irritants. One of the outstanding miracles of the eyes of birds is their remarkable "telescopic adaptability" for rapid adjustment.

The swallow, darting swiftly through the air, is able to see the tiniest insect as it swoops down through the sky. A bird of prey, even at high altitude, can perceive a small object far below and in its lightning descent (its eyes constantly change focus) so that it is able to snatch its prey without a crash landing." (P. 269, "Miracles of Science").

The robin "has unbelievably acute hearing. When a robin on your lawn stops and cocks its head to one side, it is **listening** to the soft stirring of an earthworm under the grass."

(D) **The legs and feet of birds**

The legs and feet, including claws, of birds show as much design for intended purpose as every other feature of their anatomy.

A bird cushions its landing with its legs, which consist of three single rigid bones, with joints that work in opposite directions—thus making an amazingly efficient shock absorber.

Many people wonder why a sleeping bird does not topple off its perch.

"Attached to the ligaments which operate a bird's toes, is a very long tendon which runs nearly the whole length of the leg, and broadens into a muscle on the front of the thigh. When the bird perches, its knees and ankles bend and automatically tighten the tendon, which contracts the bird's toes so that they grip tightly. The bird is then virtually locked to its perch." Who invented this natural "safety lock mechanism" so obviously designed for the benefit of the sleeping bird?

The legs and feet of birds are designed for perching, running, swimming, wading, climbing, scratching, tearing, or holding. Birds of prey have strong feet, armed with sharp, hooked talons. The foot of the climbing birds, like the parrot and the woodpecker, is equipped with two toes in front and two behind. Scratching birds, like chickens and turkeys, have short, thick toes, fitted with stout, blunt claws. The wading birds, as the crane and heron, are long and slender-legged for walking in the water of lakes and marshes. Many of the swimming birds, like ducks and swans, have webbed feet. God is very wise: He equips every creature perfectly to meet its needs—even to the "fur"-covered foot of the ptarmigan, to protect it against the extreme cold.

The thick foot of the ostrich is a weapon of defense; on the other hand the peregrine (falcon) has a hind toe like a steel spike, with which it knocks its prey senseless when it hits it in its power-dive, a lightning-like "stoop" from above.

The Remarkable Feet of the Jacana

Perhaps in no other bird's feet is "design" for an intended purpose so evident as in the Jacana.

"The jacana has most remarkable feet. It has very long spreading toes which are exceptionally slender and weak. At first glance it would seem that Nature erred in giving this creature such freakish equipment, but she didn't. The jacana spends most of its life stepping from one floating lily pad to another in search of food. Its outlandish feet distribute its weight evenly over the wide surface of the pads, enabling them to support the bird." In all seriousness, HOW could a bird with short, stubby feet EVER develop the long, slender feet necessary to walk on lily pads? Every time a bird with short stubby feet tried to walk on a lily pad, it would sink, and the poor thing would die of frustration in less than a week—if it did not drown before that!

And so the jacana, unintentionally, becomes another witness for God and divine creation, for it is clear to all that the feet of the jacana HAD to be as they are, from the very beginning, in order to do what the jacana does—walk on lily pads. FEET ANY LESS THAN OR ANY DIFFERENT FROM WHAT THE JACANA HAS WOULD NOT WORK AS THE JACANA USES THEM. No theory demanding "gradual change" by chance mutations can account for a highly specialized organ as the feet of the Jacana.

Speaking of the special equipment that a duck has, one author says, "Notice the feet of the duck: at the end of each leg he has an ingenious paddle, or oar, to drive him through the water. Either on the surface or underneath, the duck is able to proceed because of his webbed feet. The question arises, did he take to the water because he had webbed feet, or did he get webbed feet because he took to the water? It is evident that the latter **cannot be the case**—for what would the duck have done in the water while he was getting or developing those webbed feet? Also, the down on a duck that keeps him safe in the water MUST have been provided him in the very hour of his origin, or the water would have been as fatal to him as it is to a chicken today!"

(E) The bills and beaks of birds

The bills or beaks of birds are very efficient, and carefully designed devices for obtaining food, and for protection, in some cases. A beak consists of an upper and a lower mandible, or jaw. Birds which live on tough-shelled nuts usually have strong, heavy bills so they can crack the shell. Birds like hawks, owls and eagles have hooked bills so they can tear their prey apart. Scavengers, like vultures, also have hooked bills, but they are much weaker, for the flesh of a dead animal will tear off much more easily than that of a live one. Boring birds are furnished with a long, straight and pointed beak with which to dig into bark in search of grubs and insects. The goose and duck are furnished with a spoon bill, suitable to get food from the bottom of lakes and pools.

Let us mention a few of the strikingly different kinds of bills, among the many hundreds, that birds have—and note especially that they are designed for a PURPOSE. The bill of the curlew bends down, that of the avocet curves up, whereas that of the snipe is almost straight. The bill of the stork is pointed, that of the spoonbill is flat, and that of the flamingo has a sharp, right-angle bend in it. The beak of the falcon is hooked, that of the touraco is short, that of the adjutant is long, that of the toucan is enormous and that of the pelican carries a pouch underneath. Who is responsible for this variety? Consider the fact that the beak of a bird is designed for a purpose: that it might eat a particular kind of food. Consider:

The bill of the **northern shrike** is hard and used as a hammer. He kills mice and small birds by giving them a sudden blow with his bill on the back of their heads.

The **woodcock** lives largely on worms. He has a comparatively long bill that is flexible at the end; with this he probes into worm holes in search of his food. The flexible end of his bill enables him to probe readily until the worm is discovered.

The long-billed **curlew** has a beak well suited to drag crabs and worms from their holes in the sand.

The **woodpecker finch** of the Galapagos has a most curious method of obtaining its food. It picks up a long, thin cactus thorn in its short bill—admirably adapted to holding the thorn—and with it pokes out insects hiding in crevices of bark and wood! When the insect runs out of hiding the bird drops the prod and eats the insect!

With its arched, blunt beak, the flaming **ibis** dredges its food out of mud banks.

Of all bills, that of the **pelican** is one of the queerest. Diving for fish, it uses its more-than-foot-long bill like a mechanical scoop. When the fish is caught, it is stowed away in a fleshy, pouch-like sack that extends between the two sides of the lower mandible. From this reservoir the pelican swallows the fish at its leisure. But the pouch serves yet another purpose. The pelican partially digests its food, then regurgitates it into this same pouch. Then the young pelicans eat right out of this pouch, when father or mother (both parents take turns feeding their young) opens its mouth for "junior" to get his meal! What bird would "invent" such a system to feed its young? But God has plans of His own, and these are seen in nature on all sides, reflecting the fact of special design in creation.

The **tooth-billed pigeon** of the Samoan Islands has a highly specialized bill that has notches like teeth in the lower mandible. It feeds mostly on the fruit of a fig tree, and this bill is admirably designed for that purpose.

The **shoe-bill stork** has a great, broad bill, depressed in the middle and hooked at the end—suggesting a large wooden shoe. The stork, you will recall, is a voiceless bird. But the shoe-bill stork claps its mandibles together, and so expresses itself in times of danger or excitement! Such phenomena in nature are NOT the result of "survival of the fittest"—for certainly a "voice" is an asset. But this queer organ (the shoe-bill) was so designed by the Great Creator who fashioned many kinds of life to carry many lessons to the world of men.

The active **hummingbird** has a long, slender bill, that serves as a drinking straw to extract nectar from the long "throats" of flowers.

The **flamingo** has a built-in sieve in its bill with which it sifts small shellfish and other tidbits from the mud of shallow water.

The **plant cutter** birds (**Phytotomidae**) have conical bills that have fine saw cuts along the edges of their mandibles, and with these cutting edges they cut off pieces of leaves, buds and fruit for food.

The **wood hewers** of Central America have bills that curve downward and are long and slender. With these they search for insects and larvae in the cracks in bark and in tree crannies—and so the Creator of all, Who has adjusted ALL life and made it interdependent, provides a special bird as a "tree surgeon" to protect trees from the ravages of insects! Did such a provision of benevolence for trees "just happen"—or was it all in the original blueprint, in the overall plan of the Creator.

The **woodpecker** lives chiefly on insects lodged in the bodies of the trees (often in decayed parts). Its bill is straight, hard and sharp—like a chisel—so it can dig and bore after the insects.

The **gannet,** which feeds on fish, has the sides of its bill irregularly jagged in order to hold more securely its slippery victims.

The mandibles of a **heron** are long and pointed, and the beak is especially suitable for spearing small fish and frogs in shallow water.

The snipe has a long soft bill, with a nerve going to its end, giving it feeling. The tip of the bill is moveable. Because the bird cannot see down in the mud, it must depend on this type of a bill to locate worms, its food.

The Strange Bills of the Nuthatch, Bower-birds and Crossbills

Of all the hundreds of types of bills, we consider these three among the most peculiar.

The nuthatch will wedge a nut in a crack of the bark of a tree. Pivoting on its legs, it strikes the nut with the full force of its body with its beak, which serves as a hatchet—and it certainly knows how to use its hatchet bill to best advantage! While opening a nut "it almost seems to prefer to hang head downwards, probably because this position adds power to its strokes."

Two species of **bower birds** actually paint the twigs and grass stems that form the walls of their bowers. One of these, the **spotted bower,** was filmed in the act.

"The paint used was chewed-up grass mixed with saliva, and the actions of the bird when painting were . . . as follows: The bill, exuding paint, was wiped repeatedly with short jabs, first on one side and then on the other, on the stems forming the bower walls." (The Strange World of Nature, p. 109).

The **crossbill** "has one of the strangest tools (its bill) in nature, and surely one of the most specialized." Its peculiar crossed mandibles are used to pry apart the cones of certain pine trees. Here is how it is done:

"The bird inserts its opened bill under the scale of a fir cone and levers it up with a lateral movement. The lower mandible, which is applied to the body of a cone, acts as a fulcrum, while the upper part of the crossed bill does the work. While the scales are held apart in this manner, THE SINGLE SEED, with its delicate wing attached, is then removed by the tongue." (Ibid.).

If evolution, with its need for countless ages of time, had to be depended on, the poor crossbill (before it was a crossbill) would have starved to death a million times over—that is, if it had to depend only on the seeds hidden in pine cones. And if it lived on other types of seeds, it would never need its crossbill! This highly specialized organ (the bill of the crossbill) had to be made **AS IT IS,** at once, to work as it works.

Let us raise one more question relative to birds' bills. How does the evolutionist explain the outlandishly large bill of the South American **toucan?** The bill of the toucan is so large it makes the bird look ludicrous. It is about half as big as the bird itself, and seems unnecessary—except that the toucan is expert at catching fruit tossed to it. Fortunately, though the bill is gigantic and looks heavy, it is actually hollow and light, supported by an interior network of interlacing bony fibers.

How account for this awkward bill? It certainly is no advantage, in this instance, as far as eating is concerned, for the toucan lives mostly on fruit. Did the Creator want to make a bird bill **obviously not** designed to give the bird an advantage in eating habits? God is **sovereign** in His creative activities. "Natural selection" and "random mutations" leading to advantage do NOT explain the bill of the toucan—but divine creation does!

(F) Other unique features of the anatomy of birds

The heart and lungs of birds are truly phenomenal. The heart of the bird is the largest in proportion to its body size of any animal, and its rate of beating is far more rapid than man's—sometimes as high as 600 beats per minute! Through the bird's heart is pumped "the richest blood in the world"—i.e., blood with the highest count of oxygen-carrying red cells. A bird maintains a very high temperature (about ten degrees higher than man's) which assures a steady flow of energy, regardless of weather conditions. And this, incidentally, makes the bird ravenous, which in turn means that each bird will consume enormous quantities of insects, and so keep down insect pest populations.

"The lungs of man constitute about 5% of his body volume; but the respiratory system of a duck, in contrast, makes up 20% of the body volume (2% lungs and 18% air sacs). The anatomical connections of the lungs and air sacs in birds seems to provide a one-way traffic of air through most of the system, bringing in a constant stream of unmixed fresh air, whereas in the lungs of mammals stale air is mixed inefficiently with the fresh. IT SEEMS ODD THAT NATURAL SELECTION HAS NEVER PRODUCED A STALE AIR OUTLET FOR ANIMALS. The air sacs of birds apparently approach this ideal more closely than any other vertebrate adaptation." (Scientific American: article by Carl Welty, on "Birds as Flying Machines.")

Mr. Welty says that the respiratory system of birds is far superior to that of mammals—and wonders WHY evolution (natural selection) did not do as much for mammals as for birds!

This is indeed a difficult problem for the evolutionist, Mr. Welty, but it is not difficult for those who believe that all nature is the handiwork of the Sovereign, all-wise God, who gives to each form of life blessings and abilities best suited to their . . . status and function in life.

Another writer calls attention to the "super-efficiency of the bird's respiratory system." Because the air passes through the air sacs as well as through its lungs, "the bird gets oxygen when it inhales and also when it exhales, because the air passes through the lungs to the air sacs and, on its return, again passes through the lungs. The lungs consequently receive two doses of oxygen." So DIVINE DESIGN works wonders for birds that evolution is unable to achieve for more advanced forms of life—according to evolution's adherents.

A thousand and one "miracles of anatomy" could be cited that make birds among the most marvelous of all God's creatures. Some of these miracles of construction are:

The tongue of a woodpecker. To extract grubs from trees, a woodpecker has a tongue so long it curves over inside the bird's head and is actually anchored, not in the throat where one would expect, but IN FRONT OF ITS EYES, to give it more length!

The uncanny time-sense in many birds. Many coastal birds have a built-in sense so precise that after inland trips they can return to shore for feeding AT THE EXACT HOUR WHEN THE TIDE IS RIGHT.

(2) **Birds' Eggs give most convincing Evidence of Special Creative Design**

An egg looks simple enough—but it is "incredibly complex," from the air space at its end to the twisted cords that suspend the yoke in perfect tension at the egg's center. The yoke is something like a boat: it is lighter at the top where the germ cell is. No matter what way the egg is turned, the germ cell, being in the light top section, is **always** on top, near the warmth of the mother's breast! The egg shell has tiny funnel-shaped pores that let the embryo breathe. If you varnish an egg, the embryo dies because it needs oxygen that seeps through the pores of the egg.

A baby chick starts to breathe with its lungs two days before it is hatched. There is enough air in the little air space at the end of the egg to keep the chick breathing for just two days. Then, when the air runs out, the chick jerks its head, and what would seem to be its death struggle, gasping for breath, **proves to be the needed agitation of its head, with the temporary hard cone** on its soft bill, that breaks the egg shell, and lets the chick get out of its shell! SUCH AN INGENIOUS ARRANGEMENT THAT CAUSES WHAT WOULD APPEAR TO BE A DEATH STRUGGLE TO TERMINATE IN LIBERATION AND LIFE, IS THE WORK OF AN INTELLIGENT BEING OF VAST RESOURCES OF THOUGHT AND ACHIEVEMENT. To believe that such a thing could happen by "chance" is like believing one could pour a bag full of alphabet letters in a kettle of soup and from the mix pour out a "Gettysburg Address" or an Essay by Emerson.

The Master of all life has so created its various departments as to preserve what we have before referred to—"the balance of nature." This carefully DESIGNED "balance of nature" can be traced to the very origins of life: eggs; for the NUMBER of eggs wild birds lay varies from one to thirty each season. Birds which build their nests in protected places usually lay **few** eggs; on the other hand, domestic fowl, whose eggs are used by man, lay **many** eggs—obviously so designed for man's benefit! A quail, whose nest is on the ground where it is subject to more hazards, will lay up to thirty eggs; while the eagle, whose nest is on a high cliff, or in a tall tree, lays only two eggs.

Even the color of eggs is well planned by the Supreme Architect—and all with a purpose in view. Woodpeckers lay white eggs; most other birds lay eggs with colored or spotted shells. The nests of woodpeckers are in the dark hollows of trees, and white eggs are more easily seen in a dark place when the mother bird returns from the bright sunlight of the open. Birds which lay eggs in open nests on the ground usually lay eggs **with brown spots.** This makes them look much like stones and clumps of dead plants that surround the nest. Did the birds select the color of the eggs they lay? Of course not. Who did then? The Great Designer, who wisely created all things. The red-wing blackbird, which builds its nest in bushes near water or in the tall grasses in the marsh, lays three to five pale-blue eggs, streaked with purple—obviously camouflaged to make them hard to find in their natural surroundings. Did the rid-wing blackbird think up this special color design for its eggs, or did the Creator so plan it to give added protection to the species?

The Egg of the Murre

The egg of the murre is distinctly **pointed** at one end—and there is a reason for its odd shape. The egg of the murre is often laid on a narrow rock shelf high above the sea, along the coast. When the wind blows across the rocky shelf, **the egg rolls in a small, tight circle,** with pointed end inward. Even a strong wind will make the egg spin, but it will NOT roll off the ledge! Obviously, the murre did not decide what shape it wanted its eggs to be. An Intelligence outside the bird did all the designing.

(3) **Bird's Nests give convincing Evidence of Special Creative Design**

Of God's creatures, birds show more "personality" than most others. Their songs, their distinctive beauty, their very nests, reveal individuality of a high order. And the nature of birds is as different as that of individual men: some kinds of birds are "cross," some are cheery, some are lazy (e.g., the cuckoo), some are fierce, some are gentle—gentle as a dove—and some are industrious, bundles of energy, as the hummingbird. Who gave birds their distinctive personality? If you give evolution the credit, WHAT caused one kind to differ from another when both kinds live in the same environment?

Let us consider the miracle of VARIETY in bird's nests. This is a subject of surpassing interest, and causes us to fall in love even more with these fascinating creatures of God's handiwork. Each species has a characteristic nest! And the young birds need never be taught how to build their nest.* Who teaches the young bird to build a nest, according to an established pattern? And why do they never deviate from that pattern? Though there are literally thousands of types of birds' nests, some simple, some complicated, there is NEVER any deviation from established style! This is most amazing, and is evidence of Divine creation, certainly not of "evolution."

God made mention of birds' nests in the Bible. This shows the Divine interest in maintaining "balance in nature," for if mother birds are not protected, and are indiscriminately killed, the **insects will greatly multiply,** for birds keep down insect populations. Here is the Biblical reference to birds' nests:

---

*All authorities agree, "The first nest a bird ever builds is just as expertly constructed as any made thereafter."

"Four generations of weaverbirds were bred under artificial conditions in which they never saw a nest or nest material. Then the FIFTH generation of the birds were set free. **At once** they began constructing with unerring skill the complex woven nests of their ancestors!" (Alan Devoe).

"If a bird's nest chances to be before thee . . . and the dam (mother bird) sitting upon the young, or upon the eggs, thou shalt not take the dam with the young; but thou shalt in any wise let the dam go . . . that it may be well with thee" (Deut. 22:6, 7).

Nature (the work of God) has been careful to protect birds' nests, not only by obvious efforts to camouflage, but also by keeping the female birds plain-colored, drab or brown—while their mates may be dazzling red, yellow, blue or white. The reason is clear: it is usually the female bird that sits on the eggs—so nature keeps the female bird colors plain, to protect the mother, her young and the eggs. Surely the Hand of God can be seen in such obvious provisions in nature! For the same reason eggs frequently are given protective coloring.

"The speckled eggs of the piping plover, laid in a slight hollow of a beach, are almost invisible to the passerby because of their coloring and pattern, WHICH MAKE THEM APPEAR TO BE A PART OF THE SHELL-STREWN GROUND." They were planned that way, don't you think?

As we describe some of the more interesting birds' nests, note how many of them are intended to **conceal** the nest or camouflage it or merge it into its surroundings—all, of course, for the protection of the birds and their eggs and their young.

The PLACES where birds build their nests vary greatly. Most nests are built in trees or bushes and are shaped like a cup; but some choose the grass, others sand, or even rock, some dig into a clay bank, others select the eaves of houses, some build inside the trunks of trees, and still others choose dark and foreboding caves.

There is also great variety in the selection of MATERIALS out of which they build their nests. Some use sticks laid crosswise. Others use grass, stems, roots, leaves, moss, yarn, feathers, horsehair, and even mud.

The examples of peculiar birds' nests we give here will bear out this statement of Alan Devoe:

"Birds' nests are often so elaborate that it is almost impossible to believe such skill can be instinctive."

Some birds, such as the megapodes of Australia, lay their eggs in sand.

In Australia also lives a strange race of birds called "mound builders," or "brush turkeys."

"The cock Brush turkey in spring walks backward in circles, kicking the fallen leaves as he goes until he raises a mound at least six feet high, and many yards in circumference, and often weighing as much as FIVE

TONS. In this heap several hen turkeys lay their eggs, and there the eggs stay until the heat of the sun and the warmth of the rotting rubbish hatch them—just as alligator's eggs are hatched."

Who CREATED the type of egg that would hatch under such adverse conditions ? And who taught the father and the mother bird their respective duties—the one to make the huge mound, the other to lay her eggs in it? Did such a devious scheme of hatching "just happen?" Of course not; it was so "planned."

We might mention further, concerning these "mound" nests, that "the birds have to lay their eggs (in these mounds) at a time when the temperature is remarkably uniform"—otherwise, the eggs would never hatch. Who teaches the birds the need of selecting the proper TIME OF YEAR to lay their eggs in "mounds"?

Emperor penguins have the most unusual nest of all.

"The single egg rests on top of the bird's feet, tucked under a feather flap that hangs down from the lower belly. Before going to the ocean to feed the incubating bird stands close to its mate; the egg is transferred to the mate's feet and tucked under a flap there!"

Who put that flap on BOTH father and mother penguin, that they might take turns going to the ocean to feed? This is only one of a million evidences that God provides for all His creatures. He so made them that they would not lack food.

"Behold the fowls of the air: for they sow not, neither do they reap, nor gather into barns; yet your HEAVENLY FATHER FEEDETH THEM." (Matt. 6:26).

The nest of the FLYCATCHER is a beautiful, symmetrical structure: it resembles the horn of plenty.

One of the most remarkable of all nests is that of the TAILORBIRD. These birds actually sew large leaves together with fibers —using their beaks as needles!

The long-tailed TITMOUSE builds a bottle-shaped nest, skillfully woven from the cotton-like down of the willow. The PENDULUM TITMOUSE goes one better, and suspends its nest from a flexible willow branch!

The nest of the BAYA SPARROW is designed to give the parents a chance to think things over if they quarrel! The nest is built with TWO entrances, one for mother and one for father, and the nest contains separate rooms! So, when they are not on speaking terms, they can pout it out in the seclusion of their own rooms!

The CHIMNEY SWIFT builds a nest of twigs by gluing them to the inside of a hollow tree or a chimney not in use. He pastes the nest to the wall with a sticky material from his mouth. WHERE DID HE GET THE GLUE FACTORY IN HIS MOUTH? And why does the chimney swift have such a glue factory and other birds do not? To try to explain this phenomenon by "natural selection" or "random mutations" is ridiculous.

The rufus FANTAIL makes a curious nest at the fork of two tree branches; it decorates its nest by attaching a tassel that dangles from the bottom of the nest—and the next generation, and the next, and the next, WILL BUILD THE SAME TYPE OF A NEST! No one knows why, except—GOD MADE THEM SO!

The red OVENBIRD of Argentina builds nests of mud, mixed with a binding material. It builds a domed nest out of mud, with hair and rootlets added. It constructs an entrance chamber on one side, then a curved passage to an inner chamber which is lined with dry grass. Here it lays its white eggs.

The Superb Achievement of the Oriole

The Baltimore oriole builds a deep, purse-like nest that dangles lightly from the descending outermost twigs of a tree. Its nest "is among the most complicated known to man." Many thousands of shuttlelike movements of the oriole's bill are needed to produce the thousands of stitches, knots and loops found in the average oriole's nest. When completed, the nest looks like a small hammock. It is so well made that it often hangs on a tree for years without being broken by winds or storms.

But the KINGFISHER, we are told, "through thousands and thousands of years" of the "workings of evolution," came to a different conclusion as to the relative merits of nests and nesting. He decided to build a nest in the side of a bank; so he digs a tunnel **four to twelve feet into a bank along a stream.** At the end of the tunnel he builds his nest—out of fish scales and bones that he has partly digested and regurgitated!

But the saucy WOODPECKER frowns on such an idea as building his nest in mud. So he chisels out a hole in solid wood (and he has the beak to do it with) "as round as if measured with a compass." First he goes downward at an agle for about six inches, and then he goes directly down for about ten inches more. He is careful, while building his capacious home, to carry the chips away from the tree and scatter them at some distance to divert suspicion. AND THE NEXT GENERATION OF WOODPECKERS WILL FOLLOW THE SAME PATTERN! One can readily see that the woodpecker was MADE to dig a nest in wood, and the oriole's bill was MADE to weave with, and the kingfisher's beak and feet were MADE to dig with! The problem is as simple as that: each species is MADE to perform as it does. Why complicate the matter with much mystery and theorizing? Why not accept the simple obvious fact that WHERE THERE IS CLEAR EVIDENCE OF DESIGN FOR AN INTENDED PURPOSE A DESIGNER DID THE DESIGNING. The only Designer who could create such an involved system of life as we find in this world is the Almighty God.

Consider next this phenomenon: Without "overseer or master" a colony of a hundred to two hundred African WEAVERS will get together and build a gigantic, mushroom-shaped mass which turns out eventually to be a veritable "bird apartment house." Each individual nest in this huge conglomerate mass is entered from below. Future generations may add to the mass, until eventually "the whole thing may collapse from sheer weight and crash to the ground—tree and all!"

The Henpecked Female HORNBILL

We speak—that is, we men do—of the "henpecked man." But nature provides the spectacle of a "henpecked female." (The phrase doesn't sound right; but anyhow, you know what we mean).

"The male of the African HORNBILL walls up the opening to the nest in a hollow tree with mud, until ONLY THE FEMALE'S BILL CAN BE PROTRUDED. He keeps her locked up in the small nest; he then proceeds to bring her all her food, and likewise that for the young later on. And she remains a prisoner there UNTIL HER YOUNG ARE NEARLY FULL GROWN."

Most females are patient, but not THAT patient, by nature. How, in the course of a thousand million years, did that male ever persuade his wife to submit to SUCH tyranny? And why, in the name of self-preservation, did he WANT to assume the task of providing food daily for his imprisoned wife and children, when at times the task wears him out to the point of complete exhaustion, and even death? Such oddities in nature just do not make sense, until we realize that GOD MADE THEM SO, for reasons best known to Himself! "Natural evolution" could not persuade a female to permit herself to be penned in like that! Such freaks in nature as the "imprisoned female African hornbill" can not be accounted for other than on the basis of Divine Creation.

We could write page after page on such birds' nests as the swallow's nest of mud, built under the eaves of barns; of the "expanding nests" of some species of hummingbirds who use spiders' webs in the construction "so their nests will stretch with the growth of their young!" Of the nest of the Toddy bird that looks like an inverted gourd; and that of the esculent swallow that looks like a miniature canoe! Such miracles are more than strange; they are prevalent in nature to attract our attention to this fact: there is a Master Designer Who has worked in the shadow **behind** the strange workings of "nature."

(4) Peculiar Birds that Defy the Rules, and bear Witness to God and Creation

Among the "roughly 25,000 species and subspecies of birds" (Carl Welty's estimate, in the Scientific American) there are innumerable oddities and strange specimens in the world of birds that defy explanation. One must just accept them—as part of God's vast creation. There are birds that talk; bower birds that go through an intricate procedure of building a BOWER where the male courts the female; birds of great beauty, like the cockatoo, the peacock and the birds of paradise; there are birds that are scavengers and birds that are pure killers. Let us consider some of these interesting kinds of STRANGE BIRDS that bear witness to God and Creation.

(A) The Water OUZEL: "the Bird of Three Elements"

"Though the ouzel is 'the most bouyant bird in all the records of ornithology' and can float on the surface of water like grease, seeming to ride just above the water, instead of partly in it, the hidden observer is startled to see the water ouzel suddenly descend into the swift stream LIKE A LEAD SINKER! Then this strange creature of three worlds— land, air and water—WALKS about on the bottom, as though he were made of iron instead of flesh and bones."

This seemingly miraculous conduct is possible only because of special equipment that the ouzel has—and this argues for the fact of specific and special creation. The bird is provided with a special muscular apparatus that instantly exhausts the air from all its body, and gives it the weight needed to sink in swift water, and to stay down! Then when it comes to the bank it fills its body with air and so instantly regains its lost buoyancy, and floats away on the surface of the stream as though it never did anything else! To fly at will, float on water or sink like a stone into the water and walk along the bottom of a stream requires DESIGNING from some Master Artisan. Evolution would match God if it alone could achieve such marvels.

(B) The Accommodating HONEY-GUIDE of Africa

"The honey-bird of Africa, a bird no larger than a sparrow, will deliberately lead men or honeybadgers to a bees' nest, twittering loudly to them as it does so, while they respond with calls or grunts (as the case may be) to show that they are following. The association is obviously of benefit to both parties: the bird cannot break into the nest itself, but the men or badgers can, and are bound to leave enough honey and grubs around to satisfy a small bird." (The Living Sea; p. 124).

Who gave the honey-guide such wisdom? Who first taught it to do this? To say that it is "instinct" and that instinct is the "congealed actions of centuries" explains nothing—for what force first caused "actions" to become repetitive? And why did one bird turn out to be a honey-guide and another a falcon, for example? The whole mysterious realm of nature can not be explained by such trite phrases as "congealed actions" and "random mutations." In the honey-guide we find a behavior pattern that is certain, yet static, with no evidence whatever of having come about through a gradual process. The honey-guide **seems** to have intelligence above instinct; but it is an intelligence that **God** gave it.

(C) The KIWI: New Zealand's Wonder Bird

The "National Geographic" gives this vivid description of the incredible Kiwi:

"Impossible!" insisted British scientists in 1813 when they first learned of New Zealand's unique bird—the flightless kiwi. . . . Only when the skin (of a kiwi) was exhibited in a British museum would skeptics admit the existence of this strange inhabitant of the antipodes.

Little wonder that Britain's men of science at first considered the report in a class with stories of the mythical mermaid and unicorn. Who had ever heard of a bird WITH WHISKERS LIKE A CAT'S and with NOSTRILS AT THE TIP OF ITS LONG, CURVED BEAK? Where else lived a bird THAT BURROWS LIKE A GROUNDHOG and LAYS AN EGG EQUAL TO ONE-QUARTER OF ITS OWN WEIGHT? Who indeed had seen a bird with NO TAIL and with useless inch-long wings hidden beneath a coat of silky, hairlike feathers?

And yet—there it was. **Apteryx australis,** they decided to call this surprising creature. The first name means "wingless." . . .

It is a natural wonder, rivaling Australia's duck-billed platypus. The shy kiwi differs almost as much from its flightless relatives as it does from birds in general.

There are other notable features of this rare creature: the fact that the male incubates the huge egg, then turns the chick loose to fend for itself. Then too, though it has sturdy clawed feet, "it can move as silently as a rat."

Like the platypus, the kiwi defies ALL "laws" of evolution! It is impossible to trace its ancestry, according to evolutionists' concepts. Here are some of the unanswerable questions: From whence did the kiwi inherit its ability to lay such a huge egg? No other bird in the world lays an egg proportionately as large as that of the kiwi. From whence did it get its strange feathers, true feathers, but entirely different in style from those of other birds. Why does it have nostrils at the end of its long beak—and in addition, why is the end of its beak a highly sensitive organ of touch? Evolution can not trace its descent from any other animal: it is such a conglomerate creature NO DIRECT CONNECTION WITH ANY OTHER GENUS IS IN EVIDENCE.

(D) **PENGUINS: Birds of the Antarctic that Resemble Men in Dress Suits. They swim but do not fly.**

Penguins are among the world's best witnesses to how God in creation has adapted animals to a hostile environment. Living in one of the coldest sections of the world, where they have neither seeds nor insects to live on as most other birds have, penguins are PERFECTLY ADAPTED TO SURVIVE IN AN EXTREMELY HOSTILE ENVIRONMENT. Were they not equipped as they are, they could not endure the rigors of the Antarctic for a month, much less a season. IT IS IMPOSSIBLE FOR A LESS HARDY BIRD GRADUALLY TO "ADAPT" ITSELF TO SUCH RIGORS OF CLIMATE: it would die a million deaths in the "gradual" process demanded by evolution—and never attain "adaptation" through "random changes." Consider these marvels of "adaptation" that God endowed the penguins with. (The facts here quoted are from an article in December, 1957 "Scientific American," by William J. L. Sladen, p. 45.)

"The penguins almost certainly ORIGINATED in the Antarctic region, for fossil penguins found in that area of the world go back to early in the Tertiary Period (i.e., some 50 million years ago)."

This is a confession damaging to the evolutionists. In the

first place, it excludes any descent from birds who formerly lived in temperate climates. In the second place, it establishes, what we have before stated, the FIXITY of genera; for Dr. Sladen believes that fossils prove the genera to have REMAINED UNCHANGED FOR AT LEAST 50 million years! So, if there has been no evolutionary change in the last 50 million years, WHEN will evolution go to work on the penguins?

"The penguin's body is beautifully adapted to its life in cold waters. Unlike other birds, it is almost completely covered with feathers. Its dense coat of short, stiff feathers, overlapping almost like scales, gives it excellent insulation against heat loss."

And now we read further of a most interesting provision God made for the benefit of their progeny.

"The only piece of bare skin on its whole body is a very narrow strip on its abdomen which widens to about an inch and a half WHEN THE BIRD INCUBATES ITS EGG and becomes grown over with feathers again after the eggs are hatched."

Penguins (the Adelie species) seem to display INTELLIGENCE in the construction of their nests, made out of stone. This is what they do:

"They build a nest of stones. . . . The purpose of the nest, in part, IS TO KEEP THE EGGS ABOVE WATER when snow falls on the nest and melts." Did the penguins figure this out, or, DID GOD, THE CREATOR, PLAN IT SO WHEN HE MADE THEM?

Because penguins, living in the Antarctic, have no seeds, grasses, plants, worms or insects to live on, God "adapted" them for survival under such hard conditions by enabling them to LIVE ON SEA FOOD, AND YET HAVE THEIR ROOKERIES INLAND. This is a most amazing "adaptation."

"Emperor penguins arrive at their breeding quarters around the middle of March at the beginning of the Antarctic winter. It takes two months to incubate the eggs, and the male does all the sitting himself, while the female feeds at sea. At hatching time the female returns (travelling sometimes 50 to 60 miles over the trackless wilds of that frost and ice-bitten area) and finds her mate, though there is no nest, and proceeds to feed her chick. The family maintains its unity in spite of long separation in an icy wilderness without any fixed home. The dedication and endurance of the father are also quite remarkable. HE GOES WITHOUT FOOD THROUGH THE COLD ANTARCTIC WINTER FOR A PERIOD OF ABOUT THREE AND A HALF MONTHS." This is approximately 105 days. And that is while the bird is active, and NOT in a state of hibernation. As far as we know, this feat is unparalleled in nature, at least among birds, whose appetite all ornithologists know to be ravenous.

That feat—going foodless for 105 days in the extreme cold of the Antarctic—demands a miracle of construction that could be achieved ONLY by One of super-ability and super-intelligence.

We might add, for the reader's information, that penguins live on ocean krill, a small shrimp, which is also the main diet of many species of whales.

Actually, there are scores of other amazing "adaptations" that all point to the inevitable conclusion: Someone who knew what He was doing, created the penguin to MEET THE EXTREME CONDITIONS OF ITS HABITAT. The penguins HAD to be made as they are to survive in their hostile environment.

It is interesting to learn too that the parent penguin who returns to the rookery from the sea, ALSO FEEDS THE BABY PENGUIN. This it does by regurgitating food from its own supply, held in store in its body for this very purpose.

Can you think of anything more wonderful than God's amazing provision for FOOD for the penguins and their young? Can you think of an "adaptation" more amazing than that which gives the penguin the ability to go 105 days WITHOUT FOOD, and survive and remain well, in the dead of a winter of intense cold, while the bird is actively engaged in taking care of its one egg?

It is interesting that the Emperor penguin, that does NOT build a nest of stones, has its own marvelous nest, a "built-in" arrangement, that most certainly reveals Divine forethought in creation and perfection of DESIGN for an intended purpose.

"The emperor penguins incubate their single egg between their feet as they stand upright, A FLAP OF LOOSE SKIN (especially made for this purpose) COVERING AND PROTECTING THE PRECIOUS EGG FROM THE INTENSE COLD OF THE ANTARCTIC WINTER." ("Strange Animals at the Zoo"). (Caps ours.)

It is easy to see the handiwork of God in such an obvious "adaptation."

(E) **The Common, yet Uncommon, WOODPECKER—A Miracle Exhibit in perfect Adaptation for an Intended Purpose.** It has been called "Nature's Power Drill."

The woodpecker lives in a far more friendly environment than the penguin—nevertheless, the woodpecker reveals in its structure amazing "adaptation" to what the Creator designed it for.

"Woodpeckers are highly specialized for their tree-climbing and grub-hunting activities. Their feet are strong and equipped with sharp, curved claws. Two toes on each foot are directed forward, while the other two point to the rear, thus making an effective pincer for grasping the bark of the trees. (Three-toed woodpeckers have only one hind toe on each foot). The feathers of the tail are stiff and end in sharp spines. These spines are pressed against the ridges in the bark of tree trunks and branches and help prop the bird as it digs for grubs or excavates a nesting site.

"The woodpecker's head is large and its neck short and powerful, enabling the bird to deliver rapid and forceful blows with its stout beak. This beak, with its chisel-shaped tip, is an EFFECTIVE WOOD-CUTTING TOOL. With it, the bird penetrates the bark and wood of trees, where wood-boring grubs, hibernating insects and insect eggs are to be found. Once a small hole is made, the woodpecker's tongue dislodges the insect prey. The tongue is long and slender and can be protruded a considerable distance from the mouth; its tip is usually pointed and BARBED and is COVERED WITH AN ADHESIVE SECRETION."

No intelligent person can study the tongue of a woodpecker without realizing that it is well designed for an intended purpose.

This "flying power drill" has been called "nature's most baffling bird."

"How a woodpecker can violently slam its head against solid wood hundreds of times a minute without knocking its brains out, or at least getting punch-drunk, still remains a mystery. Scientists think the secret may be in the structure of the woodpecker's skull, which is constructed with a set of tiny cross braces . . . which seem to give the skull more flexibility."

One of the favorite foods of the woodpecker is the beetle. At certain times there are more beetles than he can eat—so, wise old bird that he is, he "stores" the extra beetles ALIVE in a neatly designed and constructed "prison" and so keeps a supply of FRESH food on hand! Now note the woodpecker's uncanny ability:

"The woodpecker knows how to estimate and drill EXACTLY the right size hole, so that he can squeeze the live beetle into the hole and yet not permit it to worm its way out! If he makes the hole too little, he couldn't get the insect in; and if he made it too big, the insect would be able to wiggle its way to freedom." Such a situation of course demands a very close "tolerance"—to use a machinist's expression. THE UNBELIEVABLE WOODPECKER CAN DO THAT VERY THING! And, remember, the woodpecker has to change its calculations for EACH beetle it puts into live storage—for all beetles differ some in size and shape.

Surely, every woodpecker in the world is a LIVING WITNESS to the fact that GOD MADE IT AS IT IS. Evolution can in no wise explain how the woodpecker got its unique tongue, its specially constructed tail, its designed feet, and above all, its marvelous chisel-like beak! That such amazing equipment, differing from that given to other birds, would have been perfected through long ages of "gradual change" is a preposterous assumption without valid reason for acceptance. We repeat what we have said so often before: ANY SPECIALIZED ORGAN—like the tongue, or the beak, or the tail of the woodpecker—MUST BE PERFECT

BEFORE IT SERVES ITS INTENDED PURPOSE. A beak that is only "half" developed to serve as a chisel, or a tail that is only "partially" developed to aid in climbing a tree, or a tongue only 10% long enough to reach a grub hidden inside the trunk of a tree, is absolutely USELESS. The "specialized organs" that all creatures have, had to be PERFECT from the beginning--otherwise they are worthless and impractical.

Questions: IF such "specialized organs" came to pass through the processes of gradual change, due to "random mutations," what good purpose did they serve while they were in the PROCESS of developing? And what did the poor creature do UNTIL its specialized organ was fully developed? AND WHERE IN ALL NATURE, IS THERE ONE EXAMPLE—just one—OF A PARTIALLY DEVELOPED SPECIALIZED ORGAN THAT IS NOT A USEFUL ORGAN IN ITS PRESENT STATE?

(F) **The FALCON—Nature's Great Display of "Controlled Power"**

A peregrine falcon "normally kills its prey by climbing above its victim, then 'stooping' on its quarry like a thunderbolt. A split second before the impact the talons are brought into position. AND THEN THE STRIKE. The long, needle-sharp hind claw shears through flesh and bone and, amid a puff of feathers, the prey falls to earth. . . . The peregrine falcon can strike its prey with such force that it knocks its prey clean in half." (Nature Parade, p. 209).

"Watch the master flier of them all, the peregrine falcon, also called the duck hawk. Many times I have seen one, high above me, turn its nose downward, give a mighty flap for thrust, then close its wings and plummet toward the earth like a hurled stone with incredible speed. Suddenly there is an exploding puff of feathers as the falcon strikes a bird with its large clawed fist. The prey is usually killed outright. But then comes the most amazing maneuver of all: the falcon darts under the falling bird, **flips over on its back** and catches the prey neatly in its talons! HERE IS ONE OF THE GREAT DISPLAYS OF CONTROLLED POWER IN NATURE." ("The Truth about Hawks." Audubon Magazine; article by Peter Farb).

Could such effective "streamlining" in a body, such coordination, such masterly control, such POWER, come to pass as the result of "chance mutations"—or, do we see in the peregrine falcon a perfect adaptation for a desired end, and hence, A DEMONSTRATION OF WHAT GOD HAS ACCOMPLISHED IN CREATION.

(G) **The Strange Antics of BOWER BIRDS: bizarre and "seemingly thoughtful Activities."**

In our discussions we often have called attention to actions of animals that SEEM to suggest intelligent, thoughtful activities, as though the animals had planned a course of action deliberately, as the result of careful consideration as to what was wise and best.

The male Bower bird seems to follow a course of "thoughtful activity" in its unusual procedure in its courting. The facts we present here are from an article in The Scientific American, on BOWER BIRDS, by A. J. Marshall. We quote:

"In the 1840's, a Captain Stokes came upon a peculiar object in the wild bush of western Australia. It was a neat structure consisting of two parallel walls of sticks stuck in the earth, forming a little avenue. The avenue and its entrances were paved with a scattering of white shells. The captain at first decided that it had been built by some aboriginal Australian mother to amuse her child. . . . Later in his voyage settlers showed him another 'playhouse' like the one he had seen. This one was occupied by its builder and owner—a gray, pink-crested male bird about the size of a small pigeon, called the BOWER BIRD."

"Captain Stokes' report on the incident was one of the first accounts of a phenomenon that has continued to perplex scientists to this day.

"Later, as Australia was populated, bower birds gained a fabulous reputation: their bowers were found strewn not only with shells, pebbles, bones, bits of precious opal and pieces of quartz, . . . but also bits of broken glass, nails, beer-bottle tops and brass cartridge cases!"

Some bower birds build the bowers like a maypole; others build in the form of a pyramid which may reach up to nine feet high; yet others build their bowers in the form of a conical hut; and others add a low stockade to their hut! What is even more curious, "some bower birds actually paint the inside walls of their bowers."

How complicated these procedures in the building of "bowers" may become is seen in this quotation.

"The blue-black satin bower bird paints the inner twigs of its avenue bower sometimes with charcoal. This charcoal painting is a ritual of fascinating complexity in which it uses a tool of its own making. The bird first collects charcoal (from charred trees, burned by forest fires) and grinds it up in its beak to a sticky black paste. Then it selects a fragment of bark and fashions a tiny oval wad. This is used as a stopper to keep its beak slightly open and allows the charcoal stain to ooze from the sides of the beak. The bird then vigorously smears the stain on the twigs of its bower. . . . ALL THIS HAS BEEN OBSERVED AND PHOTOGRAPHED. . . . Other bower birds plaster the inner twigs of the bower with fruit or grass mixed with saliva."

THE MALE BOWER BIRD DOES ALL THIS TO ATTRACT A FEMALE TO HIS SIDE. But whatever the factors may be that cause the female to come to him, they work out so that reproduction occurs at the time of year that is most propitious for the survival of the young!

The male actually seems to PLAN how he can influence and win a bride! And he goes to a great extreme to build an attractive, inviting "bower" where he invites the female for the solemnities of courtship. It is one of the most involved courtship rituals in all nature.

How are we to explain the apparent "thoughtful activities" involved in the construction of the bower? The "intelligence" is NOT inherent in the mind of the bird, but is given by Divine Creation so cleverly that it SEEMS to be more the result of native intelligence rather than a gift from the Creator in the form of instinct.

The point is: SUCH AN INVOLVED PROCESS CANNOT BE ATTRIBUTED MERELY TO "INSTINCT" BUT MUST BE SEEN AS A WORK OF DIVINE CREATION. Instinct that results in such involved construction and actions subsequent to construction, cannot be accounted for by any theory of evolution that we know about. It MUST be the work of a Supreme Architect who built into the very nature of the bower bird the "machinery" that causes it to build a characteristic bower generation after generation, even when not influenced by watching other birds build their bowers! EVERY BOWER BIRD IN AUSTRALIA IS A WITNESS TO GOD AND DIVINE CREATION.

(H) The HUMMINGBIRD: God's Perfect Little Helicopter

The smallest bird in the world is the "fairy hummingbird" found in Cuba. It measures only 2¼ inches from the tip of its bill to the tip of its tail, and weighs but a fraction of an ounce. The majority of the 580 species and sub-species are tiny birds under four inches in length. The hummingbird cannot walk (it uses its feet only for perching), so it has to fly to get about.

Most birds are, aerodynamically, the most perfect flying machines on earth. The hummingbird, in addition, is the only perfect HELICOPTER. His wings are attached to his shoulders in such a way that he can poise motionless in the air, and he can fly in any direction, forward, sidewise, up, down—and even backward, a feat no other bird can do. (Coronet Magazine).

"While hovering, a ruby-throated hummingbird beats its wings up to 75 times PER SECOND. In addition to backward flight, the hummingbird has achieved the aerodynamic miracle of sideways flight . . . which it does with no appearance of difficulty." ("Nature Parade," P. 206).

The hummingbird has many other unique features that set it apart as a SPECIAL WITNESS FOR GOD AND CREATION. We mention its wise way of building its nest.

By using a spider web as the framework of her nest, the mother hummingbird has a home that is strong and can be expanded easily. When she builds it, the nest is little more than half inch across on the inside. As the babies grow they push against the sides, **stretching the flexible spider web,** and the nest becomes larger with them!

Let us ask: Who gave the tiny hummingbird the ability to hover in the air without moving in any direction, and to fly in any direction, including the ability to fly sidewise and even backwards—feats no other bird can accomplish? And Who gave both the wisdom and the ability to the little hummingbird to make an expanding nest for its growing babies, so that they always have a nest that is the right size, no matter how old or young they are? And Who made each species of hummingbird **static** in its abilities, instincts and peculiar characteristics, so that generation after generation (instead of exhibiting a gradual change) they stay virtually the same? GOD, is the only answer.

(I) GERMAN WARBLERS: Sky Navigators Par Excellence

In the August, 1958, issue of the **"Scientific American"** Magazine is a revelatory article by E. G. F. Sauer, ornithologist at the University of Freiburg, Germany. These facts are quoted from Mr. Sauer's article.

Each Fall, the little German garden warbler, weighing barely three-quarters of an ounce, sets off one night on an unbelievable journey. All alone (never in the collective security of a flock) it wings its solitary way southward over Germany, France and Spain and then swings south to its distant goal in southern Africa. It flies on unerringly, covering a hundred miles or more in a single night, never once stopping in its course, certain of its goal. In the spring it takes off again and northward retraces its path to its nesting place in a German or Scandinavian thicket—there to hatch a new generation of little warblers which will grow up, and WITHOUT BEING TAUGHT, will have the self-same capacity to follow the same route across continents and oceans by the map of the stars!

To discover how they oriented themselves Prof. Sauer and his assistants experiment with warblers in cages with a glass opening at the top, so that they could see part of the sky, but nothing else of their surroundings. They also tested the birds in a cage placed in a planetarium—that is with a dome, showing an artificial replica of the natural starry sky—and they found that when the stars were hidden either in the real sky or in the planetarium by thick clouds the birds became completely disoriented and confused. Their experiments proved conclusively that these birds were guided only by the stars in their long semi-annual migrations. The behavior of the warblers, in these special studies, "leaves no doubt that the warblers have a remarkable hereditary mechanism for orienting themselves by the stars—a detailed image of the starry configuration of the sky coupled with a precise time sense which relates the heavenly canopy to the geography of the earth AT EVERY TIME AND SEASON.* At their very first glimpse of the sky the birds automatically know the right direction. WITHOUT BENEFIT OF PREVIOUS EXPERIENCE, with no cue except the stars, the birds are able to locate themselves in time and space to find their way to their destined homes."

Prof Sauer continues, giving us more of the "mystery" in this phenomenon: "Even more difficult to explain is the mystery of how the birds ever came to rely on celestial navigation and to develop their skill in the first place. . . . What evolutionary process was it that endowed these animals with the highly sophisticated ability to READ THE STARS?" (Caps ours).

No wonder Prof. Sauer questions how "evolution" could perform such a miracle in these birds and other animals, some of practically no intelligence (to wit, the crab). We, too, question the ability of "evolution" to accomplish such marvels. But we have the answer that evolutionists reject—but it is the only answer that really explains these incredible phenomena: GOD MADE THEM SO. What evolution could not possibly accomplish in a billion ages, GOD DID IN INSTANTANEOUS CREATION.

The entirely inadequate thinking of many evolutionists is set forth in the suggestion once advanced (so it is reported) by Mr. Huxley.

"Six monkeys set to strumming unintelligently on typewriters for millions of years would be bound to write in time all the books of the British Museum." He was working on the thesis that the Law of Chance will produce anything, only given time enough. But Huxley was wrong—wrong as he could be. It takes INTELLIGENCE to write books—say for example, Shakespeare's 35 plays and 154 sonnets.

---

*Mr. Sauer's amazing experiments and discoveries throw light on at least PART of the meaning of Genesis 1:14: "And God said, let there be lights in the firmament of heaven . . . and let them be for SIGNS, and for seasons . . .". "We know," quoting Sauer again, "that the warblers ARE NOT THE ONLY CREATURES possessing this gift (of being able to be guided by the stars): other birds, insects, crabs and spiders have been found by experiment to be capable OF GUIDING THEMSELVES BY THE SUN OR STARS."

And our confidence in the Bible, the Word of God, is confirmed by this startling fact: ages before the modern era of scientific investigation, that has brought these facts to light, THE BIBLE TOLD US THAT GOD PUT THE CONSTELLATIONS IN THE HEAVENS TO BE FOR "SIGNS." Birds and insects and many other animals find guidance from the stars in their local or worldwide peregrinations or flights. Men of course also use the stars as well as the sun for guidance by day and by night.

You and I know that six monkeys could pound typewriters—yea, 6,000 monkeys could pound typewriters—for all eternity AND NEVER PRODUCE ONE OF SHAKESPEARE'S SONNETS OR PLAYS, for the simple reason that the INTELLIGENCE to create his works is lacking in monkeys.

And in Huxley's illustration we find the fundamental error of most evolutionists: they believe that given sufficient time, ANY AND ALL PHENOMENA OF THE PLANT AND ANIMAL KINGDOMS PRODUCE THEMSELVES. But we all know that one can shake a barrel of printer's type for untold ages and unless guided by INTELLIGENCE they will NEVER—no NEVER—assemble themselves into the Lord's Prayer, Shakespeare's "Merchant of Venice," or any other literary creation by intelligent men. If INTELLIGENCE is lacking, the Law of Chance can produce nothing but jumbled chaos. The orderly "designs" in nature and the intelligent "adaptations" and the marvelous "instincts" and characteristics that have been given to birds, animals and insects, far beyond their limited intelligence, PROVE the fact of a Supreme Creator.

The BEAUTY of Birds Witnesses for GOD

Beauty is found in many realms of nature: consider the exquisite beauty of many flowers, the loveliness of many insects, especially butterflies and moths. No artist or photographer can possibly do justice to the striking beauty of the jewel-like tones of every shade of the rainbow as seen in various species of moths and butterflies! In some, while in flight, the most striking thing is the way the colors shift and change, varying with the angle of the light or the eye of the observer. This subtle play of color reflected from the wings of a butterfly is called "structural color," produced by the ultramicroscopic structures on the tiny scales on its wings. Consider also the many scores of kinds of brilliantly-colored tropical fish, moving about coral reefs. There is a fascinating beauty also in the grace of movement of fish and in the effortless speed with which they move through the water.

Nor should we forget the beauty in gems. No one can say of that beauty that it is there to "attract the opposite sex" or to attract insects to insure pollination as in certain flowers. In gems beauty is there for beauty's sake! Who that has seen it has not been charmed by the marvelous transformation of certain minerals and gems under the spell of black light? Under artificial light, the ruby, in the dark, will glow as though on fire. The moonstone is famed for a bluish-white play of light; the sunstone for a brilliant play of reflections. In Labrador spar appear all the colors of peacock's feathers. And who is not fascinated by the glow of a large blue-white diamond, rightly cut! Can any one deny that such displays of beauty are reflections of the Creator's love of beauty? The fact that there is beauty in nature and that men have the capacity to enjoy it, proves that nature is not a meaningless farce that just "happened," but is rather the well-planned result of the Creator's intelligent work.

But of all things beautiful in the world, what can surpass the beauty of birds? What can outdo the peacock for grandeur of display?* What can supersede, for symmetry and sheer finery, the gorgeous tail of the Australian lyrebird? Were they not real, no artist in the world could conceive such matchless beauty as is found in the amazing birds of Paradise of New Guinea! Some of the 18 species have "all the colors of the rainbow nicely blended." The Magnificent Bird of Paradise has an iridescent green vest with a sickle tail. In addition to their striking colors, in various combinations, birds of Paradise have the remarkable distinction of their peerless plumage, which take the form of delicate aigrettes, copious capes and ruffs, waving plumes and odd shaped "wires" that extend well beyond the body. Some of these "wire" adornments are really strange: for example, one species has a peculiar green rolled ornament on the ends of its long tail wires. No one knows WHY all these lovely adornments, except that GOD MADE THEM THAT WAY.

There are of course a thousand and one other beauties in the world of birds. The Red-headed Woodpecker is a showy bird. The Maryland Yellow-throat is a striking figure, with his bright

---

*One author says, "The peacock's feathers show a repeated and resplendent pattern, produced by the united effect of the combination of distinct and different tints, marked at fixed distances that are minutely fractional, on each separate spray of each feather; and each point in each spray requires a different development to produce the harmonious over-all effect of the glorious peacock's feathers." (Number in Nature). There are billions of chances to one AGAINST such symmetry, such careful, minutely accurate structural planning, with a definite design of beauty, being the result of blind force, which can neither see colors, nor take account of measured space, nor delight in the overall result. SUCH BEAUTY, SUCH SYMMETRY, SUCH UNFAILING DESIGN IN MINUTE DETAIL JUST COULDN'T HAPPEN BY CHANCE: such a result had to be planned and the plan had to be executed by One able to plan and able to perform.

and various colored feathers. The Black Swan has been described as "a royal bird of grace and beauty." The roseate Spoonbill of Texas is most impressive. Orioles and Cockatoos, Motmots and Tanagers, Jays and Warblers, Bluebirds and Goldfinches, Terns and Swallows, the Kite and the Bittern—all have a characteristic beauty all their own. WHY SUCH A VAST DISPLAY OF BEAUTY IN NATURE, especially in the world of birds? GOD MADE IT SO! To think that all this beauty, this endless variety of loveliness, came about through processes of evolution, is without foundation—it is the handiwork of the Master Artist, the One whose every movement is rhythm and whose every message is a song.

The SONGS of Birds Witness to the Creator

John Burroughs, famous naturalist, hit the nail on the head when he wrote,

"The songs of most birds have some HUMAN significance, which I think is the source of the delight we take in them.* The song of the bobolink to me expresses hilarity; the sparrow's song, faith; the bluebird's, love; the cat-bird's, pride; the white-eyed fly-catcher's, self-consciousness; that of the hermit thrush, spiritual serenity: while there is something military in the call of the robin." (Green Treasury, p. 521).

The **house wren** seems to bubble over with emotion, when it sings, for his enthusiastic song "literally shakes every feather on his small body." The "eerily descending song of the **veery** (thrush) sounds like neither voice nor instrument, but rather like a thin, vibrant whistle, faraway, wild, remote." Who can deny that the one song stirs the emotions while the other tends to put one into the hazy dreamworld of quiet reverie?

The cheery notes of the rose-breasted **grosbeak,** obviously overflowing with good fellowship, are "tossed into the sunshine far and wide in his rich, rapid warble." The **cardinal** is another vigorous, enthusiastic singer, and incidentally he continues to sing right through the winter. "His clear, vibrant whistling has all

---

*Evolutionists teach that the singing of birds is caused by "breeding activity" and the need "to protect the family food supply." This is crass materialism and hardly explains the continuous singing of the caged canary and the endless variety of song put forth by the versatile mockingbird, and the cardinal's songs in winter. John Burroughs has a much more realistic approach: birds sing, not only for their own propagation and protection, but also for MAN'S benefit! And we might suggest: no doubt the great Creator Himself takes delight in the melodious songs of His birds, and in their pleasing beauty.

the free spirit of a country boy on his way to the swimming hole," and so he breathes hope, joy and confidence to man. Who can deny that God made birds with a ministry to **man** as well as to their own families?

Wrens are in the top rank as singers: "full of trills, runs, and grace notes, the volume startling from a bird of such small size." The picturesque **bluebird** is one of the most pleasant of singers, with his warm-hearted warbling.

Famous authors and naturalists have written essays on the marvelous songs of the nightingale. Pliny, of ancient Rome, wrote (Natural History):

"Nightingales pour out a ceaseless gush of song. . . . (Think) of the consummate knowledge of music in a single bird: the sound is given out with modulations, and now is drawn out into a long note with one continuous breath, now varied by managing the breath, now made staccato by checking it, or linked together by prolonging it, or it is suddenly lowered, and at times sinks into a mere murmur, loud, low, bass, treble, with trills, with long notes . . . soprano, mezzo, baritone; briefly, it has all the devices in that tiny throat which human science has devised with all the elaborate mechanism of the flute."

Edward Thomas, of England, speaks of the swift notes of the nightingale resembling "the liquid sweetness as a grape," yet "wild and pure as mountain water in the dawn."

Undoubtedly, the mockingbird is "the most gifted of all song birds." He likes to tease, and he can imitate all kinds of noises of animals and the songs of other birds. He can even duplicate the squeak of an old wheelbarrow or the barking of a dog. Mockingbirds are brilliant singers, "having marvelous technique . . . and some observers say . . . "they excel even the nightingale and the American thrushes in their emotional outbursts." While not all mockingbirds are as adept as others, there is on record "one mockingbird which imitated 32 different species of song birds during the course of ten minutes of continuous singing."

The hermit thrush is said to emit "one of the most ethereal of all songs, a leisurely series of rising cadences so bell-like, so spiritual in tone and rendering, that they seem beyond the ability of even a bird."

Who created the tiny throat of the nightingale, the thrush and the wren, made them far more versatile than any flute devised by man, and gave them a heavenly ability to produce such consummate music? Who fashioned the throat of the mockingbird so that it can reproduce, faithfully, a hundred sounds that scarcely can be distinguished from the original? Who put the endless variety of music in the world of songbirds, giving us sounds and music all the way from the laughter of a loon in the night and the call of Canada geese, to the glorious singing of a robin or a meadowlark? Surely every bird in the world is a living witness for God its Creator. Unaided evolution could not produce the throat of a nightingale or the voice box of a mocking bird and put ten thousand miracles of song in other species of birds.

ODDITIES IN BIRD LIFE WITNESS FOR THE CREATOR

Young birds nesting in trees are so helpless that the mother bird must put food into their mouths. The bills of these young birds usually have **yellow rims** so that the mother bird can see them easily, even when it is dark. Young birds nesting on the ground, like the quail, are usually spotted or striped. When danger comes, the mother gives a cry of warning and the young will lie flat on the ground, camouflaged by their spots or stripes, so that they blend into their surroundings. Who so designed these young birds for their protection?

There are over ten thousand species of birds—and each has distinctive characteristics that mark it as a special creation of God. Some are odd specimens indeed. Why is the male of the **Phalaropes** less brightly marked than the female—contrary to the usual order? Who designed the Mexican "roadrunner" or "snake killer" so that he can easily kill rattlesnakes? Who designed the fantastic kingfisher, with his large crested head, long bill and short tail? Who created the strange three-wattled bellbird, that has mustaches like an ancient Chinese mandarin? Who first dressed the male umbrella bird in velvety black and gave it a crest that grows up and forward from the crown, shading the bill and providing a sort of umbrella for its head? This same bird also has an odd appendage called a lappet that dangles from the upper breast. When excited or disturbed, the bird can expand that appendage that hangs from its throat from a normal eight inches to thirteen. All these oddities point to special CREATION, the work of a Designer, rather than the result of blind chance mutations. Who put the collar on the ruffled grouse? Who made the saucy jay? Who decorated the redshafted flicker, with its spots and splashes? Who decorated the snowy egrets with such beautiful plumage that the desire for their plumes for women's hats almost led to their extermination in the early years of this century? Who put "the world's most extraordinary plumes" on the King of Saxony's bird of paradise? Who constructed the original of

Wallace's Standard-wing Bird of Paradise—"the bird to make you rub your eyes: the bird that differs most remarkably from every other bird of paradise." Who fashioned the gorgeous golden oriole? Who made the parrot and the black Hill Myna so they could imitate the human voice, and learn to talk? Surely, the Hand of God can be seen everywhere in His creation of such an endless variety of life, especially in the amazing world of birds.

THE MIGRATORY INSTINCT: A Witness for Divine Creation

Birds do not have much mind, but they are gifted with amazing instinct that tells the mother bird to turn her eggs, teaches the new home maker to construct a nest in harmony with that used by its species, even though she has never done it before nor even seen it done, and leads birds into amazing feats of migration that are inexplicable mysteries.

Without baggage, lunch or bedding, and with no chart, compass, map or guide—except the sun and stars—the migratory birds make unbelievably long trips, and they make them successfully, even though when first they make the trip they have never been there before!

The champion "globe trotter" and most renown of all migratory birds is the Arctic tern that summers in the far north and winters in the Antarctic. Its annual round trip may be well over 22,000 miles—for it does not make its journey in a straight line, but meanders off its course while going and coming.

The Atlantic golden plover, a robin-sized traveler, goes south by one route in the fall and returns in the spring by another route!

"Their route is in the form of a great ellipse. From Canada they strike out over the Atlantic to South America. In Spring they return by way of the Mississippi valley."

Frederick C. Lincoln, noted ornithologist and American authority on migration, says,

"The Golden Plover travelling over the ocean covers the entire distance from Novia Scotia to South American without stop. This is accomplished with the consumption of only a few ounces of fuel IN THE FORM OF BODY FAT." Such an amazing "engine" is a billion times more efficient than man's best airplane.

The feat of the Pacific plover is equally astonishing. It flies from Alaska to Hawaii. How it can traverse thousands of miles over the trackless ocean wastes and find tiny specks of islands in mid-ocean is a mystery that has baffled scientists for years. In the Spring it returns to Alaska—and in the Fall it goes back TO THE SAME TINY ISLAND IN THE PACIFIC!

A little bird called the Wheatear, no larger than the English Sparrow, travels every year all the way from AFRICA to GREENLAND. It actually crosses an ocean to go to another continent—and it makes this long trip year after year, on a regular schedule.

More than a hundred different kinds of our American birds spend the winter in Central and South America. Not so dumb, are they?

The tiny blackpoll warbler nests in Canada, then wings its way to Brazil, 4,000 miles away.

Much of the travelling of migratory birds is **at night,** and much is over the wilderness of waves of what might seem to a small bird to be boundless oceans. And here is another mystery: eastern birds mix freely with western birds and with tropical birds during the winter season in South and Central America; but when the time arrives for the northward trip **every species behaves according to its own pattern,** and goes back to its own home, flying over the same fields and alighting on the same fence posts! Miracle supreme: when their sons and daughters are old enough to migrate, they go to the same place their ancestors went, HAVING NEVER MADE THE TRIP BEFORE. And they go unguided by any other bird! The whole subject of bird migration is clothed in deep mystery. We read: "What strange power impels a tiny winged creature to leave its summer home within two or three weeks of the same day each year (long before its food supply is exhausted and before the extreme cold weather comes) and fly thousands of miles to a winter home IT HAS NEVER SEEN? What then drives it to return again to the same part of the United States or Canada where it had been raised and to arrive so punctually that the date of its return can often be predicted to within a single week? How, year after year, does it find its way back to the identical field or wood where it raised its first young?"

Birds are not the only creatures that migrate. Insects such as the monarch butterfly and the locust take long migrations. The eel, salmon and other fish also migrate, in most mysterious and unbelievable ways. The whale, the porpoise and the seal find their way through water as unerringly as the birds do in the air: and they migrate long distances.

What remains an insoluble enigma to the scientist is clear to the devout believe in a personal God. Migratory birds and other migratory animals do as they do BECAUSE GOD MADE THEM SO. That is the all sufficient and only satisfying answer to this mystery of migration. Evolution has NO answer.

So sum up: BIRDS and their characteristics are **marvelous witnesses for God and Divine Creation!** With NO intermediary stages to be found either in nature or in fossil form, how could anyone believe that heavy-boned reptiles, with scales, could evolve into light, efficient "flying machines" with feathers and wings? How can one explain the thousand and one **unique** characteristics possessed by as many birds, the ability of hummingbirds to fly backwards, the strange adaptations of flightless penguins that can live and raise their young in the frigid antarctic, on sea food only—and yet keep up their rookeries 50 to 75 miles inland? the voice box of the mockingbird and the throat of the nightingale, capable of making such wonderful music? the bill and tongue of the woodpecker?

BATS: Flying Mammals with Built-in Radar

Bats are not birds, but are the only "flying mammals" there are.* They sleep during the day and come out of their caves and other hiding places to hunt for food at night.

Their knees face outward and bend backward instead of forward, as in most mammals. The foot has a short sole and five toes, all about the same length. These features are **exactly** what a bat needs to hang on to rocks and trees, for bats hang upside down to sleep. Here again we see PERFECT ADAPTATION for the manner of life planned for it by the Creator. "Natural selection" would never develop bats into such a peculiar manner of sleeping: it requires less "change" from the conventional types, to sleep right side up as most birds and mammals do.

The outstanding peculiarity of bats, which sets them off from all other mammals, is their unique "radar" system. They fly by ear, and guide themselves by the echoes of their high-pitched squeaks!

---

*The flying lemur and flying squirrel glide through the air, but do not fly.

"The complicated flaps of skin found around the nostrils of some bats, and certain strange structures in their ears, are for the projection and reception of sound waves of ultra high frequency upon the principles we now call sonar. The bats emit bursts of sound of frequencies up to 32,000 per second,* but for intermittent periods of as little as a two-hundredth of a second each. These air-borne waves bounce back from obstacles ahead of the bat when it is in flight, and are picked up by the supersensitive ears of the animal IN THE BRIEF PERIODS OF SILENCE BETWEEN THE BURSTS OF SOUND. So sensitive and accurate is this system that the bats can alter their course in time to avoid hitting the obstacle."

Donald R. Griffin, Prof. of Zoology at Harvard, has this to say about "Bat Radar" in an article in the July, 1958, "Scientific American."

"In these days of technological triumphs it is well to remind ourselves that living mechanisms are often incomparably more efficient than their artificial imitations. There is no better illustration of this than the sonar system of bats. Ounce for ounce and watt for watt, it is BILLIONS of times more efficient and more sensitive than the radars and sonars contrived by man."

"To appreciate the precision of the bats' echo-location system we must consider the degree of their reliance upon it. Thanks to sonar, an insect-eating bat can get along perfectly well without eye-sight. . . . Bats easily find insects in the dark of night, even when the insects emit no sound that can be heard by human ears. A bat will catch hundreds of soft-bodied, silent-flying moths or gnats in a single hour (simply by the use of its sonar system).

"One highly specialized group, the horseshoe bats of the Old World, have elaborate nose leaves which act as horns to focus their orientation sounds in a sharp beam; they sweep the beam back and forth to scan their surroundings.**

From eating insects or sipping nectar to sucking blood, demands a radi-

---

*Such sound frequencies are quite imperceptible to the human ear which is usually insensitive to frequencies above 12,000 per second.

**There are about 2,000 varieties of bats, many of them with highly specialized organs. Some bats live on nectar and pollen; they have elongated muzzles and very long tongues—sometimes with a brush at the tip of the tongue, ideally adapted to their type of feeding. Unquestionably, the most dramatic "adaptation" in feeding habits is that of the vampire bats of tropical America. They live on blood drained from living animals! They tiptoe up to their sleeping or resting victims. Then with the sharp, narrow blades of their specially designed upper incisor teeth, they excavate a small segment of skin and underlying tissue, usually without waking or paining their host. The vampire then sucks up, with piston-like movements of its tongue, the blood that flows into the excavation. Its digestive tract is designed to receive and digest blood, and is reduced to a folded tube for that purpose. (Adapted from an article by William A. Wimsatt, Cornell University).

cal CHANGE in a large part of the bat's anatomy. To be useful, it HAD to be accomplished **suddenly;** gradual change from one system of eating to another, radically different, could NOT be accomplished, and have the bats survive during the period of change—for either the bat has to live on INSECTS or NECTAR or BLOOD; there is, in this case, no middle ground. During a period of "transition" from one feeding method to another, only those bats with the original feeding method would be best "adapted" — hence the transitional forms would be greatly handicapped in the "struggle for existence." Obviously, the theory of evolution, which demands a gradual change and "survival of the fittest," is theoretical nonsense.

"The most surprising of all the specialized bats are the species that feed on fish. They have a well-developed system of frequency-modulated ('FM') sonar, but since sound loses much of its energy in passing from air into water and **vice versa,** the big puzzle is: HOW CAN THESE BATS LOCATE FISH UNDER WATER BY MEANS OF THIS SYSTEM?" They do this as they fly along close to the surface of the water.

"Probably the most impressive aspect of the bats' echo-location performance is their ability to detect their targets IN SPITE OF LOUD 'NOISE' OR JAMMING. They have a truly remarkable 'discriminator,' as a radio engineer would say. Hundreds of bats will fly in and out of the same cave within range of one another's sounds. YET IN SPITE OF ALL THE CONFUSION OF SIGNALS IN THE SAME FREQUENCY BAND, EACH BAT IS ABLE TO GUIDE ITSELF BY THE ECHOES OF ITS OWN SIGNALS. . . . With an auditory system that weighs only a fraction of a gram, and in the midst of a great volume of surrounding noises, a bat can distinguish its own echoed signals, even though they are 2,000 times fainter than the background noises!" And think: the original sound that the bat emits can not be heard by the human ear!

Certainly, such a marvelous system MUST be the work of an infinitely careful and capable Workman! It is impossible for us to believe that such an involved, delicate and intricate system of sonar developed by "chance mutations." It is far more reasonable to believe that a SUPREME BEING made the bats and equipped them as He did.

And again we ask—if bats were evolved, from what animal did they evolve? Some evolutionists claim that bats evolve from mice; but if so, WHERE ARE THE INTERMEDIARY FORMS? **There are none in existence in nature today, nor are there any in the world of fossils.** It is quite a jump from mice's feet to bats' wings, and from ordinary ears and nostrils to the intricate sonar system that bats have!

The True Explanation of the Wonders in Nature

It has been proved **abundantly** that there is no such thing as "acquired characteristics" that Lamarck supposed. He taught that "an animal which acquired a characteristic would transmit it to its progeny," and so the offspring would get not only inherited characteristics but also "acquired" characteristics. That theory has been **completely demolished.** The modern science of genetics and innumerable experiments have proved that characteristics are passed on to succeeding generations **only** through the genes and chromosomes in the germ cells.

For many generations the Paduangs of Burma have, from early childhood, stretched the necks of their young girls by winding malleable brass rings around their necks. As the neck is stretched, additional brass rings of half inch diameter are added, until as many as twenty are around their necks, yet their baby girls are NOT born with long necks!

For many generations the Chinese tortured their girls by tying back their toes so they would have small feet; BUT all new Chinese babies were always born with normal feet! Weismann experimented with rats by cutting off their tails for generation after generation. He proved conclusively that by so doing one can NOT develop tailless rats.

Long ago true scientists gave up the Lamarckian theory of "acquired characters."

Darwin taught evolutionary development through "natural selection"—the "survival of the fittest." In the struggle for existence, needed and useful variations remained and variations that proved weak or useless perished. But we have stated many times that any "specialized organ" is USELESS until it is fully developed. How could a mother's breast, the eye of an eagle, the legs of a grasshopper, or the voice box of a mockingbird develop through millions of years, seeing it would be **useless** until fully developed? Nor does the more recent theory of "sudden changes" called "mutations"* solve the problem—for such chance mutations are always comparatively minor, and can in

---

*Most all mutations, whether produced by chemicals, X-rays or other outside stimuli, tend to HARM the individual. For the past fifty years scientists have experimented with the **Drosophilia** (fruit fly) and have succeeded in producing many freaks, including horrible abnormalities. They finally produced: "A strange breed of four-winged flies" at the California Institute of Technology. BUT, according to their own confession, "the accomplishment represents A STEP BACKWARD along the path of evolution to a time when (probably) all flies had four wings. Dr. Edward B. Lewis, professor of biology (California Institute of Technology) said in the Institute's **"Engineering and Science,"** that "by tampering with the genes of the tiny fruit fly we have constructed a four-winged fly. . . . But these four-winged flies cannot fly." Prof. Leroy Victor Cleveland, discussing the work done by modern scientists on the fruit flies, says, "When scientists descend to accepting LOSSES of organs (or functions) as their only proof of evolution of new organs and structures, they are supporting a lost cause."

no wise account for the tremendous gaps between major groups, such as from reptiles to birds; nor does it account for the great DIFFERENCES that exist among the genera of a family or phylum. But there **is** an explanation for all phenomena of nature: GOD MADE THEM SO.

At last scientists have come around to the conclusion—since "natural selection" is not the answer, nor is the theory of "acquired characteristics"—that it MUST be "random mutations" —or else GOD. But "random mutations" can in no wise explain the marvelous gamut of life with MILLIONS OF EVIDENCES OF "SPECIAL DESIGN FOR AN INTENDED PURPOSE" as well as MILLIONS OF PERFECT "ADAPTATIONS" that fit an individual perfectly to its environment, such as fish are adapted to water and birds to the air, and the many thousands of specialized organs that show perfect adaptations, with absolutely NO evidence whatever of intermediary forms. And since, moreover, forced "mutations" are usually destructive, we are driven to the conclusion that God's creative work is the answer.

Professors J. T. Patterson and W. S. Stone, writing in "Evolution in the Genus Drosophila" (Macmillan Co., New York; 1952), admit,

"The only alternative to evolution by selection among random mutations, is . . . DIRECTED MUTATIONS . . . POSSIBLE ONLY UNDER SUPERNATURAL GUIDANCE."

Patterson and Stone are evolutionists, but they admit that science "cannot hope, on the basis of the theory of gradual change through mutations and selection, TO DEMONSTRATE THE EVOLUTION FROM ONE SPECIES TO ANOTHER IN THE LABORATORY." (Pp. 235, 503. See also p. 1: "Organic evolution is not a repeatable experiment.")

What a confession **that** is! They do not want to admit that GOD is the Author of creation. They will not under any circumstances give up their theory of evolution, even though they know now that they CAN NOT DEMONSTRATE IT IN THE LABORATORY. But they insist on clinging to their idol and refuse consistently to admit that GOD MADE ALL THINGS AS THEY ARE. Most modern scientists admit that the theory of evolution can not be proven, but they would rather accept it as a theory than to admit GOD and His work of creation.

Chapter 10

BEES AND ANTS: THE SOCIAL INSECTS
The Phenomenon of "Community Instinct"

(1) BEES, THE "MASTERPIECES OF CREATION" (Lutz)

When viewed from either the standpoint of their physical makeup or their social instincts, bees are truly a masterly piece of creation!

A colony of bees, called a swarm, may number from 10,000 to 60,000 or more individual bees. Bees have a most amazing "economy" in their hives. The work is divided among different groups—and each group instinctively knows what it must do. There are three kinds of bees in each colony. The queen, the drones (males) and the workers (undeveloped females). The queen does not rule the colony: her function is to lay all the eggs from which the new bees develop. And that is quite a job, since from 1,000 to 1,500 bees emerge daily, in the summer season, to replace an equal number of deaths.

One section of the hive is set apart for the nursery and it is here that the queen in her daily rounds lays the tiny eggs in specially prepared cells. The eggs hatch in three days into grubs (small worm-like creatures also called maggots or larvae). They are fed in the cell until they grow to the point where they fill the cell. They are then sealed in; they pupate, and soon a new bee emerges. We will have more to say about this interesting cycle later.

The worker bees literally "work themselves to death." In a colony of 50,000 bees there are about 30,000 workers. These bees average 10 trips in a day in the summer and visit a total of 300,000 flowers. Their wings fray out from much flying—and they usually die in about two months. Those that emerge in the fall live longer, as a rule.

The Intricate Anatomy of Bees—Showing "DESIGN" for a Purpose

The whole anatomy of bees is so intricate, involved and wonderful, naturalists and biologists could write volumes on the subject. Our present purpose is to call attention to a few of the "specialized" organs that grace the body of the bee—

organs and features that are obviously THERE FOR A PURPOSE, or so constructed as to reveal special creative design for a desired end.

The honeybee has **sharp tips** on its claws, to enable it to walk along on any rough surface; between its claws it has a little pad or cushion called the **pulvillus** that enables it to walk on smooth, slippery surfaces, such as glass.

Its Pollen-collecting Legs

Not only does a bee serve a most needed function in cross-pollination, but also the pollen it gathers is part of its food. Its body and legs are clearly **designed** for that purpose. Pollen clings to the hairs on its legs and body, and is transferred to pollen "baskets" on its hind legs. These "baskets" are made by a peculiar arrangement of hairs surrounding a depression on the outer surface of the legs.

On the middle pair of legs at the knee, is a short, projecting spur, used to pack pollen in the pollen baskets. On the inner part of the hind leg are a series of side combs used to scrape together the pollen that has stuck to the hairy body of the bee; with these side combs the bee then transfers the pollen to its pollen baskets. She then packs down the pollen in the baskets with the spurs on her middle legs. Long hairs on the front pair of legs remove pollen from the area of the bee's mouth and head. The middle pair of legs are used to scrape the pollen off the thorax and front legs; the stiff hairs of the third set (hind) legs comb the abdomen and also the accumulated pollen on the second pair of legs—and then she deftly puts the accumulation into the pollen baskets on the hind legs! The whole procedure is so efficient and practical, one can not help but conclude that someone must have planned it that way! Finally—when the bee reaches its hive, it uses a spur at the tip of each front leg to push the pollen out of the pollen baskets and into the cells of the comb in the hive.*

"The walking legs of the honey bee are modified for collecting food. Each is highly specialized and quite different from the others, so that, TOGETHER, they constitute a complete set of tools FOR COLLECTING AND MANIPULATING THE POLLEN upon which the bees feed." (P. 291, "Animals Without Backbones").

---

*For a more detailed and technical description of the marvellous legs of the honeybee, see pp. 291, 292 of "Animals Without Backbones."

The Antennae

The two rodlike projections that extend in front of the bee move constantly. They are not "feelers" but "smellers"—the "nose" of the bee. On the tips of these antennae are thousands of tiny "sense plates"—from 2,000 to 30,000 of them!

The Creator has provided bees with an ingenious means to keep these sensitive sense plates at the ends of their delicate antennae clean and functioning. When the bee inserts her head into nectar-holding flowers the antennae may become coated and clogged with bee glue or other foreign substances. On the bees' front legs is a movable piece of tough tissue, which can be raised by the bee, thus creating a small opening. On the outer edges of this opening are stiff, short hairs that act like cleansing teeth. To clean her right antenna, the bee bends the antenna toward the left, opens the "gate" of her cleansing apparatus, inserts her antenna, closes the "gate," then draws her antenna back and forth between the stiff hairs until all dirt and dust are removed! She does the same thing with her other antenna until both are clean and functioning again! Clever, isn't it? Can anyone believe that such a practical, ingenious setup came to pass by chance mutations, rather than that it was so planned and made by the Almighty when first He made bees?

Bees' Wings

The bee has two pairs of amazingly efficient, powerful wings that give convincing evidence of special "design." The bee has a rather bulky body and needs large wings to fly efficiently. But large wings would, on the other hand, hinder the bee's entering the narrow six-sided cell in the hive. So the Great Designer solved this problem in "engineering" in this manner: the larger front wing has on its rear edge a ridge to which hooks on the back wing are fastened when flying. This device converts the four wings into TWO LARGE WINGS FOR FLIGHT. When not in flight, the wings are released and they overlap, greatly reducing their size! The wings, moreover, are so made that in flight they move in a figure eight design, which makes it possible for the bee to go any direction—up, down, side to side, backwards and forwards, or remain motionless while hovering before a flower—much like a hummingbird. This system of wing structure is so complicated and yet so perfectly adapted to its intended purpose that one can not help but marvel at the Genius who designed it!

Why does a Bee have Compound Eyes?

Between their two large compound eyes, having many facets, bees (like many other insects) have three tiny eyes. Taken together (the compound eyes and the small eyes) the bee must have wonderful vision. There seems, however, to be a special purpose for the bee's compound eyes. The bee is largely guided by what is called "the polarity of sunlight."

The complex eyes of bees serve as a most complex compass, built into its head. These compound, faceted eyes are sensitive to the degree of 'the polarity of sunlight.' It should be explained that the waves of light streaming from the sun in all directions travel directly outward, in a straight line, in one direction. Now as the earth revolves, an animal (including insects) on its surface views this direction of the light from a constantly changing angle as the sun rises and sets. The bee, through its intricate, compound eyes, by simply glancing at any part of the sky in daylight, can interpret this angle immediately, and thus determine the position of the sun, the time of day, and its own position relative to its hive or the place where its food is! And this makes possible its long flights from its hive and its knowledge of its way home. Not only so, it makes possible the intricate "dance of the bees" (which we discuss later in this chapter) by means of which bees communicate to other bees in the hive vital information about their newly discovered food supply.

Surely the eyes of the bee, and the use it makes of its eyes, demands a Creative Intelligence of a high order—to put such wonders in such a small insect! And by proving **creation** we automatically **disprove** evolution.

The Bee's Stinger and Sting

A worker bee has a sting at the tail end of her abdomen. The sting has little barbs at its point which turn backward and make the sting stick in the victim's skin so firmly that the bee cannot pull it out. She must literally tear herself away—and leave part of her internal organs attached to the sting. **Soon after that she dies.** Who or what made a 'mistake' like that! Certainly **evolution,** seeking ever for "the **survival** of the fittest," would not do **that.** And what bee would desire, or help evolve a sting that meant her own destruction? This is a perfect argument for Creation, and a perfect argument against evolution ("natural selection" and "survival of the fittest")—for here is a case where the price of a specialized organ is DEATH, not survival, growth, development or enlargement. But GOD made it so—and it remains so to this day.

The efficiency of the sting is assured by this special arrangement: attached muscles pump the poison into the wound even after the bee has flown away. Just a "chance arrangement" of blind evolution? No; it was **planned** that way.

BEES AS CHEMICAL FACTORIES

Bees have a special **"honey stomach"** separate from their own food-digesting stomach. The bee carries the nectar in this special honey stomach; there the sweet, natural fluid is transformed into honey! Were "survival of the fittest" the law of life, bees would not be interested in developing a second stomach: for they could readily adapt to live on nectar. But GOD planned it that way for "colony" needs and colony survival— and also for the needs of man—when He gave bees their second stomach.

The delicious **honey** that the bee makes out of nectar contains levulose, dextrose, other sugars, dextrines, gums, vitamins, proteins, mineral salts (calcium, iron, copper, zinc), iodine, several enzymes, and many other vital and nutritional substances. The little honeybee, as far as we know, is the only creature on earth that makes honey in quantities large enough to benefit man.

Bees make a **bee glue**—"propolis"—from the sticky covering on certain buds. If a mouse chances to get into their hive, they will sting him to death, and then dispose of the body by coating it from head to tail with "propolis," bee glue. This forms an airtight mausoleum for the decaying mouse: so there is no odor nor contamination of their living quarters! It was not in vain that God enabled bees to make this glue or varnish.

Bees make **wax** out of honey in four little pockets (manufacturing centers) on their abdomens. In order to start the secretion of wax special heat is needed; so the bees gather together in a large pendant mass, their wings buzzing rapidly all the while. Presently, "a strange sweat, white as snow . . . begins to break out over the swarm." These are wax scales that are removed by the bees with a pair of pincers found at one of their knee-joints. These scales are then chewed into a soft paste which can be readily molded into the delicate wax film of the cells. Even skilled chemists cannot make bees' wax as good as that made by bees!

"Bees' wax is unlike anything else. It contains a fatty acid called cerin, minute quantities of alcohol, myricin, hydro-carbons, etc. It has a higher melting point (140°) than other waxes."

They also make another grade of wax. When the larva

has grown in its cell to fill the space, worker bees seal over the cell with a special type of **porous** wax so that the larva can breathe.

Bees make a magic **"royal jelly"** that they feed to their grubs for their first 48 hours after hatching from eggs. This "royal jelly" is manufactured in the ductless glands of the nurse bees. When queens are desired, the nurse bees feed the grubs five days on royal jelly, instead of only two. If queens are not desired, at the end of forty-eight hours the grubs are taken off the royal jelly diet and fed a mixture of honey and pollen dust—mixed in EXACT proporptions! This is another instance of surprising knowledge and accuracy. This change in food brings about the birth of a neutral (female) bee—the workers with which we are familiar. How the prolonged feeding of the grub on royal jelly brings about the change from a normal neutral worker bee to a queen bee is not known.

After the grub is sealed in its wax cell, the larva spins a silk cocoon; but the larva's "silk factory" is presently discarded when the larva is transformed into its final "bee" form. How is it that the ability to make silk is present with the bee in the larval form just when needed, and no longer? God gives His creatures all they need, when they need it—but he does not load them down with excess machinery.

Needless to say, though we have touched on some of the high spots, our resume of the bee as a "chemical factory" is superficial. Were one to go into ALL of the chemical abilities of the bee, it would be a most astonishing presentation of the manufacture of proteins, enzymes, digestive juices, various and sundry types of cells, and a thousand and one molecular combinations that would startle us into rapturous astonishment.

BEES' WORK OF POLLINATING FLOWERS

Bees pollinate over fifty flowers and agricultural crops while collecting nectar. In this way they are "fifty times more valuable to society than through the honey they produce." Without their polination, orchards would produce little or no fruit, and many crops could not be grown. "No practical substitute for this pollination has been found."

When the bee goes into the flower after nectar it innocently collects the golden pollen as it rubs its body against the precious powder when it enters the flower. When the bee goes to another flower, some of the pollen is rubbed off on the second flower, and that flower is fertilized, with the bee as the unwitting agent. Then the surplus pollen is carried back to the hive by the bee in its "pollen baskets."

As a pollinator, the bee is very efficient, due to its habit of visiting only one plant species, **and no more,** at certain times of the day and at certain seasons.

Did this just happen—or did the all-wise Creator plan it so. For its own sake the bee might as well be promiscuous in visiting flowers, but for the sake of pollinating flowers—done unconsciously by the bee—it is necessary that bees go to the same species of flowers for a period of time before switching to other flowers. The pollen of one species of flower will not fertilize another species of flower. Evolution has no answer to this mystery of why the bee sticks to one species at a time in seeking nectar; but the devout, logical thinker knows that GOD MADE IT SO.

A little thought will convince one that bees and the flowers they pollinate MUST have been created at about the same time, for "the flowers need the bees for pollination and the bees need the flowers for food—for their very survival." Here again we see a wonderful partnership in nature that could only be the work of an Infinite Creator.

BEES AND HEREDITY

The queen is the mother of all the 10,000 to 100,000 bees in the hive. Fertilized during a nuptial flight by a male bee four or five days after her emergence from her cell as a queen bee, she may lay as many as 2,000 eggs a day during the nectar gathering season, and keep that up for two or three years! All from **one** mating! After the male's sperm is deposited in her body, the sperm sac is torn from him, causing his death. Then she returns to the hive and deposits one egg to a cell, so the maggots are hatched in cells.

"The baby bee, which hatches out of the egg in about three days, certainly does NOT resemble its mother. It is a fat white grub with neither wings nor legs and almost no head. Helpless, it lies waiting in its cell for nurse bees to feed it. So hungry are these youngsters that each one needs over a thousand meals a day. The greedy little creatures grow so fast that in six days each fills its cell tightly and is ready to take the next step in its life, the step that is called pupating. The nurses build a wax cover over the cell and the larva spins a silk cocoon inside. Within the larval skin wonderful changes take place. Legs and wings push out and the body changes shape to make three distinct parts—head, thorax, abdomen. The skin hardens and turns dark. After twelve days the adult worker is ready to cast off her larval skin and chew her way out of her cell."

The transformation of grub into adult bee, during the pupating process, is a MYSTERY AND A MIRACLE far beyond human comprehension. It is impossible to explain it by natural causes. It is a well known phenomenon that can only be explained by admitting a supernatural agency.

When the queen desires a worker bee, her pressure on the sperm-carrying sac in her body forces a sperm into the egg—and a female bee is conceived. If a drone or male is desired, she does NOT press on the sperm sac; the drone is thus an example of "parthenogenesis," or virgin birth: for the drone has a mother and grandparents, but NO FATHER. This complicated manner of procreation defies explanation; save on the basis of Divine Creation.

So this little creature knows how to do what man cannot do: it knows how to control the sex of offspring.

And bees have continued producing queens, workers and drones since ancient times—and honeybees are still honeybees!

Without serving an apprenticeship, twenty-four hours after emergence from her cell and cocoon, the young bee begins her duties as a nurse bee, and she performs her duties without instruction, confusion or lack of skill—the perfect example of both individual and "community" instinct. (Later on we will speak more of "community" instinct). She also is able to make royal jelly and feed her sisters who are just coming into adult life. The entire process, from the nuptial flight, to the laying of so many eggs for so long a time, through the stages of larva, pupa and adult bee, is so marvelous—as wonderful as the metamorphosis of a caterpillar into a beautiful butterfly—that it can not be accounted for by any theory of evolution. It is the work of an all-wise Creator.

VARIETIES OF BEES

Most, but not all, bees live in colonies; some are "solitary" in their habits. Nor are all bees the well-known "honeybee" variety.

There is a tiny black **"sweat bee"** that nests in the soil. It is a wild bee that is useful in helping pollinate fruit trees, alfalfa and other agricultural crops.

The **"resin bee"** builds a nest of pebbles, sticking them together with resin. Inside this nest, the bee builds cells for honey and eggs.

"The **Carpenter bee**" of Africa, unable to make wax, excavates a chamber in a pithy plant stem. She deposits one egg at the bottom of the shaft, leaves with it some beebread (pollen soaked with honey) for the young bee yet to be born, seals off the egg, lays another egg on top of the seal, provides some bee bread for that one, seals it off, lays another egg, etc., till she reaches the top of the shaft. In the subsequent emergence, strange as it may seem, the topmost bee emerges FIRST, even though that egg was laid last; then the second from the top and on down until the last one, which was the first egg laid! "The first shall be last."

One of the most interesting of all varieties of bees is the so-called **"Leaf-cutter"** bee. There is an essay by J. Henri Fabre, noted entomologist, in the book, "Green Treasury" (pp. 463, ff.) that gives this incredible information about the Leaf-cutter bees:

If one has a cooking pot that has lost its lid, no one will attempt to go to the store and buy a lid to fit it without taking an exact measure of the top of the pot. This homely illustration will help us appreciate this amazing feat of the Leaf-cutter bee.

"The Leaf-cutter has no mental picture of her 'pot' because she has never seen it; in fact, she has probably never seen any sort of a 'pot' built by her neighbors. She must, far away from home, cut out a disc from a leaf that will FIT the top of her 'jar' when she gets it made. . . . In doing her job the leaf-cutter cuts a pile of discs (from leaves), finds a vacated chamber of the Capricorn from which the nymph has departed, and in this she builds her cells. Using various sizes cut from various types of leaves she constructs cells. (Mr. Fabre counted one cell that was made of 714 pieces of leaves.) She barricades the opening into the chamber by 350 more pieces of leaves—making a total of 1064 pieces of leaves used so far in her construction job. One dauntless bee and one alone has produced the whole of this prodigious mass!"

The pieces stacked up to make lids were brought up before the cells were made. When completed, she places these round pieces of leaves—'lids'—and they fit perfectly! "When cutting these pieces for the 'lids' the bee was as sure of her scissors as a dresssmaker guided by a pattern—AND YET SHE WAS CUTTING WITHOUT A MODEL, WITHOUT HAVING IN FRONT OF HER THE MOUTH (of the cell) TO BE CLOSED. All leaf-cutters have the same talent for making lids for their 'pots' (cells)."

Amazing, isn't it. Such wisdom (without native intelligence to warrant it) must come from outside the bee, from the all-wise, Divine Creator. Who can contemplate the marvels of bees and not glorify the God who created them?

There are many other varieties of bees, all with distinctive characteristics. Some of these bees are the "Mining bee," the "Cuckoo bee," the "Giant cotton bee," etc. In this study our main interest is in the well known honeybee.

THE LANGUAGE OF BEES

Unbelievable as it sounds, honeybees actually have a "language" by which they "talk" to one another. To be sure, it is not a spoken language; however, they communicate with each other through special movements (called "dances") and through scents. The famous Austrian scientist, Prof. Karl von Frische, whose work with bees has won him international fame, has demonstrated that the honeybee indicates the presence, direction and distance of pollen and nectar food to other members of the hive, by executing **strange little geometric dances.**

A foraging bee comes home with a full load of pollen and nectar. She flies straight to the hive to share her harvest and to tell the other bees where it is. When she reaches the hive she first of all gives other bees sips of nectar from her mouth; then she begins a little dance, called the round dance. She circles to the right, then to the left, and repeats this many times. She dances energetically for about half a minute. This means, "There is good food nearby."

Other bees fly out to find the food. They know which kind of flower to look for **by the scent of the nectar** on the returning bee. When these bees get back to the hive with their loads of food, they also perform the same dance—provided there is still plenty of food left there. But when the food begins to run out the returning bees dance less and less vigorously and more briefly, so that fewer bees are stimulated to go to that place.

Now, bees sometimes fly a mile or more on foraging trips. On long trips they use a **different** dance to announce food that is more than about 100 yards away. Through this dance, the bees can actually tell **how far away the food is, and in what direction!** "The dancer makes a short straight run forward, wagging her abdomen from side to side. Then she turns left in a half circle, comes straight forward again with her tail-wagging motion, and circles to the right. She reveals the approximate distance to the food by the speed of the dancing—the faster the dance, the closer the food. She tells the direction of the food from the hive by the direction in which she makes her tail-wagging run. . . . Most of the bees that pay attention to her message will come to exactly the spot she visited!"

Von Frische was so surprised at his discovery that he said, "No competent scientist ought to believe these things"—and yet his work has been verified and proven correct.

After Von Frische had learned "the language of the bees" he got an assistant to put some bee food at some distance from a hive—but Von Frische knew not where. When a bee found it, it was distinctly marked, and then Von Frische watched carefully its dance when it returned to the hive. He interpreted the message the bee danced to its fellows at the hive; and then the naturalist said to his assistant: "The food you placed is 320 meters from the hive"—and he also gave the direction. Checking showed that actually the food had been placed 332 meters from the hive, and in a direction he had estimated correctly to within four degrees!

Ronald Ribbands, University of Cambridge naturalist, has discovered **that** in addition to the "dance" of the bees, their "taste and smell" also play an important role in their methods of communicating with each other.*

THE PHENOMENON OF "COMMUNITY INSTINCT"

The extraordinary abilities of bees is explained by all—creationists and evolutionists—as due to "instinct." Bees truly have amazing instincts. Take for example their striking **"architectural"** abilities, which are far beyond what the bees' limited intelligence warrants. Bees, as you know, construct a perfect hexagonal cell— "the cleanest, most appropriate, most enduring receptacle for honey that can be conceived." Bees construct their cells in double tiers—directly opposite each other—with one bottom serving for both cells. And since the cells are six-sided, **each one of the six sides of each cell is also the side for another cell next to it**—the best possible shape for the prevention of waste. These cells, called "one of the wonders of the natural world," are made of thin plates of pliable wax. ALL OF THIS CONSTRUCTION WORK IS DONE IN TOTAL DARKNESS. When occasion requires, such as in odd corners, the cells are shaped square, triangular, or just to fit the space. The cells are tipped up slightly from the horizontal, to hold the honey better. They are always filled before being capped. The cells are geometrically accurate, having "a precision that baffles description." This amazing ability is of course for the benefit of the COLONY, and is called "colony instinct."

Colony instinct is further seen in the realm of **community protection.** A stranger approaching a hive may notice bees circling around in a wide sweep. These are guards constantly on the lookout—seeking the protection of the colony, not especially their own—and if an enemy threatens the hive one of the guards notifies the colony and a large detachment of "soldiers" goes forth, ready to attack if any attempt is made against the hive: and the courage of the bees knows no limit in defence of their home and their treasure.

The bees' plan to generate heat when needed. As is true of all insects, honeybees are cold-blooded creatures: they cannot regulate their body temperature to a specific degree as people can. However, differing from most other insects, they can and do pro-

---

*For those who wish further information on this subject, we suggest the pamphlet, "THE SCENT LANGUAGE OF HONEY BEES," by Ronald Ribbands. Published by The Smithsonian Institute (1956); (Publication Number 4243). The pamphlet will be sent free to those engaged in research.

duce considerable heat by the activity of their bodies. In a cold hive bees begin a muscular activity that resembles shivering. The colony forms a compact cluster. The bees on the outside of this cluster crowd closely together and turn their heads inward, thus forming a sort of shell. The bees in the center of the cluster move rapidly, shake their bodies, and fan their wings in a lively manner. In this way they produce a summer temperature inside that cluster though the thermometer may show freezing on the outside of the hive. In the summer when it is too hot inside the hive, bees air-condition the hive by gathering outside, beating their wings vigorously, so blowing air into their hive. These activities of course give more evidence of "community instinct."

Bees are Guided by INSTINCT—not by Intelligence

J. Henri Fabre, the great French entomologist, says, "the bees' instinct is **fixed,** unchanging, limited and non-progressive as the law of gravity."

Fabre placed a piece of straw in some cells in a hive and the bees extracted these straws as often as placed there UNTIL THE HONEY-GATHERING PERIOD HAD PASSED and the egg-laying season took its place. Then the bees would ignore the straws in the cells. Even the queen would lay her eggs in cells with straws in them, as she did in the perfect cells having no straws. The workers would then seal them up at the proper time, as they did the untouched cells. **If these bees had the faintest degree of intelligence they would know no young bee could develop in the abnormal condition with a straw in the cell.** On the passing of the honey-gathering season, their 'instinctive disposition' had changed and they were helpless to recall the departed impulse. Fabre tells of other tests, and the insects' failure to adjust themselves, and so he concludes that bees are "hopelessly non-progressive, and non-intelligent."

Fred Kohler, prominent evolutionist, writing on "The Societal Organism," (P. 59, "Evolution and Human Destiny"), concedes that the individuals in a beehive do NOT show evidence of the intelligence that the whole colony shows.

"In its functioning the (bee) colony acts as if it 'knew' what it was doing. This appears particularly remarkable when one considers that the individual insect apparently does NOT possess by itself the degree of intelligence evident in the functioning of the colony. Despite the apparent lack of consciousness of the individual insect, **the colony shows a rational behaviour—a behaviour that is directed to assure the survival of the colony.** HOW IS SUCH A SITUATION POSSIBLE?" (Caps, bold face, ours).

Well might Fred Kohler ask, "HOW IS SUCH A SITUATION POSSIBLE?" He further states, "As true instincts are neither taught nor transmitted by example from one generation to the next, they must, as there is no other possibility, be part of the genetic code determining the species." He then suggests that instincts are subject to "mutations" just as much as physical characteristics. We would like to ask Dr. Kohler—HOW CAN AN INSTINCT NECESSARY FOR THE SURVIVAL OF THE COLONY COME ABOUT "GRADUALLY"? If the "instinct" to build wax cells is only partly there, the colony will not survive.

And how can he possibly explain a "colony instinct"—geared to the survival of the colony rather than the individual and yet it is a "part of the genetic code" of each individual—as the result of evolution, when evolution teaches the SURVIVAL OF THE FITTEST: i. e., each individual seeks FIRST for its own survival?

We repeat: "a gradually evolving instinct" is IMPOSSIBLE—for **the first bees that ever lived** had to know as much as their modern mates about cell construction, wax making, royal jelly, the secrets of feeding, the way to predetermine sex, nectar gathering and honey making—otherwise the entire colony would have perished before evolution had a chance to get it started! To believe that the abilities, characteristics, physical make-up, specialized "organs" (such as the stinger, the antennae cleaner, and the pollen basket) and instincts came about through gradual development is utterly impossible to the logical, open mind. Such a theory can be received only by those who have been brainwashed into blind adherence to a dogma, believing because "others (with 'authority' or 'scientific standing') believe it." IF there ever was a time when the colony instincts of bees were only partially developed, there never could have been bee swarms that survived! BEES FROM THE VERY BEGINNING HAD TO BE AS THEY ARE NOW, or there would be no bees today.

No wonder that Charles Darwin, in one of his books, found in the common honey bee a problem that baffled him **"more than any other he encountered."**

"It looks as though God Almighty," says H. Gracey (in, **Evolution and the Honeybee**) "in this little insect (the bee) prepared a trap to catch and baffle the ablest men that ever tried to support the evolutionary theory. In the honey bee we have a highly endowed little creature with instincts that seem to rival reasoning powers more closely than the instincts of any other creature—and yet there is no door left open for the entrance or the transmission of these wonderful peculiarities. The parents of the bee have none of these instincts to transmit; and the honeybee itself can transmit nothing. Mr. Darwin's theory of transmission is closed at both ends. We must admit: THIS IS THE HAND OF GOD."

(2) Ubiquitous ANTS: Energetic Witnesses for God and Creation

"Go to the ant, thou sluggard; consider her ways, and be wise: which having no guide, overseer, or ruler, provideth her meat in the summer, and gathereth her food in the harvest" (Prov. 6:6-8).

"The ants are a people not strong, yet they prepare their meat in the summer" (Prov. 30:25).

GO TO THE ANT, thou mental sluggard—and learn lessons of God's great work of creation! Ants are to be found most everywhere: in cities, in the fields, in deserts, in forests and in dense jungles of the earth. Ants belong to the insect order Hymenoptera, to which bees and wasps also belong. Some ants are winged and some are wingless; some are noticeably large (i.e., for ants) and some are so small they can scarcely be recognized as ants. In color they may be red, black, red and black, brown or yellow. They can be readily distinguished from other insects by their slender waist (called the petiole) on the top of which rise one or two bumps called nodes. ALL ants live in colonies; there are no solitary kinds. Each colony consists of three kinds or castes: queens, males and workers (undeveloped females). Queen ants lay eggs. In some ant colonies, the workers come in different sizes—and each does the work best suited to its size. Some of the workers are called "soldiers" because of their unusually large jaws which enable them to defend the nest. Some ants can sting; others bite; and still others squirt a bad-smelling, irritating fluid on their enemies. Some ants, like the famous army ants, are blind, and are guided by the senses of touch and smell. There are over 2500 species of ants—each a marvel of creation.

Stages in the Development of Ants from Eggs

With most species, only the queens and males have wings; and they have them only for the mating flight. After mating, the male ant generally soon dies. A lone queen can start a new colony.

Upon her descent to earth, the queen either bites or breaks off her own wings. (If other ants are with her, they often tear off her wings for her). She eats her fill of food which the workers bring in. The workers take the eggs as she lays them, place them in nurseries, carry them from day to day from one gallery to another, even bringing them out of the nest into the sunshine, then restoring them to an underground gallery at night, so that their eggs get both the necessary heat and moisture. The eggs hatch (according to the season) in from fourteen to forty days.

Small, white fleshy grubs emerge—legless and conical in shape. They are helpless and have to be fed by the workers with a special, semi-predigested food. The larvae may reach the chrysalis form in a month or six weeks, but some species live through the winter in the larval stage, requiring the unremitting attention of their nurses throughout that long period!

The chrysalis may be either naked, or be incased in a neat silken cocoon. Now here is one of the **fascinating miracles of** "colony instinct" with which the Creator has gifted some species of ants:

"When a larva is ready to spin its cocoon, worker ants bury it in the ground. Burying is necessary, because the chubby, legless larva needs something on which to fasten its silken thread as it begins to spin. The silk flows from the larva's mouth, not from the abdomen, as in spiders. SOMEHOW OR OTHER THE WORKERS SEEM TO KNOW WHEN THE COCOON IS FINISHED. They then dig it up, clean it off, and carry it to a suitable place in the nest. The workers again know when to help, when the new adult ant within the cocoon is ready to emerge. THEY THEN CUT OPEN THE COCOON and free the new adult ant, which is so feeble it can not open the cocoon by itself! How can one account for this amazing "colony instinct" that causes the worker ants to look so thoroughly after the larva cocoon—doing exactly as they should, at exactly the right time—even knowing when to exhume the buried cocoon? Evolution has no satisfactory answer; such an intricate procedure, INVOLVING THE WELFARE OF THE COLONY, and taking no thought for the welfare of the individual ant, proves that a Master Mind created both the little animals and their instincts that make such an involved, altruistic system work!

The Strange Case of the Tents made by Baby Ants

Dr. William Mann first observed the strange family life of the curious ant, **Polyrhachis simplex,** when he explored the Kerak region of Palestine in 1914.

"Small silk-and-leaf structures that the explorer found on bushes near his tent were the tip-off that **Polyrhachis** was living in the vicinity. Each of these structures sheltered leaf hoppers, which exude a kind of nectar that the ants feed on. The ants were sheltering the hoppers, as humans keep milk cows, to furnish food for the colony."

Later Dr. Mann discovered HOW the aerial "cow barns" for the food-producing leaf hoppers were built.

"Worker ants were carrying Polyrhachis' newly hatched larvae to the building site. THEN THESE INFANTS (LARVAE) SET TO WORK SPINNING THE SILK TO MAKE SHELTERS FOR THE FAMILY'S MILK COWS!"

This is a phenomenon of the first order: silk-making larvae, before they made their own cocoons, are used by the worker ants to make shelters out of silk, to protect the "cows" (leaf hoppers) used by the ants as their source of food! Remember, this achievement is NOT the result of native "intelligence" in the ant—for they have practically no intelligence; but it is the result of an OUTSIDE INTELLIGENCE WHO EQUIPPED THESE ANTS WITH THE "INSTINCT" TO KNOW HOW TO DO THIS AND TO DO IT!

True Symbiosis: The Case of the Blue Butterfly and the Ants

If the case of the Ants who make Shelters for their Cows is interesting, this one is even more so.

"In June the larva of the large blue butterfly of England hatches on the wild thyme bushes. It feeds for about twenty days, then moults. After moulting the larva stops eating and wanders about aimlessly. At this point ants gather about the larva and (1) stroke its honey gland with their antennae and drink the sweet droplets it gives off. Finally an ant picks up the larva in its powerful jaws and carries it underground to its nest. (2) Here the larva is permitted to feed on ANT GRUBS, and it continues to yield 'honey' whenever the ants stroke it. By winter the larva has become four times its original size and has gone into hibernation. The following spring it becomes active again and soon encases itself in a cocoon. (3) In May a lovely butterfly emerges from the cocoon, makes its way above ground, where it flies off to lay its eggs!"

Here is a case in which ants sacrifice some of their own grubs, in order to keep their "cows" (butterfly larvae) alive and willing to be "milked" of their "honey" on which the ants feed! In this manner two different orders of creatures help each other—true symbiosis. One is dependent on the other. Explain this phenomenon? GOD MADE IT SO!

More Evidences of the "Colony Instinct" in Ants

When building their homes, they divide into troops. One troop does the excavating. Their habitations are well-planned, and include a central assembly, or "club house," where they gather in large numbers. They provide numerous entrances to their underground dwellings. At night, when most of them are resting, they keep a few sentinels on duty. They also make outside roads leading up to their hill.

The colony instinct in ants results in a well-organized community life. After completing their home, with its many chambers and tunnels, all well-planned and constructed, "they gather food, feed their young, and tend to their domestic animals—and EACH MEMBER OF THE COLONY FULFILLS ITS DUTIES WITHOUT HESITATION OR CONFUSION. . . . "Certainly there is no counterpart among other living creatures to the military, food-gathering, cattle-keeping, and slave-making activities of the ants or to the perfectly ordered system of the beehive."

How Ants Maintain their Food Supply

Some ants keep "cows," others run "farms," some make "biscuits," and still others store their food in living vats! Other means of gathering and storing food are too numerous to elaborate, but all are extremely interesting and instructive.

HARVESTER ANTS, large, long-legged red or black ants, eat dry foods, especially seeds. They gather and then store the seeds in their nests. After long periods of wet weather, the harvesters bring their seeds out into the sunshine to dry. Sometimes, when the seeds sprout in the nests, the workers remove the growing plants before they clog the passageways!

Some ants steal their food. Dr. Charles D. Michener says,

"Some species of tiny ants nest near or actually inside the nests of larger ants. These small ants creep into the passages of the big ants and steal their food. When they are chased, they escape into the tiny passages they built that are too small for the big ants to enter."

The small "ARGENTINE" ANTS, when outdoors, eat dead insects, flower nectar and honeydew—the sweet juice excreted by plant lice and other small leaf-sucking insects. To be sure of a good supply of honeydew, the Argentine ants—and many other species—care for these plant lice as farmers care for their cows. They stroke their little "cows" to coax them to give droplets of honeydew. They even carry their insect "livestock" to different parts of the plant, or to different plants in the garden, to make sure they have plenty to eat. Some ants take their plant lice into their own nests for the winter. And they will dig tunnels in the soil for the convenience of root-sucking plant lice.

Ants that make biscuits! Studies of the common Mediterranean ant, **Aphaenogaster barbarus,** have revealed that these ants actually make biscuits from the seeds they collect!

The seeds are collected and dried, and later put out in the rain till germination begins. After germination the seeds again are dried . . . and later chewed into a kind of dough that is then dried into biscuits! (Nature Parade, p. 27).

The famous "LEAF-CUTTER" or "PARASOL" ANTS THAT MAKE GARDENS AND RAISE THEIR OWN CROPS! These amazing creatures are called "Leaf-cutter" or "Parasol" ants because they may be seen in processions, each one bearing above his head a bit of green leaf! These bits of green leaf are NOT for food, but are taken to their nests and made into compost—for these ants are actually FARMERS (perhaps the only "farmers" in the animal kingdom, with the exception of certain termites). They deliberately sow, prune, manure, weed, and harvest their crops, which are different kinds of fungi. (Some seem to be related to the mushrooms we raise.) In the Bronx Zoo in New York City a colony of these farmer ants is on display, started there in 1950 when a queen and her attendants where shipped to the Zoo from Trinidad.

The British naturalist, Thomas Belt, published the results of some special investigations he made of these "farmer" or "parasol" ("leaf-cutter") ants.

He discovered that the ants do not eat the cut leaves but hash them up into a compost, on which they sow the spores of certain fungi. The ant farmers weed and cultivate these fungi as carefully as any gardener tends his cabbages. The little plants are not permitted to reach the fruiting or "toadstool" stage; instead the ants constantly prune them back—with a purpose!

The repeatedly pruned fungus forms tiny knots, about the size of a pinhead, called "kohlrabies." These are eaten by the ants. The kohlrabi we eat is really a greatly thickened stalk of cauliflower; it is not found in nature, but is the result of human horticulture. The kohlrabi of the ants is just as clear a case of horticultural know-how! THIS IS TRULY "SCIENTIFIC" FARMING!

They use this cultivated food for another purpose. By rationing the amount of kohlrabi eaten, these ants produce four or five different sizes of ants, that they put into different "castes" of workers. Those fed on minimum amounts never grow up to be more than "minims," tiny workers who tend the fungus garden and feed the larvae. A medium-rich diet develops the "mediae," workers who do most of the leaf-stripping. More food develops the big, fierce soldiers who defend the nest; they can bite so savagely that they can draw blood. And a still richer diet produces the idle males and the virgin "princesses," both winged in preparation for the nuptial flight.

(The account goes on to tell of the queen wrenching off her own wings and starting a new nest). (Condensed from the Reader's Digest article, "WONDERLAND OF ANTS").

WHO MADE "SCIENTIFIC" FARMERS OUT OF THESE TINY INSECTS that have no intelligence—that is, no power to think through a problem—but are guided entirely by "blind, tyrant instinct." When one sees such a miracle as this in nature, he can but "bow and worship."

Ants that store Honey in their own Bodies

Some ants eat nothing but honeydew. These so-called "HONEY ANTS" store collected sweets for dry, needy periods by making some of their own ants into LIVING STORAGE TANKS. While these "repletes" are yet young they are fed such enormous quantities of honey dew that their skin stretches and they swell into round balls as big as peas, full of the precious honeydew. They cannot move; but merely hang from the ceiling of the cavernous nest for the rest of their lives, receiving and dispensing honeydew as required. There is one thing about it—they probably never get hungry, so they might be satisfied in their overstuffed existence. Honey ants of one species, found in Australia, fill certain of their numbers full of honeydew, then put them in cells and keep them in prison for the rest of their lives! The other ants then feed from these living storage tanks during the dry season.

Experiments Prove Ants Act through INSTINCT, not through Intelligence

A writer in the June, '57 "Scientific American" says,

"Here is an ant. It exhibits an extremely complex pattern of behavior. Does this signify intelligence or is the behavior purely automatic? We observe, for example, that the animal attacks every foreign ant that enters its territory. Does it recognize the newcomer as a stranger and anticipate a potential danger to its own group? Or is it merely acting automatically to a strange odor from the newcomer? As a test we extract some juice from a strange ant and smear a little of it on a member of the ant nest. When the ant returns to its own nest, its nest-mates become greatly excited; they quickly attack and kill it. . . . The ants are NOT acting with intelligence but simply as automatons, responding blindly to an odor in accordance with mechanisms (instincts) which nature has built in them."

The obvious deduction is, since ants do NOT have intelligence, but are motivated solely by instincts, that these instincts were GIVEN TO THEM BY AN OUTSIDE FORCE to preserve them as a species and properly equip them to live successfully in their environment. That being the case, ants are the work of Creation and were made in the beginning as they are now. A SLOW PROCESS OF EVOLUTION, GRADUAL CHANGE, BRINGING TO PASS DECIDED INSTINCTS AND SPECIALIZED PHYSICAL CHARACTERISTICS THROUGH LONG ERAS OF TIME WOULD NEVER PERMIT THE SURVIVAL OF THE SPECIES DURING THE PERIOD OF DEVELOPMENT.

ANTS TOO HAVE A "LANGUAGE"

Though not as astounding as the language of bees, ants too have a means of communicating to other ants.

Ants rely almost entirely upon odor trails. When an ant "scout" finds a lot of good food she becomes excited and hurries back to the nest, with her abdomen lowered until it nearly touches the ground. A faint scent comes from her abdomen and clings to the surface on which she runs. When she gets home she gives some of her food to the other ants. They get excited and follow her trail to the food supply. Each ant going over the trail strengthens the odor and soon a steady stream of ants flows to and from the nest. By some mysterious way (still unknown to science) trailing ants can tell from the odor WHAT DIRECTION the scout took on the path; that is, they know if they are going or coming toward the nest on the path. If you let ants trail over a piece of paper and then turn the paper so that the path is reversed, the ants will be confused and MAY GO BACK THE WAY THEY CAME. . . . "An ant is not really intelligent. It is guided by instinct. Ants can learn quickly to follow a trail (by the scent), but if a trail is made to follow a circle, ANTS MAY FOLLOW IT ROUND AND ROUND UNTIL THEY FALL OVER DEAD."

VARIETIES OF ANTS

The largest of our common ants are the shiny black CARPENTER ANTS. Some species are nearly half an inch long. These large, awkward-in-movement creatures may get into the woodwork of a house and riddle it with their galleries. They make their nests by digging in the soil or by chewing galleries in wood. They do not eat the wood, but remove it to make space for their nests.

HARVESTER ANTS dig nests in the soil and live on the seeds they gather and store for winter.

HONEY ANTS live on nectar from flowers or honeydew from aphids ("cows"), as we have described above.

KIDNAPER ANTS hide in the walls of other species' nests and then steal their babies.

SLAVE-RAIDING ANTS tear open the nests of other ants, seize their hapless young and carry them away to make slaves of them.

AMAZON ANTS cannot live without the help of slaves of other species. Amazon ants are large, bright red or partly black ants with peculiar, long, sickle-shaped jaws. These jaws make excellent weapons for fighting, but they are very poor tools for eating, digging, feeding or carrying babies. So these ants **must** capture slaves to keep themselves from starving to death. An Amazon queen will enter a nest of black ants to start her colony. She will probably murder the black queen with her sharp jaws. Soon she is accepted by the black ants as their new queen. Later, her brood enslave the black ants and make them do their work.

In Kenya, Africa, is a species of DRIVER ANTS in which each colony has three queens that turn out no fewer than 11 million eggs annually! The workers of this species reportedly can kill a wounded elephant and then pick his bones clean. The workers in this species have TWO stomachs: one for their own use, the other, a "social" stomach for food for the non-working members of the nest! Who designed **that** phenomenon? It **had** to come from an outside Intelligence.

CORNFIELD ANTS, very numerous and widely distributed, eat the sweet secretions of corn-plot aphids. Aphids lay their eggs in the ant burrows. When these hatch in the spring, the ants place the aphids on weed roots till the corn is planted and growing. Then the ants transfer the aphids to the corn roots, thus insuring a constant, desirable food supply!

MOUND-BUILDER ANTS construct great cities in the soil, carrying up dirt and sand bit by bit until they have mounds three feet high and ten feet in diameter, filled with hundreds of tunnels, rooms and storage vaults. These large ants, also called the ALLEGHENY ANTS, often build on wooded slopes, among pine trees. They mix pine needles with twigs, straw and other debris, in making their vast honeycomb of intercommunicating passages and chambers in the mounds. Some students estimate that perhaps 100,000,000 ants may occupy one of these larger nests, and that the nests may remain tenanted for 20, 40 to even 80 years, if left undisturbed.

One of the most fascinating of all ants is the **Oecophylla smaragdina;** the word "Oecophylla" meaning "leaf house." The "leaf house" these odd ants build is high up in trees! And the leaves are the living leaves of the tree which are woven together by silky threads. About a half century ago a traveling zoologist, Franz Doflein, observed what actually took place up there in the boughs.

"Worker ants, working in gangs, held the leaves together, clinging to the edge of one with all six legs and holding the edge of the other leaf with their mandibles. If the distance between the leaves was too great, an ant chain was made by one ant holding another in its mandibles, until the chain was long enough to span the distance from one leaf to the other. Sometimes chains of seven or eight ants are necessary to reach from one leaf to the other. Once this has been accomplished, the whole chain slowly retreats to the leaf on which the supporting ant was standing, until the two leaves have been pulled together sufficiently so that they could be held in place by a single row of ants. Then another gang of workers appears, EACH ONE CARRYING AN ANT LARVA IN ITS MOUTH." Franz Doflein then observed that, while the adult ant can not spin a thread, the larva can. Using their own larvae like upholstery needles—or like shuttles, to use Doflein's term—THE ANTS WOULD THEN WEAVE, OR SEW, THE LEAVES TOGETHER until they had their nest completed!

What "chance mutation", pray tell, first led the ground-loving ants to venture into the business of building nests in high trees, out of LEAVES and not out of sand and dirt and pine needles? And how many million years did it take to make the change? AND HOW IS IT THAT THEY DID NOT STARVE TO DEATH WHEN THEY WERE, SAY, HALF WAY ALONG FROM GROUND-LOVING ANTS TO TREE CLIMBERS AND TREE BUILDERS??? Obviously, the "leaf house" ants WERE THAT WAY FROM THE VERY BEGINNING: no evolution could ever negotiate, gradually, such changes in ant habits!

And who taught this non-intelligent little animal to sew leaves together? And who gave them the acrobatic skill to build an "ant chain" from one leaf to another, and draw distant leaves together, to serve their purposes? Who first suggested to these ants—if at one time they were accustomed to life on earth, let us say, as ordinary Harvester ants—that they might use their own larvae to make silk threads for them? And Who taught them, after they got used to the idea that they could use their larvae to make silk threads, how to sew leaves together with these threads? The very asking of these questions shows how preposterous is the idea of "gradual change by slow, evolutionary processes." To teach an intelligent dog tricks is one thing; to get a non-intelligent ant, bound by "the tyranny of instinct," to make revolutionary changes, is quite another thing.

THE FEROCIOUS "ARMY ANTS"

The Army ants live in the tropical climates of Africa, South America and Mexico. They are large, fierce, expert hunters. They eat only meat, which they find and kill on regular hunting raids. They never make a nest, but often when not on the march they will hang together in great masses on trees, like a swarm of bees. They are nomads, going from place to place. They are blind and are guided by "feelers" instead of sight.

"On the march, workers carry the larvae in their jaws. The queen army ant does not lay her eggs continuously as do queens of other ants. She lays them in huge batches at regular intervals. When the queen is swollen with eggs, the army camps. At this time the larvae of the previous brood make their cocoons. Within one week the queen may lay 25,000 eggs. In two or three weeks, when the older brood have all emerged from their cocoons, and the new batch of eggs have hatched into tiny larvae, the colony starts marching again—this time to make vigorous raids for food in all directions. Every night the entire colony moves to a new location. Workers carry their young ones while on the march. . . . The queens of army ants never have wings." (Charles D. Michener).

They have instincts that enable them to achieve some remarkable things, such as crossing water. Carveth Wells tells about their spectacular method of crossing water. He witnessed this scene in the Malay jungle.

"Rivers do not stop a marching column of army ants; on reaching a river, the main body waits while scouts look for the best place to cross. The scouts find a bend in the stream, where the current is shunted diagonally across the river-bed to the other side. Next, the ants form heaps and slowly wriggle themselves into a solid ball, about the size of a coconut. Then, in some inexplicable manner, enough momentum is obtained to carry the ball of squirming insects down the slope to the water's edge, where it falls in with a splash.

"Here the ball rolls about, so that an ant may be on top one second and entirely submerged the next. . . . The current keeps the ball rolling, so that each receives only a momentary ducking. The instant the ball touches the bank on the other side, the ants unscramble, toddle ashore, and continue their march!" (P. 222, "Nature Parade").

Because of INSTINCT given by the Creator, the feat becomes believable; but were one forced to believe that they GRADUALLY ACQUIRED this amazing ability, it is too much to give credence to—for they would have died a thousand deaths while learning to cross streams, and never would have developed the proper technique, even in a billion years! They were originally MADE to do this; such a feat can NOT be acquired gradually.

Dr. T. C. Schneirla's Testimony

T. C. Schneirla, of the American Museum of Natural History, wrote on THE ARMY ANTS. It is published in the SMITHSONIAN REPORT FOR 1955 (Publication No. 4244).* We quote:

(1) Speaking of the **Eciton** (army) ants, he says (p. 391): "It can be said that there are NO LEADERS in these swarms except in a very temporary and limited sense, and that not in the sense of human leadership; but the swarm at any stage is 'directed' COLLECTIVELY in a complex manner through the activities of all ants participating in the raid."

This bears out the statement in the Bible, Proverbs 6:6-8: "Go to the ant . . . consider her ways, and be wise: which HAVING NO GUIDE, OVERSEER OR RULER, provideth her meat in the summer", etc.

(2) Dr. Schneirla infers that there has been NO EVOLUTION in the Army ants for the past 65 million years.

"The Dorylines, one of the eight major subfamilies of ants (of which the army ants are a species) . . . have survived very successfully from early Tertiary times, or at least 65 million years, on the basis of the unique combination of a nomadic behavior with a fully carnivorous way of life."

Other Witnesses say, ANTS ARE NOT EVOLVING

The noted anthropologist, Loren C. Eiseley, writing in a recent issue of "The Scientific American," flatly states that **ants are not evolving.**

---

*For another interesting article on ARMY ANTS, see the essay by Thomas Belt, on "Army Ants," p. 450, "Green Treasury."

"Ants have led their present lives for more than 80 million years, while man's civilization is scarcely more than 7,000 years old. . . . They have changed very little, if at all. They are one of the small 'immortals'. They attained their present relatively high biological specialization very long ago and have since been marking time or evolving so slowly that the modifications are extremely minor."

What a confession for an evolutionist! We agree: Ants are NOT evolving—nor have they ever evolved. They were CREATED as perfectly adapted to their environment as they are now; otherwise they never could have survived.

Another authority, an evolutionist, says,

"Insects appeared ages ago, before the first vertebrates, the true fishes, the snakes, the lizards and the birds. They most certainly have had time to develop along higher lines (but) THEY SEEM TO HAVE REMAINED IN EQUILIBRIUM WITH THEIR ENVIRONMENT and have to a certain extent marked time" (Book of Knowledge).

Maurice Maeterlinck, noted naturalist, says (in "The Life of the Ant"):

"The ants are the most abundant of all insects in the Tertiary deposits. We find them in the Eocene, the most ancient of these deposits. . . . Eleven thousand, seven hundred and eleven specimens contained in the Baltic amber have been examined, as well as hundreds of other specimens found in the Sicilian amber of the middle Miocene. But here is a most disconcerting fact (i.e., to evolutionists); contrary to expectation, we find that the more ancient ants are NOT more primitive than those found in fossil amber, and that the latter, despite the millions of years which divide them from the ants of today, are almost as fully specialized, almost as civilized. Many of them, Wheeler tells us, had learned to seek out plant lice (and use them as their 'cows'). . . . Now the rearing of 'cattle' . . . mark the culminating point of their present civilization. What then are we to conclude? Well, if we choose, we may draw very strange conclusions—as, for example, THAT EVOLUTION IS LESS PROVEN, LESS CERTAIN THAN IS GENERALLY ASSERTED, that all the species, with their diverse degrees of civilization, DATE FROM THE SAME MOMENT, and were, as the Bible declares, CREATED ON THE SAME DAY, and consequently, that tradition is nearer to the truth than science."

Chapter 11
THE MARVELOUS MYSTERY OF MAN

"SO GOD CREATED man in His own image" (Gen. 1:26, 27).
"I am fearfully and wonderfully made" (Ps. 139:13).

MAN—a tripartite being of body, mind and soul—created in the image and likeness of God, is the supreme miracle and mystery of life on earth.

Part I. THE BODY OF MAN

"Your body," says Dr. Peter J. Steincrohn, "is the WORLD'S MOST INCREDIBLE PIECE OF MACHINERY. It manufactures, improves and repairs itself. It has illimitable reserves." The doctor is right: the average man can get along without his gall bladder, spleen, tonsils and appendix. He can dispense with one lung, one kidney, two-fifths of his liver, part of his brain, most of his stomach, both eyes, ears, arms and legs—and still live!

As an "incredible piece of machinery," the body is highly complicated and efficient. It is an intricate assembly of thousands of mechanisms working together in synchronized obedience to direction. It contains chemical factories that process chemicals of countless varieties. The body consists of trillions of living cells.

In the body are **"a hundred thousand different kinds of protein molecules!"** These protein molecules are highly complicated.* Typical proteins are **collagen**, which provides framework material; **hemoglobin**, which carries oxygen in the blood cells; **keratin**, contained in the protective cells of hair, skin, and fingernails; **myosin**, which converts energy from chemical to mechanical form in muscle cells; and the innumerable specialized **enzymes** and **hormones** so important in regulating body activity.

Each of the trillions of cells in the body is a LIVING ORGANISM—living protoplasm—and in order to carry on life processes they must convert food, air and water into energy and tissue and food for tissue. These elements must be changed into such a form that they can be absorbed as food and be carried to all parts of the body. This process is called DIGESTION. Air must enter the body to oxidize foods; this is called RESPIRATION. Altered food and oxygen are dissolved in the blood and

---

*To illustrate how **highly** complex protein molecules are: one molecule of hemoglobin, in the blood cell, may have in it 758 atoms of carbon, 1203 of hydrogen, 195 of nitrogen, 3 of sulphur, 1 of iron, and 218 of oxygen, making a total of 2378 atoms. All of the other 100,000 protein molecules are highly complex and ALL are different!

carried all through the body to hungry cells. The heart pumps the blood to all parts of the body; this is called CIRCULATION. Waste products, like the ashes of a furnace, must be removed; this operation is known as EXCRETION. All of this highly complicated performance must have directing intelligence, and this is in the **Central Nervous System,** assisted by the **Autonomic Nervous System.** Working closely with the nervous system are the DUCTLESS GLANDS which pour hormones into the blood stream when needed for the control of various activities.

The Miracle of Conception

It is incredible but true that this amazing human body comes originally from just TWO TINY CELLS: the female egg (ovum) and the male sperm. Conception—the instant when new life is created—takes place the moment a male sperm penetrates and fertilizes a female ovum. Starting with the union of these two cells, as a tiny bit of living protoplasm, and sheltered in the mother's womb, this minute bit of life grows, divides and redivides, and develops—until finally it grows into the amazingly complex being called man.

Beyond doubt, this is the greatest miracle in nature, that "a SINGLE FERTILIZED CELL SHOULD HAVE THE POWER TO DRAW **seemingly from nowhere** THE 30 TRILLION CELLS OF THE VIVIPAROUS HUMAN, having all the necessary structures and organs that make up the body of man!"

"The wonder of this miracle is deepened by the fact that while all the cells in the body started from ONE fertilized cell, the ovum, which was itself neither muscle, nerve, blood, or bone cell, each separate colony of cells produced by the process of division becomes an organ whose cells all have a ponderable and demonstrable DIFFERENCE between them and the cells of any other organ in the same body"—and different from the two parent cells! ("Theory of Evolution and the Facts of Science," P. 52).

Remember too that the male spermatozoa are incredibly small. In one discharge of seminal fluid there may be from 200,000,000 to 500,000,000 spermatozoa—enough potentially to populate the entire north American continent! Each spermatozoon is composed of two main parts: the flagellum, or tail, and the cell proper, which is the head end. It was PLANNED that way, for the flagellum is the "motor" that drives the sperm on its way to find and contact the ovum. When contact is made, the flagellum dissolves and the sperm head (the true cell) unites with the ovum. The two fuse into one—and conception has taken place!

"The sperm with . . . admirable FORESIGHT has with it a minute quantity of an enzyme, **hyaluronidase,** which has the ability to loosen the egg's protective sheath, and so permit penetration. . . . At the same time the egg's wall (also exhibiting admirable FORESIGHT) goes through a drastic change, and it becomes **no longer penetrable by another sperm assault."** ("Wonders of Conception," Reader's Digest).

The ovum (egg) is 35 to 40 times the size of the head of the sperm; and yet the ovum is smaller than a small period at the end of a sentence! Though this ovum is a single cell, it is highly complex. In it is the all-important nucleus, and in the nucleus are wonders untold, including 24 chromosomes*—worm-shaped structures horizontally striped with bands of light and dark. And— wonder of wonders—these chromosomes contain thousands of genes (estimates vary from 3,000 to 30,000), the "seeds of inheritance." So infinitesimally small are they that

"even the most powerful peering eye of the electron microscope cannot see inside the unbelievably minute chromosomes" and the far smaller genes. ("Secrets of the Human Cell").

The male cell is even more infinitesimal. It is only one-fortieth as big as the ovum.

The Miracle of the Growth of the Embryo

"No mind," said Dr. Arthur I. Brown, "can grasp the wonder of the growth of the embryo. One who has watched this under the microscope or by examination of the very tiny early embryos almost feels as if it cannot be real—such miracles simply are not possible!"

Two cells have united: the ovum and the sperm. They soon develop THREE distinct types of cells (different from each other and radically different from either the original ovum or sperm). These three—the primary germ layers—eventually produce ALL THE ORGANS, MUSCLES, BONES AND VARIOUS PARTS OF THE BODY in the bewildering process of embryological development!

At the end of a month, the growing embryo has its own circulatory system, at the end of the second month the fetus is only an inch long; at the end of the third month a bony structure is beginning to form; at the end of the sixth month the fetus is 10 to 14 inches long—but if born prematurely at this point it will not be able to survive. During the seventh month the baby begins to move about, and two more months of prenatal life are added to complete the maturing process before birth.

The moment of birth is at hand. The mother's body reacts automatically to this great event, and "this fantastically complicated series of events

---

*The traditional figure of 24 chromosomes for each sperm and ovum is still given in most texts, though recently geneticists discovered that some contain only 23.

have interacted, each perfectly timed, to produce the most superb of all achievements: a new human life." ("Miracle of Birth," by J. D. Ratcliff).

The Fallacy of the "Recapitulation Theory"

Evolutionists have developed a theory to explain certain structures and changes that take place in the embryo. It is called "The Recapitulation Theory." It states that "every creature passes through stages in its embryonic development similar to those which its remote ancestors passed through in evolving upwards." Their slogan is—"Ontogeny (the development of the individual) recapitulates Phylogeny" (the development of the race). Many people have been deceived into believing in evolution through these subtle, though false arguments. Their principal arguments are:

(1) "Human life begins as a protozoan." This is untrue. We all know that human life begins with the union of TWO cells that are NOT ordinary protozoa, but are the specific reproductive cells of man.

(2) "The human embryo at one stage of development has gills like a fish." This is entirely fallacious. The so-called "gill slits" in the human embryo are NOT gill slits at all, but pharyngeal arches. They have grooves, but NO perforations, as in gills. Douglas Dewar, English naturalist, said, "In the embryo of a reptile, bird or mammal (including man) **no clefts form between arches,** which never assume the characteristics of gills. It is clearly incorrect to call them gill arches. **The embryo of a higher animal never passes through a "fish" stage** . . . Embryology lends no countenance to the view that the higher vertebrates evolved from a fishlike ancestor. It is only by putting far fetched and artificial interpretations on embryological phenomena that they can be made to fit in with the evolution hypothesis." (P. 49, "Difficulties of the Evolutionary Theory.")*

(3) The human embryo in one stage of its development is said to have a tail like a puppy. This so-called "tail" is simply the coccyx, or end of man's spine. As the embryo grows, the coccyx is covered with tissues and muscles, and the "tail" no longer shows, though it is still there as the end of the spine. The coccyx serves as an anchor for useful muscles. So this argument fades into nothingness.

(4) "The human embryo bears a confusing and close resemblance to the embryos of other animals." This "resemblance" is purely superficial. Close examination at any stage of development always reveals "striking differences." NO TWO EMBRYOS OF ANY TYPES OF LIFE ARE EVER EXACTLY ALIKE: all bear characterstics of their own family or genus.

Many students are deceived by the "schematized" drawings of overly-zealous evolutionists. Some advocates of the Recapitulation Theory have not hesitated to forge 'embryonic connecting links.' Prof. Haeckel, one of the earliest advocates of Embryology as "proof" of Evolution actually

---

***Gray's Anatomy,** a medical authority, says "perforation does **not** occur in the pharyngleal arches in birds or mammals."

forged some features of his drawings. When tried by the Jena University Court, and convicted, he confessed,

"A small per cent of my embryonic drawings are forgeries; those, namely, for which the observed material is so incomplete or insufficient as to compel us to fill in and reconstruct the missing links by hypothesis and comparative synthesis." Then follows this startling indictment of other embryologists: "I should feel utterly condemned . . . were it not that hundreds of the best observers, and biologists lie under the same charge."

And **this** is called **"Science"?** We are under the impression that Science should be honest, and deal with FACTS—not falsifications intended to support a theory that can NOT be supported by facts!

Many drawings of evolutionists are "schematized," "doctored," and changed, and many plaster of Paris casts of so-called "missing links" are "reconstructed" to make them APPEAR AS THE EVOLUTIONIST THINKS THEY OUGHT TO APPEAR, with no thought of what they actually are or were in nature! They twist the facts to support their hypothesis.

At the British Association Meeting at Edinburgh, August 10, 1951, Prof. T. S. Westoll called the Recapitulation Theory "Sheer nonsense!"

(2) The Miracle of the Growth of the Body

Many changes take place quickly after birth. Before birth the baby lives in water—in the amniotic fluid, in a sac—a planned device to protect the growing embryo. At that stage, the lungs of the unborn child are practically solid flesh; but as soon as the new born infant gasps for his first breath, accompanied by vigorous crying, the air sacs in his lungs expand, and never again, as long as he lives, will they be entirely devoid of air.

His heart gradually slows down from its furious pre-natal rate to a still rapid 140 to 150 beats per minute.

He is born with a sucking instinct; for his very life depends on it. And now a great mystery begins: the tiny 6 pound baby grows, and grows and grows—until he reaches maturity. WHAT MYSTICAL FORCE MAKES THE BABY GRADUALLY TURN INTO AN ADULT? No one knows; the only answer is, GOD MADE IT SO. And why does the marvelous process of growth STOP at just the right time? Why doesn't the baby grow to be 30 feet tall? And what keeps the growing process symmetrical and at a uniform rate? That is, why is it that one arm of the boy doesn't grow to be five feet long, while the other one, let us say, stops at 10 inches? Why is it that the nose on the boy's face always turns down, as it should; what if, occasionally, it should turn UP, or be twisted SIDEWISE? What a mess **that** would be! If his nose turned up, instead of down, he would be in trouble in every rainstorm! Fortunately, God has regulated the "growth mechanism," so that none of these theoretical monstrosities actually happen. The growth of the baby into manhood is uniform at all times.

"Nobody can say why all parts of the body grow in such beautifully regulated proportions. Look at the skin of a six-year-old, for example. Why doesn't it grow too much, giving him folds like a hippopotamus, or so little that he looks like a balloon about to burst. This is one of the INCREDIBLY PRECISE COORDINATIONS OF NATURE."

"During its first twenty years, the human body grows to 'eight heads high' with a fantastic series of speed-ups and slow-downs, and the infinitely complex influences which change the tiny infant into the full-grown man or woman are interwoven with such astounding precision that even the most cynical skeptic must call it miraculous." ("The Miracle of Growth," by Herbert H. Kenny).

(3) SECRETS OF THE HUMAN CELL*

The body, essentially, is made up of trillions of cells—most of them ALIVE, and able to reproduce themselves.

In a previous chapter we discussed the marvels of cell construction and cell life. Human cells are similar in construction to all cells, having three principal parts: outer membrane, cytoplasm and nucleus. A gossamer-like membrane encloses the entire cell; the cytoplasm is the jelly-like substance in which the nucleus floats; and the all-important nucleus is the "executive" part of the cell.

"Each human cell nucleus contains 48 chromosomes, with the exception of egg and sperm cells, which contain half as many. Even the powerful peering eye of the electron miscroscope cannot see inside the unbelievably minute chromosomes. But indirect evidence indicates that, small as they are, they are still large enough to contain 30,000 genes—the seeds of inheritance."

These cells are so small, it takes 8,000 of them to make an inch, and 64,000,000 of them can be put in a square inch!

"These 'bricks' (cells) from which all living matter is made, are able to perform chemical transformations that baffle the world's cleverest chemists, producing infinitely complex vitamins, hormones, proteins. They perform striking feats of 'biological engineering'—the outstanding example being the formation of the human ovum and sperm.

There are five general types of cells in the human body—All coming originally from the first two specialized cells: the ovum and the sperm. These cells are (1) the nerve cells, (2) epithelial cells, (3) connective tissue, (4) muscle cells and (5) blood cells.

In general, cells in the body have two main responsibilities:

---

*These facts are taken from the chapter on "Secrets of the Human Cell" in the book, "Family Doctor," published by The British Medical Ass'n., 6-58.

their own sustenance and reproduction, and their community responsibilities. The first includes such responsibilities as eating and waste disposal, the second includes the responsibilities of each cell to all others. Tiny cells in the pancreas, for example, produce minute amounts of insulin which control sugar use by all other cells. Fat cells store tiny droplets of oil to be used for energy for the rest of the body.

Researchers have "bumped into a number of problems which have so far proved baffling." Here is one of the most mysterious:

"Since cells generally show a remarkable specificity—lung cells always divide into lung cells, white blood cells always divide into white blood cells, kidney cells always divide into kidney cells, and so on, WHY DOESN'T THE ORIGINAL FERTILIZED OVUM DUPLICATE ITSELF, instead of going through an amazingly intricate series of divisions and differentiations to produce a mouse, a whale or a man? Cell students have found no satisfactory answer."

Evolution has no "satisfactory answer" to this miracle; neither has modern science nor philosophy. But there is an answer: GOD MADE IT SO, and its secret is wrapped up in the miracle of life.

(4) YOUR SKIN: The Largest Organ of Your Body

Your tender skin is far more than a protective covering; it is an organ "ranking with the brain, heart and lungs in its importance to human life." It is the largest organ of your body, and one of the most important. It has an average area of from 16 to 20 square feet.

There are about 2,000,000 sweat glands scattered over the surface of the skin—500 to every square inch, except on your palms and soles, where there are about 2,000 to every square inch! Thousands of MILES of very small capillaries are in the skin, along with many thousands of sensory nerves. The skin also has about 2,000,000 sebaceous glands that produce sebum to lubricate and waterproof the skin. And the skin has an intricate system of countless ELASTIC FIBERS that keep it smooth and firm, close-fitting, yet pliable. ALL of these parts are highly "specialized"—created for a purpose.

Your skin has three major levels: the EPIDERMIS (outer), the DERMIS (middle) and the SUBDERMAL (inner). This "subdermal" is a subcutaneous fatty tissue, having a cushion effect. In the lower part of the epidermis is a thin sheet of cells where most skin growth occurs. These cells divide and form new cells constantly, which are crowded slowly upward to the surface. The trip may take weeks. On the trip to the surface each cell dies, and so the exterior of the skin disintegrates into microscopic layers of scales.

The result is, "some twenty or more layers of scales form the outer surface of your skin—and lend toughness to it." These invisible fragments of dead cells are constantly being rubbed off and replaced.

Your skin was PLANNED to serve your body in these ways:

(A) **Your skin helps regulate the temperature of your body.** By "sweating" your body cools off. Furthermore, when it is warm, the blood vessels in and near the skin enlarge and let the heat escape. When it is cold, the blood vessels contract and conserve the heat of the body.

So precise is this automatic mechanism that a person in good health maintains a steady 98.6° temperature whether hunting in the arctic or vacationing in the South Pacific. So amazingly efficient is the body's cooling system, centering around the action of the skin, that if the air is dry and enough water is consumed, human beings actually can tolerate temperatures of from 240° to 260° for short periods of time! This is hot enough "to grill a steak."

(B) **Your skin, with the layer of fat underneath it, is in itself a remarkable INSULATOR.** Like the asbestos around a hot water pipe, it keeps the heat in and the cold out; or, if the surrounding atmosphere is hotter than your body, the skin tends to keep the heat out of your body.

(C) **Your skin, studded with nerve endings, is the principal organ of touch, giving at least six different sensations:** light touch, pressure, heat, cold, pain, and tickling. So the skin is a highly complicated ALARM SYSTEM, constantly keeping your mind informed, by its contacts with the outside world, of sensations of pleasure and pain, enemies to the body, and friends. Each of its millions of tiny nerve receptors is a combined "receiver" and "transmitter."

(D) **Your skin helps in ELIMINATING POISONS, and throws off waste materials through the sweat glands.**

(E) **Your skin is a manufacturing plant: it makes new hair, and nails and new cells.** It continuously REBUILDS the surface layer.

(F) **Your skin makes a special pigment—MELANIN— that shields you from over-exposure to the sun**—and it automatically provides the amount needed. An extra amount shows up in a "tanned" skin.

(G) **Your skin is a combination of a sort of "leather jacket" and a "raincoat" to serve you and protect you from the elements.** Your skin is an "armor" of "overlapping fish-like scales, to protect the tissues of your body." On the other hand, it is FULL OF HOLES (sweat pores)—and yet it doesn't leak! Consider the miracle: your skin is so ingeniously made that it will exude sweat, but will NOT permit water to enter your body even though you are immersed for a long time!

(H) **Your skin also maintains a delicate system of fine hairs that cover most of the body,** though they often are too fine to be visible. These hairs reveal to us how infinitely careful the Creator was in planning for every **minute detail** for the body of man.

"EACH HAIR has a small blood vessel to nourish it, and several sacs of oil to lubricate it, a nerve to give the alarm if it is pulled, and a tiny muscle to draw it up if one is chilled or scared." ("Know Your Skin," by Dr. Frank C. Combes, Director of Dermatology, Bellevue Hospital, N. Y.).

(I) **Your skin is an indicator of your emotions,** as we all know. We turn "red" with embarrassment or pale with fright.

(J) **Your skin is an indicator of your health:** undue redness may indicate fever, a sallow "complexion" and paleness have diagnostic value to your doctor. Other skin changes and symptoms indicate diseases to the trained diagnostician.

(K) **Your skin is a remarkably roomy "storehouse,"** for salts, sugars, fats, water, and other materials, when more is taken into the body than is needed. Later, when needed, the skin returns these to the blood stream for transportation to the organs.

(L) **Your skin is a first-class BARRIER to germs**—until it is punctured or cut. Your body is covered at all times with countless bacteria, waiting to enter your body and do their evil work. YOUR SKIN KEEPS THEM OUT.

(M) **Your skin also has a marvelous chemical called KERATIN** (a chemical similar to gelatin) **that protects your body from infiltration by many inimical liquids,** such as many oils, diluted acids, alkalies and other chemical enemies of the body.

(N) **Your skin is UNIQUELY YOURS; it carries a telltale identification pattern: your fingerprints.** They are never duplicated in any one else. Your skin is so much a part of yourself that though it can be transplanted from one part of your body to another, it can NOT be successfully transplanted to another person (except an identical twin).

(O) **Your skin is an excellent doctor!** It repairs itself. If cut or bruised, it begins immediately to mend itself.

"After injury, blood gushes immediately from the tiny blood vessels of the skin, washing out dirt and germs. Then the blood vessels constrict so that the flow diminishes and soon a clot of rapidly hardening blood fills the gash. Like glue, it firmly attaches itself to all sides of the wound, then gradually shrinks, drawing the sides together. As the hours pass, thread-like connective-tissue cells, called fibroblasts, invade the clot from all sides, and gradually build up new tissue. The gap is filled and then surface skin cells grow in—and the repair job is completed!" ("What Your Skin Does For You," Journal of Lifetime Living, 8-'56).

Who can not see that the skin was TAUGHT to do that by an outside Intelligence—the Creator! That such an involved process should develop from "random mutations" is more unlikely than that a cyclone should blow desert debris together, form it into an art gallery and fill it with ancient, medieval and modern art treasures!

We have gone to considerable detail to show that what **seems** to be merely a covering for the body is in reality a highly complicated and efficient organ, that gives overwhelming evidence that it was DESIGNED and PLANNED and MADE to function as it does.

(5) **BLOOD: The Stream of Life**

"For the life of the flesh is in the blood" (Lev. 17:11). Blood is not the simple fluid it was once thought to be. In the last fifteen years modern research has discovered that "more than 70 different proteins" have been identified in it, not to speak of the various other constituents—the red cells, plasma, white cells and the platelets—of which it is composed.* And the body is constantly manufacturing new supplies of these products to replace the old. A human adult produces about 140 million NEW RED BLOOD CELLS per minute! ("Blood," by Douglas M. Surgenor, "Scientific American").

The average human body contains 5 liters of blood, which consists of 2.75 liters of plasma, 2.22 liters of red blood cells, .02 liter of platelets and .08 liter of white cells.

Blood has these functions in the body:

(1) It transports oxygen from the lungs to all the cells of the body.

(2) It carries food elements (glucose, amino acids, **proteins, fats,** etc.) from the alimentary canal to the cells.

(3) It assists in the elimination of waste products such as carbon dioxide, urea, uric acid and creatinine.

(4) It carries hormones, the ductless gland secretions that regulate many important functions of the body.

(5) It maintains a more or less constant temperature in the body.

(6) It plays an important part in resistance to disease.

(7) The blood "maintains its own composition and integrity."

The Amazing Red Blood Cells

The red blood cell was once thought to be a "dead cell" because, when it reaches the blood (it is made in the red marrow of bones) it at once **loses its nucleus.** But actually, it stays ALIVE—WITHOUT A NUCLEUS! That is a miracle of Divine design, for to rob a cell of its nucleus is like robbing a man of his heart. But the **chief function** of the red blood cells is to carry hemoglobin (which in turn carries oxygen to the cells); and since red blood cells can CARRY MUCH MORE WITHOUT A NUCLEUS TO TAKE UP SPACE, GOD MADE THEM INTO LIVING "GHOST CELLS".** Moreover, this living "ghost cell" is specially "engineered" to carry a maximum load.

"The red cells . . . are a particularly excellent piece of biological engineering. They are BICONCAVE in shape (like a doughnut, with a thin section in the middle instead of a hole), and this facilitates QUICK ENTRY of oxygen and other supplies to all parts of the cell. If red blood cells were spherical instead of biconcave we would need about NINE TIMES

---

*The blood also contains "antibodies" against various infections, at least ten substances involved in clotting, twenty different enzymes, fat-containing proteins, carbohydrate-containing proteins, metal-containing proteins, hormones, albumins, and doubtless other still unidentified substances. (See article on "BLOOD," by Douglas M. Surgenor, in "Scientic American").

**Red blood cells, without their nucleus, are called "ghost cells."

AS MANY OF THEM to distribute oxygen in the body with the same speed." (Eric Ponder, in "THE RED BLOOD CELL," Scientific American).

Let every fair-minded person consider this double miracle and see that every red blood cell in man's body virtually says,

"The Supreme Architect designed me so I can stay alive without a nucleus, and He gave me such an efficient shape, that I can DO NINE TIMES AS MUCH WORK as if I were made like a conventional cell."

This, my friends, is THE FINGER OF GOD! Evolution could work ten thousand million years and never "evolve" a cell without a nucleus that would stay alive.

There are many other phenomena, "mysterious (features), which testify to the complex structure and the vitality of the red cell. One of the most striking appears when the cell is examined under a phase contact microscope. The surface of the cell seems to move, as if wind were blowing over a field of wheat. This so-called 'scintillation phenomenon' is believed to be connected with the cell's metabolism." (Ibid.)

The deeper we look into God's creation the more marvelous it becomes. One miracle is linked to another. If the red blood cell is a miracle of construction, so is hemoglobin. We have already called attention to the fact that hemoglobin is one of the most intricate protein molecules in nature.

"The manufacture of hemoglobin is a great chemical feat. When a new red cell is made, its hemoglobin is made also. THIS SUBSTANCE IS THE MOST COMPLEX KNOWN TO CHEMISTRY. It is not only the most complex, it is one of the largest molecules that chemistry knows. . . . The chemical skill of the red-bone marrow that makes the hemoglobin is transcendent!" (Ibid.) Who gave red-bone marrow such marvelous ability?

Hemoglobin (in the red blood cell) carries oxygen **to** each body cell and carbon dioxide **from** each body cell. This involves another miracle and is accomplished by "methods of varying intricacy." We quote.

Oxygen is carried by the red cells; i. e., by the hemoglobin that the red cells carry. The iron-containing protein combines readily with oxygen —and then releases it to the hungry cells. "Whether the hemoglobin molecule will take up oxygen or release it, depends on the oxygen gas pressure in the place where it happens to be. In a place of high oxygen concentration, as in the lungs, hemoglobin attaches oxygen to itself; but when this hemoglobin molecule reaches the hungry tissues, where oxygen concentration is low, it releases the oxygen! . . . In a similar manner, the hemoglobin carries carbon dioxide away from the tissues where it has delivered oxygen." ("BLOOD," by Douglas M. Surgenor, Scientific American Magazine).

This is a MIRACLE in nature, that a protein molecule should have and lose its affinity for oxygen, at just the right time to feed oxygen to cells—and turn right around and pick up excess carbon dioxide from cells, and then release it to the plasma. This is an astonishing performance that can be explained only by admitting that GOD MADE IT SO.

Without exaggeration VOLUMES could be written on MIRACLES OF DESIGN AND FUNCTION in the blood.* Before we pass on to tell of some of the other miracles in the human body, let us consider two more miracles in the blood. (1) The marvel of the manufacture and presence in the blood of "ANTIBODIES."

Of extraordinary interest is the story of how the blood makes antibodies to fight disease. Antibodies are "those substances in the blood which are protective agents formed to fight infection by an invading organism . . . An antibody in the blood is a modified soluble protein with properties that make it stick to the type of molecule or microorganism against which it was developed. After an attack of yellow fever, for example, antibodies against the yellow fever virus are formed. These antibody molecules will immediately coat any new yellow fever viruses that happen to enter the body and WILL EFFECTIVELY PREVENT THEM from causing an attack of the disease." ("HOW ANTIBODIES ARE MADE," By Sir Macfarlane Burnet, in The Scientific American Magazine). Science and Medicine know these facts—but can not explain them.

Let us say a few words on How the Blood MAINTAINS ITS OWN COMPOSITION AND INTEGRITY. The blood works with the liver on this.

"For this nature has developed most ingenious and elaborate mechanisms. The blood halts its own escape from the body by a selfstarting series of reactions, still imperfectly understood, which involves calcium, the platelets and a number of plasma proteins, all in trace amounts. The process leads to the formation of THROMBIN, which in turn converts the protein fibrinogen to the blood-clot material FIBRIN. The internal composition and viscosity of the blood are controlled mainly by osmosis, which regulates its water content. This is no simple matter, for parts of the circulatory network, notably the capillaries, are permeable to water. The control of the water balance between the blood and the tissues with which it is in contact is exerted to a large extent by the concentrations of the large protein molecules on the two sides of the capillary wall." ("BLOOD," by Douglas M. Surgenor, in The Scientic American Magazine).

And so, when the body is cut and its life-blood would ooze out or flow away, THE BLOOD ITSELF TAKES IMMEDIATE STEPS TO STOP THAT FLOW BY CLOTTING. Truly, the Blood is a wonderful fluid—THE STREAM OF LIFE. The Divine Designer made it as it is.

---

*Volumes HAVE been written on WHITE BLOOD CELLS, ON BLOOD PLASMA, ON ANTIBODIES, ON THE CIRCULATION OF THE BLOOD, etc.

(6) THE HUMAN EYE: "The Wonder of Wonders"

The phenomenon of vision was a great enigma to the ancients —and modern man is still wondering about the marvels of it.

Aristotle cried out in wonder: "Who would believe that so small a space (as the eye) could contain the images of all the universe? What skill can penetrate such a wonderful process? This it is that leads human discourse to the consideration of divine things!" (Quoted in the 1954 Smithsonian Report).

Sir Charles Scott Sherrington, famous English physiologist, was one of the most highly honored scientists of our day. (He died a few years ago, at the age of 95). On writing a classic essay on THE EYE, he said, "Behind the intricate mechanism of the human eye lie breath-taking glimpses of a Master Plan."

Of the eye, we read:

"No scientific instrument is as sensitive to light as your eye. In the dark, its sensitivity increases 100,000 times, and you can detect a faint glow, less than a thousandth as bright as a candle's flame. You can see in brilliant, blazing light, too: in light brighter than the radiance of a billion candles. You can see light from the stars, and the nearest of all the stars is 24 trillion miles away.

"The human eye responds to light waves, which are very short— 40,000 to 60,000 of them per inch. (There are many kinds of energy waves: heat, light, radio, X rays).

"The image focused by the lens of the eye falls on the 'screen' at the back of the eyeball, the retina. The retina is a kind of carpet, made up of tiny light-sensitive cells. Each retina has about 130,000,000 cells. There are two types of cells in the retina—'rods' and 'cones,' so-called because they look something like rods and cones under a microscope. The 'rods' are used for general perception of light; the 'cones' are used to see color and fine detail." (Our Amazing Eyes, by John Perry).

It is unbelievable with what rapidity the eye works. It has been estimated that from the vast panorama presented by your eyes, each eye can send a thousand million impulses per second to the brain—and then your mind chooses significant details. You can stare at a sign without becoming aware of its message, while on the other hand a fragmentary glimpse of some familiar object attracts your immediate attention. ("Your Remarkable Eyes," Science Digest).

To bring before us the fact that the human eye is indeed "THE WONDER OF WONDERS" we quote again from Sir Charles Scott Sherrington's classic essay on **The Eye.**

How does a pinhead-sized ball of cells (the tiny human embryo) in the course of so many weeks become a child? Consider the story of just one individual part: THE EYE.

The many cells which make the human eye have first executed correctly a (process) . . . engaging millions of performers in hundreds of sequences. . . . To picture the complexity and the precision of this performance beggars any imagery I have. It suggests PURPOSIVE BEHAVIOR—not only by individual cells, but also by colonies of cells.

The eyeball is a little camera. Its smallness is part of its perfection. But this is a SPHEROID camera which focuses itself AUTOMATICALLY, according to the distance of the picture interesting it. It turns itself in the direction of the view required. Indeed our eyes are TWO CAMERAS finished to one standard so that the mind can read their two pictures as one. And the eye is contrived as though WITH FORETHOUGHT OF SELF-PRESERVATION. Should danger threaten it, in a trice its skin shutters close, protecting its transparent window. Working only with albumen, salt and water, the starting embryo proceeds, though it is only a little pin's-head bud of multiplying cells, NOT ONE-TEN-THOUSANDTH PART the size of the eyeball it makes. The whole structure, with its prescience and all its efficiency, is produced by and out of specks of granular skin cells arranging themselves as of their own accord in sheets and layers, and acting seemingly on an agreed plan! The magic juices that make the eye, go by the chemical names protein, sugar, fat, salts, water. It all sounds like a tale that challenges belief; but so it is. There is more yet.

The biconvex lens is made of cells like those of the skin but modified to be glass-clear, and FREE FROM BLOOD VESSELS which would throw shadows within the eye. It is delicately slung with accurate centering across the path of the light which will some months later enter the eye. IT IS BEING PREPARED IN DARKNESS FOR USE IN LIGHT. In front of it a circular screen controls, like the iris-stop of a camera, the width of the beam and is adjustable so that in poor light more is taken in for the image. In a camera this adjustment is made by the observer; in the eye this adjustment is **automatic, triggered by the image itself!**

Not only must the lens be glass-clear, but also its shape must be optically right. Its two curved surfaces, back and front, MUST BE TRULY CENTERED ON ONE AXIS, AND EACH OF THE CURVATURES MUST BE CURVED TO THE RIGHT DEGREE, so that light is brought to a focus on the retina and gives there an accurately shaped image. The optician skillfully grinds his glass curvatures in accordance with mathematical formulae. In the formation of the lens of the eye a batch of granular skin cells are told to travel from the skin, to which they strictly belong, and to settle down in the mouth of the optic cup and arrange themselves in a compact and suitable ball. NEXT THEY ARE TOLD TO TURN INTO TRANSPARENT FIBERS, and to make themselves into a subsphere—a lens of the RIGHT size, set at the RIGHT distance between the transparent window of the eye in front and the sensitive seeing screen of the retina behind. In short, they behave as if fairly possessed!

Furthermore, the lens of the eye, compassing what no glass lens can, CHANGES ITS CURVATURE to focus near objects as well as distant, when wanted; and not merely the lens, but the pupil—the camera stop—is self-adjusting. ALL THIS HAPPENS WITHOUT OUR HAVING EVEN TO WISH IT, without our even knowing anything about it, beyond that we are seeing satisfactorily.

The skin shutter outside, above and below this window, grows into movable flaps, dry outside like ordinary skin, but MOIST INSIDE, which wipe the window clean every minute or so by painting fresh tear water over it.

The eye's key structure is the light-sensitive screen at the back. It receives, takes and records a CONTINUALLY CHANGING MOVING PICTURE, lifelong, without change of "plate," through every waking day. And it signals its shifting exposures to the brain. It (the retina) is a nine-fold layer of great complexity. It is, strictly speaking, A PIECE OF THE BRAIN lying within the eyeball.

The cells that are at the bottom of the cup become a photosensitive layer—the sensitive film of the camera. The nerve lines connecting the photosensitive layer with the brain are not simple. The human eye has about 137 million separate "seeing" elements spread out in the sheet of the retina. The number of nerve lines leading from them to the brain gradually condenses down to little over a million. They are in series of relays, each resembling a little brain, and each so shaped and connected as to transmit duly to the right points of the brain each light picture momentarily formed and "taken." On the sense-cell layer the image has, picture-like, two dimensions. BUT THE STEP FROM THIS TO THE MENTAL EXPERIENCE IS A MYSTERY. For it is the MIND which adds the third dimension when interpreting the two-dimensional picture. AND IT IS THE MIND WHICH ADDS COLOR.

The chief wonder of all we have not touched on yet. The eye sends into the cell-and-fiber forest of the brain throughout the waking day continual rythmic streams of tiny, individually evanescent electrical potentials. This throbbing, streaming crowd of electrified shifting points in the spongework of the brain bears no obvious semblance in space pattern to the tiny two-dimensional upside down picture of the outside world which the eyeball paints on the beginnings of its nerve fibers to the brain. BUT THAT LITTLE PICTURE SETS UP AN ELECTRICAL STORM. And that electrical storm affects a whole population of brain cells. Electrical charges have in themselves not THE FAINTEST ELEMENTS OF THE VISUAL— they have nothing of "distance," nor "vertical," nor "horizontal," nor "color," nor "brightness," nor "shadow," nor "contour," nor "near," nor "far," nor visual anything—YET THEY CONJURE UP ALL THESE! A shower of little electrical leaks conjures up for me, when I look, the landscape, the castle on the height or my friend's face, and how distant he is from me!

How are we to explain the building and shaping of the eyeball, and the establishing of its nerve connections with the right points of the brain— and all starting from the tiny pinhead-sized embryo? And how EXPLAIN not the eye but the "seeing" by the brain behind the eye? THIS IS THE WONDER OF WONDERS. (Condensed; caps ours).

No evolutionist on earth can explain such miracles by his theory. The only possible explanation is, GOD MADE IT WORK THE WAY IT DOES.

Can anyone, who gives careful thought to the problem of HOW TO EXPLAIN THE HUMAN EYE, believe that it all started ages ago with a freckle or "pigment spot" that gradually developed, through countless ages, by random mutations, into the marvelous human eye? And by the way; if one eye should develop that way, why is it that there are TWO eyes, well spaced? And why is it that a third eye did not develop on the side of the head—and a fourth in the back of the head? In these days of auto-driving, it would be so practical to have an eye at the back of your head!

Many years ago, Dr. William Paley, in his book, **Natural Theology,** said,

"Were there no example in the world of contrivance except that of the **eye,** it would be alone sufficient to support the conclusion which we draw from it, as to the necessity of an intelligent Creator. . . . The marvelous makeup of this optical instrument, the further provision for its defence, for its constant lubricity and moisture which we see in its sockets and lids, in its glands for the secretion of tears, its outlet or communication with the nose for carrying off the liquid after the eye is washed with it; these provisions compose altogether an apparatus, a system of parts, so manifest in their DESIGN, so exquisite in their CONTRIVANCE, so successful in their issue and so infinitely beneficial in their use, as, in my opinion, to rule out all doubt (as to their divine origin.)"

A Miscellany of Miracles

Which of the marvels of the body would you like to consider next?

TONSILS

Shall we take a look at our mysterious TONSILS, once considered by evolutionists to be "vestigial organs," having no practical value to the welfare of the body? Learned doctors now assure us that tonsils are important—very important—to the welfare of the body.

Assuring us that "in the overwhelming majority of cases tonsilectomy is useless," Dr. Harry Bakwin, of the New York University and Bellevue Medical Center, says, "Tonsils act as a 'trap' to catch certain germs before they become widely spread through the bloodstream. In addition, tonsils appear to play a vital role in the formation of antibodies against bacterial and virus diseases."

TONGUE

Or, shall we spend some time in a minute examination of the TONGUE, with its many muscles "twining and intertwining," bound together in marvelous complexity, in a most astonishing way, which arrangement makes possible the many and varied motions that the agile tongue is capable of. Shall we look into the mysteries of its 3000 **taste buds** with their complicated structure —and each with a nerve connection to the brain? Shall we wonder WHY we taste "bitter" at the back of the tongue, "sweet" at the tip, and "salt" and "sour" tastes at the sides of the tongue?

DIGESTIVE SYSTEM

Someone will say, we **must** spend some time with the DIGESTIVE SYSTEM, rightly called "one of the supreme wonders of the body." ("Your Body's Wizardry with Food," by J. D. Ratcliff). Quoting more from this article we learn,

"Your digestive system is rugged and durable, completely automatic, and so complex that its workings are still not fully understood. . . .

"Digestion starts in the mouth, with the chemical action of one of the body's 20-odd ENZYMES—master chemists which promote reactions without themselves taking part in the reactions! . . .

The stomach is also a secretory organ, having the incredible total of 35 million glands! . . .

"The small intestine is one of the true wonders of the human body. It is absolutely essential to life. It performs the ultimate task of the digestive process. It has an elaborate muscular system. . . . The inside of the intestines is rough, folded, and contains approximately five million 'villi,' minute, hairlike protuberances—necessary in the digestive process." (Condensed from the Reader's Digest, 1953).

TEETH

Another reader will say: "Don't forget the TEETH. Wouldn't it be fascinating to trace the work of the tiny cellular 'teeth carpenters' who start before birth on their complex task?"

These "teeth carpenters" must find their chemicals for their job in the blood stream, "so the root of the tooth has a hole bored into it to carry blood vessels which enter from the deeper parts of the mouth. How do these expert "carpenters" KNOW HOW TO MAKE ENAMEL, DENTINE and CEMENT, laying the materials down bit by bit, shaping the teeth, placing the enamel over the chewing portion where it is needed, and performing all the intricate operations, right on time, EVEN HOLDING BACK THE TEETH UNTIL BABY HAS ALMOST FINISHED HIS NURSING?" (Dr. A. I. Brown, in "GOD'S MASTERPIECE: MAN'S BODY").

MUSCLE

Another will say, "you ought to discuss MAN'S MUSCLES—so complex in their action that 'When you scratch your nose with your forefinger the musuclar action involved dwarfs in complexity the workings of a hydrogen bomb" ("THE MIRACLE OF MUSCLE," By J. D. Ratcliff.)

More than half the human body is muscle—"the most remarkable stuff in nature's curiosity shop." "We speak of 'muscles of iron,' yet the working or contractile element in muscle is soft as jelly. HOW THIS JELLY CONTRACTS TO LIFT 1000 TIMES ITS OWN WEIGHT IS ONE OF THE SUPREME MIRACLES OF THE UNIVERSE. An elaborate series of chemical and electrical events, which would require hours or days to duplicate in the laboratory, occurs almost instantaneously when a muscle contracts—the twitch of an eyelid, for example." ("The Miracle of Muscle;" condensed from "Today's Health," Jan. 1956.).

Of course we SHOULD go on to write "pages" more about these "Miracle Muscles" of ours—but space is limited.

THE NERVOUS SYSTEM

A doctor among our readers says, "Do say something about Our NERVOUS SYSTEM: Our incredibly efficient and complex communications service that dwarfs any made by man."

"We are awed by the complicated tangle of wires in a large telephone cable; wonder-struck by the efficiency of a communications system that completes a call halfway around the world in a few minutes. But we are inclined to take for granted a communications system that is far more extensive, infinitely more complex—our own nervous system. Day and night millions of messages pour through its billions of cells, telling the heart when to beat faster, limbs when to move, lungs when to suck air. But for the links which it provides, our bodies would be mere masses of chaotic individual cells. . . .

"The ears have 100,000 auditory cells. Minute nerve ends in the inner ear pick up a particular sound frequency and start vibrating—waving like wheat in the wind. A current is generated. It may be so feeble that IT MUST BE AMPLIFIED THOUSANDS OF TIMES before it can be detected. Fed into the brain, by the nervous system, it is identified as a musical note. . . .

"Each eye has 139 million light receptors which send group impressions to the brain. . . .

"The skin contains a vast network of nerve receptors. If a hot object is pressed against the skin, some of the 30,000 'heat spots' will warn of the danger. It has 250,000 cold receptors and something like half a million tactile (touch) spots. . . .

"When we reach a complete understanding of the nervous system, we shall be close to understanding A SUPREME RIDDLE OF THE UNIVERSE: how that mass of cells known as man manages to behave like a human being." (Condensed from, "How Your Nervous System Works," in the May, 1956, "Today's Health").

THE HEART

Another eager inquirer then raises the question as to why the heart muscle beats a lifetime without tiring excessively.

Bruno Kisch, research associate in cardiology at New York's Mount Sinai Hospital, believes he has the answer.

"Heart muscle, like other striated muscle, is made of slender fibers. These in turn are composed of tiny fibrils. At regular intervals the fibrils are crossed by bands. . . .

"Micrographs show that heart muscle differs from other muscle in two important respects. The first is that the capillaries that carry blood to the heart muscle actually PENETRATE the muscle fibers; in other muscle the capillaries have only been observed on the surface of the fibers. The second is that among the fibrils of heart muscle are an unusually large number of granules called SARCOSOMES, which in other cells are known to contain enzymes."

Dr. Kisch believes that these sarcosomes give special feedings of rich enzymes to the heart muscle, and that, together with the deep-seated capillaries, is the SECRET of the untiring work of the heart! (See, "HEART MUSCLE," Scientific American Magazine).

The heart is the most efficient PUMP in all the world.

HUMAN HAIR

To satisfy the curiosity of one questioner, we should take a peek at human hair—under the microscope.

"Hair surface consists of overlapping scales resembling shingles on a roof. Underneath the scale is the cortex—overlapping spindleshaped cells that contain pigment. In or near the center of every hair is usually a medulla made up of cube-like cells which may contain pigment cells and air spaces." In each individual in the world his HAIR FOLLOWS A DISTINCTIVE PATTERN, set by the scales, the cortex and the medulla! One's hair is as distinctive as one's **fingerprints!** (See, "Detective by a Hair," telling of the work of Dr. Milton W. Eddy, who has perfected a system of identification by one's hair.)

THE "DUCTLESS" GLANDS

Each of us has in his body two kinds of GLANDS: Duct glands, such as the liver, that pour out their secretions through a duct to a designated place. We also have in our bodies pieces of highly specialized tissue that secrete chemicals that are poured DIRECTLY INTO THE BLOODSTREAM. They are called the Endocrine or ductless, glands. The most important ones are the Pituitary, the Thyroid, the Parathyroids, the Islet cells of the Pancreas, the Adrenals and the Gonads. These glands which together weigh only about two ounces, pour out some of the most powerful drugs known!

The Pituitary Gland

The PITUITARY is perhaps "the most remarkable gland in the human body."

"Because of its influence on other glands, it has been compared to the conductor of a great symphony. About the size of a large pea, it resides in a bony cavern on the underside of the brain, approximately in the center of the head. Something like 50,000 nerve fibers enter this fragment of tissue. Some of its hormones act as stimulants on specific targets. Thus one jogs the thyroid into activity, while others activate the adrenals, pancreas and sex glands. One helps to govern salt balance in the body. Another acts as a brake on the kidneys. One of the most fascinating of pituitary chemicals is the growth hormone: the circus midget has too little, the giant too much. Etc., etc." (See, "Your Amazing Glands," in **Your Family Doctor,** 11-'58).

Can anyone believe that this astonishing gland, no larger than a pea, "just happened" or was formed by "chance mutations"?

"The thyroid gland in your neck produces no more than a teaspoonful of hormone in a year. But if the teaspoon is only partially filled, a newborn baby can develop into a cretin—a malformed idiot.

"The adrenal glands produce only a teaspoonful of hormone in a **lifetime.** But let their hairline balance be upset and we are prey to a host of crippling and disabling diseases." (Ibid.)

THE HUMAN HAND

Some artist or musician raises the question, "How about discussing the HAND?" Gladly, dear friend, for the hand is one of the most marvelous of all God's gifts to man. The hand is unique with man.

"Nowhere in the animal kingdom is there anything comparable with the human hand. . . . the organ that has specialized in remaining unspecialized. It is an almost perfect tool-holder." "The HAND, the BRAIN and human SPEECH are the three features that distinguish man from the animals."

"The wrist has 8 pieces of bone, all wonderfully jointed. Our fingers have 19 bones, and are webbed part way up on the palm side, making a big hand that prevents things slipping through the fingers. . . . The human hand is a very wonderful thing and one of the greatest of its wonders is that the thumb is 'opposable' and CAN TOUCH ALL THE FINGERS—and so the hand can grasp tools, a pen, etc."

The hand has strength, lightness and dexterity. With the hands one can play the piano, write, paint and perform a thousand and one other actions.

W. BELL DAWSON, M.A., F.R.S.C., Canadian scientist, wrote this instructive statement about the hand:

"The design of the hand is remarkable; for the large and strong muscles which bend the fingers are NOT IN THE FINGERS THEMSELVES, but in the forearm near the elbow; and the fingers are bent by tendons or cords attached to these muscles, which pass through the wrist and across the palm of the hand to the fingers. What a remarkable plan it is on which our hands are constructed. How dreadful it would be if our fingers were like thick sausages, as they would be IF their powerful bending muscles had been placed in the fingers themselves! As it is, the hand, though so wonderfully strong, is able to do the most delicate work.

"Those who wish us to believe that man has descended from some animal, are MORE PUZZLED ABOUT THE HAND than almost anything else in the human body. For the evolutionist, who tries to prove his theory, cannot imagine, nor explain, HOW THE HAND WITH ALL ITS SKILL, could have developed from the PAW OF A BEAST. For it is only a highly intelligent being that can make use of an appliance which is so remarkably made as the hand. How then did its skill begin? The animals had no need of it whatever because their paws were already thoroughly fitted for every purpose in their animal life."

MAN'S INCREDIBLE LIVER

A physician has been listening carefully; now he wants us to suggest a few facts about "MAN'S INCREDIBLE LIVER." Take a look at a shapeless glob of liver in a butcher shop; transfer your thoughts to your own liver, the largest internal organ in your body. It may weigh almost four pounds.

"The liver is becoming recognized as one of the greatest mysteries of the scientific world. It does 19 different jobs!"

"Your liver," says Herbert D. Benjamin, M.D., "is probably THE MOST EFFICIENT AND COMPLICATED DEVICE ON THE FACE OF THE EARTH."

"This mysterious gland is the master laboratory of the body. . . . The liver's cells brew a vast and varied chemistry essential to the smooth functioning of all our organs. Some examples: our kidneys couldn't dispose of waste nitrogen if the liver didn't turn it into urea for excretion. The liver stores vitamins necessary to the birth of blood in the marrow of our bones. The liver builds amino acids into the albumin that regulates the balance of salt and water without which we could not live. And the liver manufactures bile which influences intestinal activity, so we're not poisoned by products of our own digestion." ("Your Liver is Your Life," by Paul deKruif, in Reader's Digest).

"The liver is the most fantastically complex and efficient organ in the human body. Among other things, the liver regulates the exact consistency of the bloodstream. It does this by producing three different substances and delivering them to the blood in just the proportions needed. HOW it does this nobody can even begin to guess. . . . One of these substances is FIBRINOGEN, which causes clotting when the blood is exposed to the air. This substance is so complicated that scientists have been unable to duplicate it in the laboratory.

"The second is PROTHROMBIN; it keeps the bloodstream thick enough so that internal hemorrhages through the walls of the blood vessels and organs cannot develop.

"And the third is HEPARIN that counteracts any tendency of the other two to thicken the blood too much."

Yes, "the liver is a combination factory, laboratory and refinery that forms dozens of fantastically complicated chemical operations that no man-made factory in the world can duplicate." (Dr. Glen R. Shepherd).

And the liver does all its amazing feats through one type of cell. "Reddish-brown in color, the liver contains millions of minute cells arranged into working units known as LOBULES. Each lobule is a chemical factory or storehouse." (See "Man's Incredible Liver," By Leroy Thorpe.)

Think a moment: Could this reddish-brown, shapeless mass of cells do all these miracles, without being activated and made to so perform by the Supreme Intelligence? That the human liver could evolve through random mutations is as farfetched as to believe that a modern jet plane could make itself.

THE EAR

The specialist suggests that we consider the EAR: AUDITORY MARVEL.

"Even in our era of technological wonders, the performances of our most amazing machines are still put in the shade by the sense organs of the human body. Consider the accomplishments of the ear. It is so sensitive that it can almost hear the random rain of air molecules bouncing against the eardrum; yet in spite of its extraordinary sensitivity the ear can withstand the pounding of sound waves strong enough to set the body vibrating. The ear is equipped, moreover, with a truly impressive SELECTIVITY. In a room crowded with people talking, it can suppress most of the noise and concentrate on one speaker. . . . At some sound frequencies the vibrations of the eardrum are as small as one billionth of a centimeter—about one-tenth the diameter of the hydrogen atom! AND THE VIBRATIONS OF THE VERY FINE MEMBRANE IN THE INNER EAR WHICH TRANSMITS THIS STIMULATION TO THE AUDITORY NERVE ARE NEARLY 100 TIMES SMALLER IN AMPLITUDE. This fact alone is enough to explain why hearing has so long been one of the MYSTERIES OF PHYSIOLOGY. Even today WE DO NOT KNOW HOW THESE MINUTE VIBRATIONS STIMULATE THE NERVE ENDINGS." (See article on "THE EAR," By Georg von Bekesy, in the Scientific American).

Let us consider briefly part of the marvelous structure of the ear. "To understand how the ear achieves its sensitivity, we must take a look at the anatomy of the middle and the inner ear. When sound waves start the ear drum (tympanic membrane) vibrating, the vibrations are transmitted via certain small bones (ossicles) to the fluid of the inner ear. One of the ossicles, the tiny stirrup (weighing only about 1.2 milligrams), acts on the fluid like a piston, driving it back and forth in the rhythm of the sound pressure. These movements of the fluid force into vibration a thin membrane. called the basilar membrane. The latter in turn finally transmits the stimulus to the organ of Corti, a complex structure* which contains the endings of the auditory nerves. The question

---

*The organ of Corti is so complex, its workings are almost beyond human comprehension. "The cochlea has developed within itself the extraordinary structure known as the ORGAN OF CORTI. The cochlea consists of two and a half spiral turns around a central supporting pillar. But this spiral canal is subdivided by plates of bone and membrane into three staircases. Upon this membranous partition is the special structure called the organ of Corti. THIS IS A VERY COMPLICATED ARRANGEMENT OF CELLS, placed upon a vast number of parallel fibers rather like piano wires; . . . it has been supposed that these fibers act like the wires of a piano, responding to various vibrations that reach them. . . . In this organ itself the essential elements are the hair-cells of Corti, many thousands in number, having fibers of the auditory nerve encircling their bases,

immediately comes up: WHY IS THE LONG AND COMPLICATED CHAIN OF TRANSMISSION NECESSARY?

"The reason is that we have a formidable mechanical problem if we are to extract the utmost energy from the sound waves striking the eardrum. Usually when a sound hits a solid surface, most of its energy is reflected away. THE PROBLEM THE EAR HAS TO SOLVE is to absorb this energy. To do so, it has to act as a kind of mechanical transformer, converting the large amplitude of the sound pressure waves in the air into more forceful vibrations of smaller amplitude." (Ibid.) (Caps ours).

You may have to read the above paragraphs at least THREE TIMES to follow the description; but note well what we capitalized: "WHY IS THIS LONG AND COMPLICATED CHAIN OF TRANSMISSION NECESSARY?" And the author tells us why—"to extract the utmost efficiency from the sound waves." Note that the author suggests that the EAR had a problem to solve. Now the EAR is utterly unable to solve ANY problem. To suggest that the "ear" had a problem to solve is like suggesting that the propeller of an airplane, not yet properly designed, had a problem to solve. It is not the MECHANISM that solves the problems, but the MECHANIC. GOD, the Creator, had a problem to solve, when He constructed the ear—and He did a magnificent job of it! Why not give HIM the credit for what He did?

The author of the article in "THE BOOK OF POPULAR SCIENCE" who wrote on THE EAR, had this significant thing to say (P. 2762):

"From the external ear a canal leads inwards until it is closed by a definite and unmistakable drum-head. The drum of the ear, or tympan, is no more figuratively so named than the lens of the eye. It is what it is named, AND THOSE WHO CAN CREDIT ITS ORIGIN BY THE NATURAL SELECTION OF CHANCE VARIATIONS CAN CREDIT ANYTHING." (Caps ours). This author, whoever he is, is saying exactly what we say: It is IMPOSSIBLE to believe that such an intricate and involved mechanism as the ear came into being by "chance variations" in the theoretical workings of "evolution." Since the "chance variations" of evolution are ruled out by all reason and logic, we MUST account for such miracles of construction by acknowledging that GOD MADE THEM.

"OUR LIVING BONES"

We would like to enlarge on the structure and workings of

---

while the upper ends are provided with several short stiff hairs, bathed in the fluid that fills the cochlea. These are undoubtedly the all-important cells of the inner ear." (Book of Popular Science). The fact is, the workings of the INNER EAR are so complicated no one fully understands them.

the KIDNEYS that "are among the most miraculous organs in the body"* but we are forced to stop, by reason of limitations of space. But we do want to speak of yet one more marvel of the body: "OUR LIVING BONES."

"Actually, our bones are not 'dead' but they contain thousands of small blood vessels and are quite as much ALIVE as one's stomach. Active little cells called osteoblasts work night and day creating new bone, while house-wrecking cells known as osteoclasts labor just as hard tearing down material tagged for the scrap heap. In addition, the red marrow acts as a blood-cell factory, and the bones themselves as calcium storage vaults." (See "Our Living Bones," in the May, '48 "Hygeia.").

Every bone in the body and every combination of bones was especially DESIGNED to serve a certain purpose. The skull was **designed** to house and protect the brain. The spinal column was especially **designed** with three curves in it, instead of being straight.

"The curves protect the spine against fracture. The presence of the three curves divides the weight of the three masses of organs in the body. The upper sector carries the head, the middle one, the thoracic viscera, and the lower one the abdominal organs. In a straight backbone there would be entirely too much weight concentrated at the bottom. Remember also these other features about the spine: (1) The astonishing articulations by which the several bones of the spine touch one another in many places, but invariably FIT absolutely accurately; (2) the amazing system of LIGAMENTS which bind each pair of bones together; and (3) the wholly wonderful perforation of the individual bones, making a neatly fitting conduit to carry the spinal cord without injury of the slighest degree, although it (the spinal cord) is one of the most delicate structures of the body; (4) the strange curvature, perfectly engineered; and (5) the large number of bones and joints, giving FLEXIBILITY—all these reveal to us the wisdom of the Creator. (Adapted from Dr. Brown's treatise on "GOD'S MASTERPIECE: MAN'S BODY").

Merged into the skull, is the facial portion of the head, with its 14 bones, so as to make the well-known contour for jaws, cheeks, eye-sockets and nostrils.

"The leg bones are hollow, in keeping with the engineering principle that a hollow column is stronger than a solid one of equal weight. On a weight-for-weight basis BONES ARE STRONGER THAN STEEL. Bone construction is comparable to reinforced concrete." (See "Our Busy Bones," Today's Health, Nov. '55).

But the hollow space within the bones is NOT WASTED. In some bones

---

*The kidneys contain a total of approximately 280 MILES of tiny tubules whose function is to filter impurities from the blood. In the course of a day they filter something like 185 quarts of water from the blood, purify it, and return it to circulation: THE MOST AMAZING PURIFYING SYSTEM IN THE WORLD. Who designed this system?

it is filled with the blood-forming bone marrow. God is an Expert at conservation and efficiency.

"The human skeleton represents a MASTERPIECE OF ENGINEERING DESIGN, with each component part tailored to a specific job." (Ibid.).

All authorities on bones emphasize the fact that they are DESIGNED for a specific job—but few of them call attention to WHO DESIGNED THEM. Surely, the bones themselves could not do that; it is impossible for reasoning people to believe that such "marvels of engineering," such obvious DESIGN for a desired end, just "happened" through the processes of "natural selection" and "chance variations." Who or what then DESIGNED our Living Bones? The only possible answer is: GOD, THE CREATOR, DESIGNED THEM.

We ask further—Who DESIGNED the marvelous hinge joint in the knee? Who DESIGNED the amazingly efficient ball-and-socket joints at shoulders and hips, making movements of arms and legs possible in almost every direction! ALL joints in the body are "true wonders of mechanical art." (Dr. Brown).

"BONE is a busy and in many ways quite amazing tissue. It houses the factory (the bone marrow) that produces most of the cells in the blood; it stores minerals and doles them out as needed to other parts of the body; it repairs itself after an injury; it grows, like any other living tissue, until the body reaches adulthood. Not the least of its wonderful properties is the fact that while it is growing and constantly building itself, it also serves as the rigid structural support for the body, like the steel framework of a building." ("BONE," By Franklin C. McLean, in Scientific American).

Consider now this fact: the world is ASTONISHED at the wonders of the human body—but they fail to give the Creator the credit due HIM. We list some of the titles of the articles in secular and scientific journals from which we have quoted; and note how all of them speak with amazement about the human body:

"Wonders of Conception"
"Secrets of the Human Cell"
"The Miracle of Birth"
"The Miracle of Growth"
"Your Remarkable Eyes"
"BLOOD: The Stream of Life"
"Your Body's Wizardry with Food"
"Powerful Pinheads"
"Your Amazing Glands"
"Your Liver Is Your Life"
"Man's Incredible Liver"
"Our Living Bones"

To sum up: the human body is a MIRACLE in construction and function. We have not listed one-thousandth part of its wonders. It would take a million words—and then the subject would not be fully covered. From head to toes, from skin to bone marrow, ALL parts of the body give evidence of careful DESIGN—engineering perfection. Only God the Creator could achieve such a masterpiece of construction!

Part II—THE MIND OF MAN

We must distinguish between the "mind"—man's mental ability—and the "brain"—the physical organ that is the seat of the mind. As Norman L. Munn says,

"No one has ever seen a mind. A surgeon cutting in the brain sees only nerves and blood vessels; to know what is going on in the brain he must ask the patient. . . . Only through language can we get any sort of direct picture of the working of the mind." (The Evolution of Mind; June, '57, "Scientific American").

Men in all ages have marvelled at the miracles of man's mind.

"Of all wonders," wrote Sophocles, "none is more wonderful than man who has learned the arts of thought and speech."

Dr. Henry Fairfield Osborn, noted modern anthropologist, wrote, "To my mind, the human brain is the most marvelous and mysterious object in the whole universe."

"The brain is a mystery," said Sir Charles Sherrington, of London; "it has been and still will be. HOW DOES THE BRAIN PRODUCE THOUGHTS? That is the central question and we have still no answer to it. . . . Despite significant advances, we have not yet accumulated more than fragmentary insight into what goes on in the brain . . . which is of HARDLY IMAGINABLE COMPLEXITY."

Another scientist bears witness to the same fact. "Today, even though we are awed and even frightened by the intellectual achievements of man's mind, the mechanisms that make it possible are still unknown. Knowledge of the outward form of the brain is well advanced. But what of the neurone mechanisms involved in CONSCIOUSNESS, THOUGHT, PERCEPTION, BEHAVIOUR, MEMORY? . . . They are unknown." (Wilder Penfield, Prof. of Neurology, McGill University, in an article in the 1955 Smithsonian Institute Annual Report).

The Physical Nature of the Brain

The brain, spinal cord and nerves comprise the "Central Nervous System." The brain of the average man weighs about 3.3 pounds—about 1500 grams—"a mass of pink-gray jelly-like substance composed of some ten to fifteen billion nerve cells." There are three important parts to the brain:

(1) The **Cerebrum**. It is the main part of the brain. It is in two

halves; each is heavily convoluted. About 65% of the surface area of the cerebrum is buried in the folds of the fissures that form the convolutions; so, by using massive convolutions, the Creator increased the efficiency of the brain **threefold**. The cerebrum, with its covering of "gray matter" (the cerebral cortex), which is only .1 inch thick and 400 square inches in area, is the seat "of consciousness, memory, imagination and reason."

"The ten to fifteen billion nerve cells in this cerebral cortex are the center of operations. Each of the sense organs reports on its own lines to specific, well defined regions of operations. Those for the eyes are at the back of the brain, those for the ears are well down on each side, etc.

"All messages to the brain are sifted and decoded, decisions are made and orders relayed to appropriate stations of the body. It STAGGERS THE IMAGINATION HOW EFFICIENTLY IT DOES THE JOB. With five or six times as many workers (the microscopic brain cells) as there are people in the entire world, and WITHOUT SUPERVISORS . . . it can handle equally well the information conveyed by the cells' call 'soup's on!' or 'fire!' or 'play ball!', by the sight of an object in a store window, a page of differential equations, a beautiful woman, the tiny spot of light from a distant star, or the blinking of a traffic signal.

"It is a far reaction," Sir Charles Sherrington, the Oxford Scientist, reminds us, "from an electrical reaction in the brain to suddenly seeing the world around one, with all its distances, its color and its variations of light and shade." (The "electrical reaction" he refers to is real, but unbelievably minute. And the neurons—brain cells in the cortex—are inconceivably complex in their composition and action.)

"The study of the densely packed fine structure of the cortex has generated an immense literature on the various neuron types and their arrangement and interconnection in the half-dozen layers in which the cortex is divided. WE DO NOT EVEN BEGIN TO COMPREHEND the functional significance of this richly complex design. . . . Each of the ten billion neurons receives connections from perhaps 100 other neurons and these connect to still 100 more! THE PROFUSION OF INTERCONNECTIONS AMONG THE CELLS OF THE GRAY MATTER IS BEYOND ALL IMAGINATION. It is ultimately so comprehensive that the whole cortex can be thought of as one great unit of integrated activity. If we now persist in regarding the brain as a machine, then we must say that IT IS BY FAR THE MOST COMPLICATED MACHINE IN EXISTENCE. . . . It is infinitely more complicated than the most complicated man-made machines— the electrical computers.

"Each of the 10 billion neurons is an independent, living unit. It receives impulses from other cells through intricately branching **dendrite fibers** which sprout from its central body; it discharges impulses to other cells via a single slender fiber, the AXON, which branches profusely to make contact with numerous receiving cells via their DENDRITES. . . . Connections between cells are established by the SYNAPSES—specialized junctions. . . . At these synapses the transmitting cell secretes highly specific chemical substances, whose high-speed reaction carries the signal from one cell to the next. THE WHOLE OF THIS ALL IMPORTANT PROCESS OCCURS ON AN EXQUISITELY SMALL SCALE. A neuron operates on a power of about a thousand-millionth of a watt!" (Condensed, from "THE PHYSIOLOGY OF IMAGINATION," by John C. Eccles, President, Australian Academy of Sciences, in "The Scientific American," Sept., '58).

The above is a highly technical discussion of the workings of the cerebral cortex, but we include it to alert the reader to some appreciation of how inconceivably complex are the workings of the brain. To help us better understand how truly remarkable the human brain is, we offer this contrast:

"It takes 500 TONS of equipment for just ONE telephone exchange. Sixty-two thousand man-hours of work are needed to install this one 10,000 line Dial Exchange. Eighty miles of cables and 2,600,000 soldered connections are needed for this apparatus." The human brain, of course, does INFINITELY more than a 500 ton mass of telephone equipment which merely carries messages, but gives no commands, can not remember, reason, speak or imagine.

(2) The **Cerebellum**. It is beneath the back part of the Cerebrum—and much smaller. The cerebellum controls the voluntary actions of our muscles, partly on orders from the cerebrum. It is that part of the brain concerned with our control of the body, as "balance, the body's movements, and its muscular habits and aptitudes." With his 'incomparable cerebrum,' and his likewise 'incomparable cerebellum,' man can attain special techniques like those of the pianist, surgeon or artist.

The ability to preserve our balance "is an enormously involved process"; so much so that "probably 1,000,000,000 nerve cells take part in every move we make to keep our bodies straight."

(3) The **Medulla Oblongata.** It is the enlarged bulb at the top of the spinal column. It is a center to control the automatic functions of the body, as breathing, and the pumping of the heart.

Volumes by the score have been written on both the **Cerebellum** and the **Medulla Oblongata,** and an incredible mass of wonders and miracles have been discovered about these portions of the brain—pertaining to both their composition and functions. We must leave to others the privilege of enumerating them, for the wonder of reverent minds.

Having said a little about the marvels of the cerebrum—the main portion of the brain—and its incomparable cortex, we now want to point out the great

GULF that Exists Between the Brain of an Ape and the Brain of Man.

There is a vast gulf when SIZE is considered, between the brain of an ape and the brain of man.

Prof. C. Judson Herrick, distinguished Chicago neurologist, says, "The brain of the most incompetent normal man is TWICE as big as the most accomplished ape." And the average size brain of man is "THREE times as large as the average size brain of an ape," relative to body sizes.

Moreover "the human skull contains a brain some FIVE or SIX times the weight of that of any of the existing apes—in proportion to their respective body sizes." (Book of Popular Science, p. 3210).

Not only is man's brain much larger, but also it has certain areas and lobes that the brain of an ape DOES NOT HAVE.

The portions of the brain of an ape that receive sense impressions and bring about muscular movements are about as large, relatively speaking, as those of man. But in man there is a tremendous development in the "association areas" where the process of reasoning, and other higher functions of the mind, go on. The "association areas" also enable man to read, speak, understand a language, and learn by experience.

Another authority says, "A certain small area of the cortex, in the frontal lobe, called 'Broca's convolution,' is man's SPEECH center; monkeys and apes DO NOT HAVE THIS AREA AT ALL."

And here is an additional weighty testimony: "The most remarkable change in brain form, passing up the scale from monkey to man, is the comparative ENLARGEMENT of the frontal and anterior lobes, and there can be little doubt that this (enlargement) is associated with man's supremacy in the intellectual sphere." (P. 436, Annual Report for 1955, Smithsonian Institute.)

HOW DID MAN GET THIS ENLARGED BRAIN? And how are we to account for the fact that only man has the brain areas that give the powers of speech (using organized language), reasoning, the ability to learn from experience, and the miracle of a creative imagination? One can easily see that "between the brain of the higher apes and man is a VAST GULF" for which evolution has no satisfactory explanation.

WHAT THE MIND OF MAN DOES

The mind of man has capabilities that set it apart as God's Masterpiece of creation. Not only does the mind of man control and direct the muscular activity of his body (both voluntary and automatic), but also it has the power to remember the past, plan for the future, make decisions and carry them out; the mind of man can reason, imagine and dream; the mind of man makes him self-conscious and world-conscious. There are many more activities of the mind, all marvelous, all mysterious. Let us list some of the more wonderful of the capabilities of the mind.

(1) **It is the Mind that makes our sense organs work.** An authority reminds us:

"We know something about how the sense organs in the eye, the ear, the nose and the skin transmit stimuli of light, sound, floating particles and touch into afferent streams of nerve impulses, but without the subtle powers of MIND, working in and through the brain, all these sense perceptions would be lost. Neither the highly complex ear nor eye would work at all WITHOUT THE MIND BEHIND THEM."

(2) **It is the Mind of Man that makes real to him the Outside World.** Sir Charles Sherrington, Oxford scientist, seeks to impress on us this miracle.

"I am seated in my room and turn to look through the window. I perceive sun and sky, field and road, the approaching friend. There is 'near' and 'far' in the scene; there is color; some objects move—others do not. WHERE IS THIS SCENE? Out beyond the window? Yes; but also, IT IS IN MY BRAIN. It is a transmutation, a wonderful transition, as inexplicable as fairy magic. ... It is the work of the brain! ...

"To glimpse the marvel of this, we must strip from our minds the habituation of lifelong routine. Then only can it appear to us in its NAKED WONDER—the more intriguing, because still inexplicable. ... And there is another strange thing about it, namely, that this unexplained phenomenon DOES NOT EXCITE ANY WONDER IN US. WHY?"

(3) **The Mind of Man makes Man CONSCIOUS of himself and the world around him.** This has been called "the most celebrated achievement of the human mind." (Fred Kohler).

Fred Kohler, ardent evolutionist, in his book "Evolution and Human Destiny" says, "One of the most remarkable achievements of the human mind IS ITS ABILITY TO BE CONSCIOUS OF ITSELF AND ITS FUNCTIONING. 'Consciousness' represents actually a situation in which matter has become aware of its own existence. This is indeed such a remarkable circumstance that it has been considered to be beyond the possibility of any scientific understanding . . . and has led to the belief that MAN IS IN SOME FUNDAMENTAL ASPECTS DIFFERENT FROM ALL OTHER CREATURES, and that his mind must contain some principle not explicable in terms of it being an extremely complex aggregate of matter." (Pp. 95, 96, 97; caps ours).

In his book, "The Nature of the Universe," Fred Hoyle, noted astronomer and evolutionist, comments on the miracle and mystery of CONSCIOUSNESS. "How," he asks, "is a machine produced that can THINK ABOUT ITSELF AND THE UNIVERSE AS A WHOLE? At just what stage in the evolution of living creatures did INDIVIDUAL CONSCIOUSNESS ARISE? I do not say that questions such as these are unanswerable, but I do say that it will not be simple to answer them." (P. 123).

This astounding miracle of the mind of man that makes man SELF-CONSCIOUS and WORLD-CONSCIOUS, though it is such an inexplicable mystery to the scientist, is readily understood by the person who believes in special Creation. It is explained by the fact that when God created man, He not only made his body, but He made him a LIVING SOUL (see Gen. 2:7), self-conscious, world-conscious, universe-conscious—and conscious of his fellow-men. The Bible speaks of this miracle in I Corinthians 2:11.

"For what man knoweth the things of a man, save the spirit of man which is in him?"

(4) **The mind of Man can Learn to do most anything: It can create Machines.** In its inventive or creative ability the mind of man is so far ahead of the animal creation that one can not speak accurately of a "missing link"—one must say "a missing CHAIN."

"In this respect the mind of man OVERWHELMINGLY TRANSCENDS all its predecessors. Though the ape can learn, its powers of learning are strictly limited. But the evident fact of man is, HE CAN LEARN TO DO ANYTHING. He can make any kind of object, any kind of a machine, or a book. Only man's brain can create machines outside his body. The difference then between man's mind and that of his nearest rivals is the difference between the unlimited and limited . . . The brain of man is the supreme organ of life." (P. 1969, Book of Popular Science).

(5) **The Mind of Man has the power to REMEMBER, and to learn from past Experiences.** What is this strange power of mind that enables the brain to store away in its secret repositories beyond the immediate sphere of consciousness MILLIONS OF FACTS AND OBSERVATIONS that can be brought forth to active recollection on demand or when needed? This is a miracle so stupendous as to be totally incredible did we not all know it to be true! How can a man conjure at will the image of his long-dead grandmother? or the old homestead where he spent his childhood, and the thousands of precious memories associated with it?

In some persons memory is truly phenomenal. The late Charles Evan Hughes could write a thirty-minute speech, and then deliver it from memory —word-for-word—the same day.

John von Neuman, reputed to be "one of the world's greatest scientists," had "mastered college calculus by the age of eight and could memorize on sight a column in a telephone book and repeat back the names, addresses and numbers without error." (See Life Magazine, issue of Feb. 25, '57).

So prodigious is man's ability to memorize and remember that we are told "the brain's ten billion cells can hold, potentially, more information than is contained in the nine million volumes in the Library of Congress." Another authority says, "in the 70 years of his life a person can receive into his mind 15 TRILLION SEPARATE BITS OF INFORMATION."

And think of this miracle: Recent research has "located the recording mechanism for remembered experiences "in an area of the brain having a "total surface of about 25 square inches—and only a tenth of an inch thick—in the cerebral cortex." (See "The Tape Recorder in Your Brain," 9-'58 Coronet, p. 57).

"The brain's tissue (the seat of memory) include some thousand billion billion protein molecules. THAT number is believed ample for the purposes of human memory." (**Ibid.**)

(6) **The Mind of Man has the power to FORGET some things.** We do not actually recall physical pain.

"Unconsciously certain self-protective devices in the mind erase from memory some painful memories, unpleasant appointments and embarrassing situations"—though not entirely—as though to save us from an overload of painful memories. The keen edge of grief and loss soon wear off. Many frightening dreams are forgotten before even the sleeper awakes—"hidden from view by the self-protective unconscious mind that is always on guard." (Popular Medicine).

(7) **The Miracle of the Unconscious or Subconscious Mind.** Much of the prodigious amount of information stored in the mind is in the storehouse of the unconscious mind. The subconscious mind seems to retain EVERYTHING it has ever received! And though it is difficult, seemingly impossible, to bring much of it up to the surface, modern psychologists are discovering new ways to induce the unconscious mind to dislodge its long hidden secrets.

"The most wonderful part of your mind is undoubtedly the unconscious, which lies below the immediately recoverable memory and is thousands of times larger. We do not yet know much about the unconscious mind, but we are learning. . . . By means of several devices we now know how to bring back lost memories." ("Your Brain's Unrealized Powers," Reader's Digest).

(8) **Man's Mind is "Educable"—it has the power to Learn from Experience, History and Teachers.**

Unlike the beasts, who never advance, from generation to generation, **man learns** from history, from his own experience and the experience of others. He builds on the inherited knowledge of the past. This accumulated knowledge has rapidly increased, especially since the invention of printing. In recent years accumulated knowledge has snowballed, so that in our times man has accomplished truly amazing things!

Mankind has developed such marvels as highspeed photography, airplanes that travel faster than sound, television, radio, the X-ray, nuclear fission, and a thousand and one other miracles that previous generations could never have realized for the simple reason that ACCUMULATED KNOWLEDGE had not yet advanced to the point that permitted the intricate inventions of our modern age.

(9) **Man's Mind alone is created with the ability to learn and use LANGUAGE.** Beyond all question, this is one of the greatest gifts God gave to man. The ability to learn a language and use it distinguishes him from all lower animals, and enables him to communicate knowledge to others and to learn from others.

"A child learns the use of words not merely as mechanical signs and signals (as an animal does) but as AN INSTRUMENT OF THOUGHT. This insight is beyond the reach of any animal. Even the most intelligent sub-human animal reacts only to SOUNDS and **not to the meaning of words.** The late Edward Lee Thorndike demonstrated this with an experiment on his cats. He trained them to dash to the food box when he said, 'I must feed those cats'; but then when he later exclaimed at the cats' meal time, 'Today is Tuesday' (using the same intonations as they were accustomed to) the cats sped to the box. And the words 'My name is Thorndike' evoked the same response. . . . A chimpanzee, intelligent as it is, SIMPLY CANNOT MASTER LANGUAGE. It can be taught to speak a few words, but each word takes months and eight words seem to be about its limit. . . . With a body as large and complex as ours, the chimpanzee has only ONE-THIRD as much brain to manage it. Furthermore, in the ape's brain the frontal area, which is concerned with associations and symbolic functions, IS MUCH SMALLER IN RELATION TO THE REST OF THE BRAIN than ours. It seems to lack entirely the section in the left frontal lobe known as 'Broca's area,' a part of the human brain known to be involved in speech."

"Through the use of LANGUAGE man is able to gain mastery of his environment, learn from the past, penetrate the future . . . and man's future is largely under his own control, thanks to the gift of LANGUAGE." (Norman L. Munn, in "The Evolution of Mind," Scientific American, 6, '57).

Fred Kohler, evolutionist, senses the vast importance of LANGUAGE. He says, in "Evolution and Human Destiny,"

There is little doubt that some several million years ago the ancestors of present-day man consisted of one or several species of primates. . . . Now it is known that the other primates (apes and their relatives) CANNOT BE TAUGHT LANGUAGE to any appreciable extent. . . . (They) lack the necessary organs and nervous equipment. . . . Even if animals are brought up among human beings they develop NO LANGUAGE WHATSOEVER. . . . Animal learning is based upon observation by the individual animal and the trial and error response to such observation. It is NOT based on any appreciable interchange of information with other members of the species. . . . The entirely new method of learning that is consequently available to man as a result of the development of language, increases the effective experience of each individual . . . and has achieved the developing intelligence of society."

"It is probable that the circumstance of the dawning of language in man was an extremely important factor responsible for making him what he is today."

Contemplating on the ability of man to use LANGUAGE, he deducts, "When the ratio of brain development in man, to that in the higher animals, is compared with corresponding ratio of their mental achievements, one finds that man's accomplishments SEEM VASTLY OUT OF PROPORTION. How then is one to explain that man's ability is so very much greater than one would expect on the basis of his brain development alone, basing the comparison on animal standards?" (See Pp. 73-82).

HOW ARE WE TO ACCOUNT FOR MAN'S VASTLY SUPERIOR MENTAL POWERS AND ACHIEVEMENTS over the animals, out of all proportion to his increased brain size? The answer is, when God created man, HE endowed him with these privileges, responsibilities and abilities. How deep is the thought suggested in the Scripture, "God CREATED MAN IN HIS OWN IMAGE AND LIKENESS" (Gen. 1:26, 27)!

Would any one be so foolish as to suggest but a "missing link" between the chattering monkey or the guttural growls of the Gargantuan gorilla and the smooth-flowing speech of a silver-tongued orator such as William Jennings Bryan? No, my friends, that gulf is NOT a mere "missing link" it is a gorge as wide and as deep as the ocean!

(10) **The Mind of Man is capable of THINKING: REASONING.** He holds debates, courts, parliaments—and "bull sessions." He argues, defends, accuses, tries and condemns. From childhood on up he for ever wants to know WHY? He is inquisitive, logical (and at times very illogical) and inventive. He creates books, machines, paintings and poems. This ability to **think—reason**—has given us an Einstein, standing before the blackboard, writing down the equation that unlocked the secrets of the atom! Between the gibbon in the treetop, clutching a banana, and the engineer at his desk, using trigonometry and the slide rule, is more than a "missing link"—it is a chasm as wide as a hundred Grand Canyons!

(11) **The Mind of Man is capable of Creative IMAGINATION.** Ludwig von Beethoven's "Symphony No. 5 in C-Minor" was a "triumphant masterpiece."

A contemporary composer, hearing it for the first time, was so moved by it he ejaculated, "Unbelievable! . . . Marvelous! It has so bewildered me (by its charm) that when I wanted to put on my hat, I could not find my head!" And yet this marvelous, almost matchless, creation ORIGINATED IN BEETHOVEN'S IMAGINATION. Though he was growing deaf, and could not hear his own playing, "he heard the full richness of this immortal composition FIRST IN HIS OWN IMAGINATION—and then he set it down in notes. "The fateful opening chords and the glorious theme have made it the most popular symphony of all time."

The mind's ability in the field of "creative imagination" has produced much of the literature of the world—dramas, melodramas, tragedies, sonnets, poems, stories and novels; it also has produced many of the inventions of the world, which were FIRST created in the imagination of the inventor, and then developed into practical realities.

Frank Barron, Research Psychologist at the University of California, gave this tribute to the "Charm of Imagination."

"The sorcery and charm of imagination, and the power it gives to the individual to transform his world into a new world of order and delight makes it one of the most treasured of all human capabilities." (See "The Psychology of Imagination," by Frank Barron, Scientific American, 9-'58).

Picture a William Shakespeare at his desk, writing one of his immortal sonnets or dramas. Between a William Shakespeare or a Ludwig von Beethoven and the five simple words of "Viki," the world's "most educated chimpanzee,"*—words spoken with no understanding of their meaning, but mere words, to obtain a reward, as a dog barks to get a bit of food—there is an expanse that separates as wide as stellar spaces! No use looking for a missing link; scores of chains are missing!

Beside these capabilities of the mind of man that we have listed, there are scores more, such as "perception," "apperception," "intuition," "analytical capacity," "cognition," etc.

The greatness of the mind of man can be measured somewhat by his mental ACHIEVEMENTS. The mind of man has conceived and created great telescopes, with which he has surveyed and mapped the heavens, peering two billion light years out into space. He has made microscopes with which now he can actually see the atom! He is able now to study the miscroscopic world as never before, and he is discovering wonders undreamed of a century ago. He has "cyclotrons" and "accelerators" that cost

---

*"Viki" is the name of a chimpanzee taken in their home when three days old, by Prof. and Mrs. Keith Hayes, at the Yerkes Laboratories of Primate Biology, at Orange Park, Fla. "It was a three-year experiment in bringing up a baby chimpanzee as one would bring up a child." The story of the experiment is told in the book—"THE APE IN OUR HOUSE," by Kathy Hayes.

a hundred million dollars to construct. He has scaled the highest peaks, travelled over both poles, and is discovering marvels in the abysmal undersea depths. He has developed a highspeed camera that can shoot FIFTEEN MILLION MOTION PICTURE FRAMES PER SECOND. He has discovered the secrets of nuclear fission and nuclear fusion—and he has harnessed the power of the atom. He has thrust satellites into orbit in space—and is planning now to go to the moon. He has discovered and is using many "wonder drugs" and is able to travel in his airplanes faster than the speed of sound.

Though the mind of man had not attained the measure of mastery over nature as it has now,* Dr. Alfred Russell Wallace knew the vast superiority of the mind of man over the mind of apes—and so he was led to ask Charles Darwin, "HOW DID MAN GET HIS WONDERFUL BRAIN?" Wallace's question and the modern perplexity engendered by that question are discussed in the Nov. '55 issue of **Harper's** Magazine, in an article on "WAS DARWIN WRONG ABOUT THE HUMAN BRAIN?" by Prof. Loren C. Eiseley, anthropologist. We quote from this very frank discussion.

To explain "the rise of man through the slow, incremental gains of natural selection, Darwin had to assume a long struggle of tribe with tribe . . . for man had far outpaced his animal associates. To ignore 'The Life Struggle' would have left no explanation as to how humanity by natural selection alone managed to attain an intellectual status so far beyond that of the animals with which it had begun its competition for survival."

But Wallace pointed out that "MEN WITH SIMPLE CULTURES (such as 'stone age men' in the world today) POSSESS THE SAME BASIC INTELLECTUAL POWERS which the Darwinians maintained could only be elaborated by competitive struggle."

"Man and his rise now appear short in time—EXPLOSIVELY SHORT. There is every reason to believe that whatever the . . . forces involved in the production of the human brain, a long, slow competition of human groups would NOT result in such similar mental potentialities among all

---

*The generation-by-generation INCREASE in knowledge that has characterized the history of the human race is neither evolution nor proof of evolution. Man's brain from the beginning of his creation had the same potential it has now. Loren C. Eiseley, Prof. of Anthropology at the University of Pennsylvania, says, "There is evidence that **Homo sapiens** has NOT altered markedly for hundreds of thousands of years." (Scientific American, 11-'50). Man's brain capacity and his IQ have not increased; the ancient Greeks and more ancient Egyptians and dwellers in the Mesopotamian valley of 5,000 years ago were as intelligent as our generation, though they did not have the vast store of accumulated KNOWLEDGE to draw on that our generation has.

peoples everywhere. SOMETHING—some other factor—HAS ESCAPED OUR SCIENTIFIC ATTENTION.*

Loren C. Eiseley continues, "There are certain strange bodily characters which mark man as being MORE than the product of a dog-eat-dog competition with his fellows:

(1) He possesses a peculiar larval nakedness, difficult to explain on 'survival' principles;

(2) His period of helpless infancy and childhood are prolonged;

(3) He has aesthetic impulses;

(4) He is totally dependent, in the achievement of human status, upon the careful training he receives in human society.

(5) Unlike solitary species of animals, he cannot develop alone—he has suffered a major loss of precise instinctive controls of behavior. In place of this biological lack, society and parents condition the infant and promote his long-drawn training;

(6) We are now in a position to see the wonder and terror of the human predicament: MAN IS TOTALLY DEPENDENT ON SOCIETY.

(7) The profound shock of the leap from animal to human status . . . involved the growth of prolonged bonds of affection . . . because otherwise its naked helpless offspring would perish.

(8) Modern science would go on to add that many of the characters of man, such as his LACK OF FUR, THIN SKULL, and GLOBULAR HEAD suggest mysterious changes in growth rates . . . which hint that the forces creating man drew him fantastically out of the very childhood of his brutal forerunners. Once more the words of Wallace come back to haunt us: 'WE MAY SAFELY INFER THAT THE SAVAGE POSSESSES A BRAIN CAPABLE, if cultivated, and developed, of performing work of a kind and degree FAR BEYOND WHAT HE EVER REQUIRES TO DO.!**

---

*Yes, Dr. Eiseley, something HAS escaped the "scientific attention" of most of our scientists today—and that "something" obviously is the fact that GOD CREATED MAN AND HIS VASTLY SUPERIOR BRAIN.

**We now know Wallace was right. Modern communications and transportation, plus societal adjustments, have abundantly proven the point. "Men with black skins, whose fathers lived in grass huts and hunted lions with spears, NOW STAND BEWIGGED IN COURT, pleading cases with an Oxford accent. Men with yellow skins, whose fathers used their wives as beasts of burden to plow their fields NOW CROUCH IN COCKPITS OF MIG-17's, as they roar over the Formosa strait." (See "Races in Turmoil," Newsweek, 9-15-58).

Even Fred Hoyle admits (in "Evolution and Human Destiny") that "members of primitive cultures possess a brain potential EQUIVALENT TO THAT OF CIVILIZED MAN. . . . An infant removed from such a group (having a primitive or 'stone age' culture) will on the average share the conceptual characteristics of the society he is raised in."

Obviously, all men everywhere, whether occidental, oriental, "civilized" or in a primitive 'stone age culture', HAVE ABOUT THE SAME RELATIVE BRAIN SIZE, and the same MENTAL POTENTIAL. In other words, "civilization" is built on ACCUMULATED KNOWLEDGE rather than on superior mental powers or brain potential.

Then Prof. Eiseley makes this startling admission and naive confession—

"Today we can make a partial answer to Wallace's question, SINCE THE EXPOSURE OF THE PILTDOWN HOAX all of the evidence at our command—and it is considerable—points to man in his present form, AS BEING ONE OF THE YOUNGEST AND NEWEST OF ALL EARTH'S SWARMING INHABITANTS. . . . Most of our knowledge of him (man) . . . is now confined, since the loss of Piltdown to the last half of the Ice Age. If we pass backward beyond this point . . . it is (to the scientist) like peering into the mists floating over an unknown landscape. Here and there through the swirling vapor one catches a glimpse of a shambling figure, or a half-wild primordial face stares back at one from some momentary opening in the fog. Then, JUST AS ONE GRASPS AT A CLUE, the long gray twilight settles in and the WRAITHS AND THE HALF-HEARD VOICES PASS AWAY . . .

"Ironically enough, science, which can show us the flints and the broken skulls of our dead fathers, HAS YET TO EXPLAIN HOW WE HAVE COME SO FAR SO FAST, nor has it any completely satisfactory answer to the question asked by Wallace long ago. Those who would revile us by pointing to an ape at the foot of our family tree grasp litte of the awe with which the modern scientist now puzzles over man's lonely and SUPREME ASCENT. The true secret of Piltdown, though thought by the public to be merely the revelation of an unscrupulous forgery, lies in the fact that it has forced science to re-examine the history OF THE MOST MIRACULOUS CREATION IN THE WORLD—the human brain. (Caps ours; condensed from "Was Darwin Wrong about the Human Brain?" in Harpers Magazine, 11-'55).

Prof. Eiseley still clings to the empty shell of evolution; but he virtually admits that Wallace must be right: **The amazing Brain of Man demands a special Creation.** In fact, Prof. Eiseley uses these words, directly quoted—"The most miraculous **creation** in the world—the human brain."

Part III. MAN ALSO HAS AN EMOTIONAL, MORAL AND SPIRITUAL NATURE

In addition to his superb brain and mind, man has a soul, a "psyche,"—a moral, spiritual and emotional nature—that distinguishes him from the lower animals.

After creating man, God breathed into his (man's) nostrils "the breathe of life" and man became a LIVING SOUL. See Gen. 2:7.

(1) **Man's Emotional Nature.** In addition to man's powerful INSTINCTS*—far higher and different from what animals have—

---

*Psychologists speak of the "parental instinct," "the fighting instinct," "the fear instinct," "the instinct of self sacrifice," the instincts of "sympathy, suggestion, curiosity, wonder, and imitation," as well as other instincts of lesser importance.

man has a highly complex EMOTIONAL nature, far above and beyond what animals have.

In animals, such emotions and qualities as fear and hatred are largely instinctive. But in man we see the full development of a broad gamut of emotions and moral qualities—both good and bad—that includes love and hate, joy and sorrow, peace and anxiety, trust and unbelief, hope and despair, satisfaction and frustration, elation and despondence, mercy and vengeance, pride and humility, courage and cowardice, approval and disgust, ambition and apathy, perseverance and pliability, nobility and meanness, gratitude and ingratitude, envy and good will, steadfastness and vacillation, zeal and lethargy, righteousness and unrighteousness, independence and dependence,—and a hundred more emotions and moral qualities that have enriched or degraded our lives and our literature.*

Consider this brief list, then think of the paucity of the emotional life in animals. The few emotions animals seem to have are more instinctive and are far more simple than man's vast domain of rich emotions. How can one account for the GREAT GULF between the emotional life of animals and that of man? The only answer is, Man is a special creation, with God-like gifts and qualities of mind and soul.

(2) **Man's Esthetic Nature.** Inherent in the soul of man are such abilities and qualities as an appreciation of things beautiful and harmonious, as expressed in music, the arts, song and poetry. We can enjoy a symphony and a sonata, a poem, a lovely woman and a beautiful picture, a gorgeous sunset, a majestic snowcapped peak, a delicate fern frond, an exquisite lily or orchid, as well as the scintillating grandeur of the distant nebulae. We are inspired by Michelangelo's "David," or "Moses;" we are enchanted by the ceiling of the Sistine chapel or Raffael's Madonna; we revel in the art gallery as well as in the pristine beauty of the everglades and the rapturous song of the nightingale. We look with awe at a sunset over lake Victoria Nyanza, and are enthralled by the vista from the top of Pike's peak. The soul of an ape, a

---

*The richness of our language and our literature depends largely on the wealth of emotional experiences of men: patriotism, romance, love, hate, nobility, meanness, fidelity, ambition, wonder, curiosity, etc. The intriguing fabric of human experiences, made possible by a wide range of deep emotions, is utterly unknown to any animal. We refer to "the patience of Job," "the courage of Richard the lion-hearted," "the zeal of the evangelist," etc.

monkey, or a dog, has NO RESPONSE WHATEVER to a poem, a sunset or a beautiful painting, but finds its satisfaction in a banana, a bag of nuts, or a bone. How can one account for this tremendous gulf? It becomes increasingly obvious that man is a special creation —created in the very image and likeness of God. (See Gen. 1:26, 27).

(3) **Man's Moral Nature.** When God created man He made him a free moral agent, with the power of CHOICE. This great gift involved a paralleling **responsibility:** blessing would follow a **right** choice, judgment would follow a **wrong** choice. Unquestionably, this gift of a **free will** is man's highest endowment. After his creation man (Adam and Eve) were put to the test—and they failed, and rebelled gainst God's government. So our original parents **sinned** (Gen. 3:1-7), and through them SIN and DEATH entered the race (see Rom. 5:12). We now observe in the world the double phenomena: universal SIN and universal DEATH. Animals, in the same world judged by sin, also suffer the experience of death. See Romans 8:20-23.

Why evil is present in mankind, without ANY EVIDENCE WHATEVER OF THE GRADUAL MORAL BETTERMENT OF THE RACE* (save that which results from the influence of the Bible) is also an inexplicable phenomenon that utterly refutes the theory of evolution. According to evolution the morals of mankind should gradually improve—but they do not.

Herbert Spencer glibly spoke (as do modern evolutionists and communists) of "the universal law of progress which ordains that surely evil and immorality disappear . . . surely must man become perfect."

History tells us that the horrors of wars have been getting steadily WORSE, especially the wars of our century.

The Gospel facts of (1) Man's origin by special creation in the image of God (Gen. 1:26, 27); (2) His "Fall" and the entrance of

---

*Whatever moral betterment the race has experienced has always been the result of the acceptance of the teachings of the Bible. When so-called "civilized" nations like modern Germany DEPART from historic Christianity, a Hitlerian barbarism takes over, resulting in atrocities of the worst sort, such as the attempted genocide of the Jews, and the horrors of brainwashing. When men reject Christianity and turn to communism— dialectic materialism—similar evils follow: mass murders, the enslavement of large segments of society and torturous brainwashings. To the extent that our own country has departed from Bible teachings and Bible standards, the evils of juvenile delinquency, alcoholism and crime have proportionately increased.

sin (Gen. 3:1-7 and (3) His consequent judgment—death and separation from God; and (4) His "redemption" as a race through the atonement of Christ (Rom. 5:12-19; I Cor. 15:3, 4); and (5) the "salvation" of all individuals in the race who repent and accept Christ and His Gospel (John 3:16; Rom. 10:9, 10; Acts 2:28; Acts 16:31) are fully unfolded in the New Testament and are the subjects of DIVINE REVELATION in the Bible. The amazing thing is, the Divine Revelation of the facts of the special creation of man, and his subsequent fall and ruin are abundantly attested to by the FACTS OF HUMAN EXPERIENCE. Evolution has no acceptable way of explaining either the phenomena of SIN and DEATH in human experience, or the REGENERATION of believers in Christ.

Man has a **conscience** that enables him to distinguish between right and wrong, good and evil. Man has a **will** that enables him to resist evil and say no to wrong—but it also enables him to yield to sin and do wrong. Man's moral accountability to his Creator is unique in all creation; no animal is so gifted and honored by the Creator.

(4) **Man's Spiritual Nature.** God not only gave man a superior soul, He also gave him a **spirit** that makes him GOD-CONSCIOUS. Men everywhere instinctively KNOW there is a Higher Power, and that they must give account to Him. The base religions of the world seek to placate evil spirits, and so seek to avoid retribution for their failures. But though the religion of many cultures is base, yet all men everywhere have some sort of "religion," the perfect witness to the fact that all men are created with a SPIRIT as well as a soul.

Scattered over the world are several hundred tribes without a written language, living in various types of primitive cultures. Among them are the **Negritos** of the Philippines, scores of wild Indian tribes of Central and South America, the aborigines of Australia, the **Ituri** pigmies of Africa, the **Onges,** one of the most primitive of all cultures, who live on the Andaman Islands in the Bay of Bengal—so degraded they do not even know how to make a fire; the **Wobangs,** white pigmies of the interior of New Guinea; native Papuans; **Samongs,** the tree people who live in remote forests in the East Indian Archipelago; the **Botocudos,** who live in interior Brazil, and "have not yet reached the old Stone Age level of development." They slit the lower lip and ear lobes and insert large wooden plugs in them. Then there are also the **Kubus,** who rove the jungles of Southern Sumatra, said to be "the most primitive tribe on earth;" and the **Troglodytes** of Southern Tunisia who live in caves. ALL OF THESE PRIMITIVE PEOPLE HAVE SOME FORM OF RELIGION—proof that they are "God-conscious" and that they believe they must give account to a Superior Being. True, their worship is perverted—but the SPIRIT that makes the knowledge of God possible is there, even though it is depraved.

It is a miracle of God's grace that He **did** endow man with the ability to KNOW HIM and to worship Him. To man alone, as far as life on earth is concerned, eternal life is offered through Christ. We read that "Christ tasted death for every MAN"—not animals (see Heb. 2:9). When God created man, He brought upon the theater of the world a being capable of knowing Him; his was a capacity and a privilege the lower animals were not given. True, man is now under the condemnation of sin, but he is the object of God's solicitous grace—and those who yield to the Gospel of Christ are RENEWED into the image of the very likeness of God, re-created "in righteousness and true holiness" (see Eph. 4:23,24; II Cor. 5:17). The Creator, through Christ, recreates men into His perfect likeness when they repent and believe Christ's gospel. This whole realm of the spiritual, made possible only to mankind, and made actual to all who are willing, through Christ, to submit to God's government, is a realm entirely beyond the ken of all animals. The spirit of man, through which he is GOD-CONSCIOUS and through which he may come to know God and be restored to His perfect image in righteousness and true holiness —is the ultimate witness that man is a special creation, infinitely higher than any and all animals.

Further Evidence that Man did not Descend from Monkeys, Apes or Their Relatives

Darwin taught in the DESCENT OF MAN that the early ancestors of man must have been "more or less monkey-like animals" of the anthropoid group. Many attempts have been made to discover the "MISSING LINK" between man and the apes, or between man and monkeys, but "NO MISSING LINK HAS EVER BEEN FOUND" (Popular Science). Because of this fact—truly terrifying to the evolutionists—many biologists today "no longer believe that man descended directly from . . . the apes." Most evolutionists now teach that man descended from a "common ancestor" of apes, monkeys and men. " 'Proconsul' is probably the common ancestor of apes and man" (Life). Actually, no one seems to know what this "common ancestor"—called

"Proconsul"—was like;* it exists only in the imagination of evoultionists. There is no such creature, nor was there ever.

Since no MISSING LINK between man and apes has been found, science now theorizes that "man and the anthropoid apes (gorillas, chimpanzees, gibbons and the orangutans) sprang from the same limb of a common family tree. One branch is represented by the apes, the other by man." (Scientific American).

This is a favorite trick of modern evolutionists. "They assume that the actual 'ancestors' of all 'specialized types' are some 'generalized ancestors' for which no fossils can be found." (W. Henning).

This modern evasion actually **greatly weakens** their argument and focuses attention to the "VAST GULF" that actually does exist between apes and men. And by further removing the original stock from which "man and apes descended" **they themselves create a vastly greater gulf to bridge!** If the step is great between apes and men—and Huxley said that the gap between monkeys and apes is **greater** than that between apes and men—IT IS FAR GREATER BETWEEN SOME SUPPOSED 'PROCONSUL' (lower on the scale then apes) AND MEN. In other words, if we can prove there is an "unbridgeable chasm" between apes and men, we also demonstrate that there is a far greater chasm between man and the supposed, lower-than-monkeys "Proconsul."

Argument for Evolution from "Comparative Anatomy"

Evolutionists argue that "all elements in the human skeleton are 'readily comparable' to the similar bones in apes: both have a cranium, both have ribs, both have the femur (leg) bones, etc. They further call attention to similarities in the main body organs: heart, lungs, eyes, ears, etc., and that "the brain of the apes . . . has the same convolutions as man's."

But similarity of anatomy is accounted for by the fact that

---

*Prof. Julian Huxley says, "Ten or twenty million years ago man's ancestral stock branched off from the rest of the anthropoids, and these relatives of man have been forced into their own lines of specialization." (Evolution: The Modern Synthesis.)

Frances Vere, criticising Prof. Huxley's statement, says, "This assertion is totally unwarranted; there is not one tittle of evidence to support it. 'Ten to twenty million years' give Prof. Huxley a good slice of time in which to indulge his fancy." But remember the words of Sir Wilfred Le Gros Clark, F.R.S., in a lecture at Oxford, "We must admit that WE HAVE NO FACTUAL EVIDENCE on which to base an answer to the question, 'When did the "hominid" and "anthropoid" lines separate?' THE FOSSIL RECORD OF THE EARLIEST STAGES OF HOMINID EVOLUTION IS COMPLETELY BLANK. ("Requiem For Evolution," by Francis Vere.)

apes, monkeys and men **live in a similar environment**—our world, with its atmosphere, water, chemical elements and foods types—and so all apes, monkeys and men NEED a heart, lungs, skull, brain, legs, eyes, etc., to live in the common environment. Similarity of anatomy does not prove evolution any more than the similarity between a Lincoln and a Ford means that the Lincoln "evolved" from the Ford. It simply means that THE SAME INTELLIGENCE MADE BOTH.

Instead of being misled by the similarities into accepting the evolution dogma, let us note well the striking DIFFERENCES between apes and men—differences so great as to create a chasm unbridged and unbridgeable by any theory of "natural selection." Between these two realms—apes and men—there are NO CON- Between these two realms—apes and men—there are NO MISSING LINKS,* either in the world of living animals or in the

There are Many Kinds of Monkeys and Apes

Before we list some of the many differences between apes and men, we call attention to the fact that there are scores of kinds of monkeys and several kinds of apes (gorillas, gibbons, chimpanzees and orangutans) all **radically** different from each other; whereas **men, the world over, are built on practically the same pattern!**

It is worthy of mention that man is the **only** species in his genus and the only species in his family! This is a powerful argument for Creation: for man is UNRELATED.

Apes are tropical animals ranging from the three-foot Gibbon, a treedweller, to the six-foot, terrifying Gorilla, with the fiendish features, who may be several times heavier than man. Science agrees, "Neither Gibbon nor Gorilla can in any way be considered a Missing Link." When an ungainly Orangutan seeks to stand upright, its arms reach almost to the ground; it has comparatively short legs.

The male Proboscis monkey has a grotesquely elongated "Roman" style nose (snout). Why? Shall we ape Darwin's line of argument and suggest that this odd nasal appendage developed by "natural selection" because the female of the species preferred males with such outlandish protuberances? Nonsense!

There is another natural "clown" among the monkeys—the male mandrill. It is a short-tailed baboon, with a swollen blue face, a red nose and red lips, a yellow beard, and with odd swellings below its eyes. It is a

---

*"Missing link" is used here as it is commonly used. It has come to indicate a fossil or animal that bridges gaps between various major or lesser groups.

268

rare caricature! If the female of the species were responsible for this apparition, she surely did a good job of it!

The pigmy marmoset monkey is as small as a squirrel. The sacred monkeys of India are "whitish animals with blackish faces," having great projecting eyebrows.

From which of these species did man descend? And from whence came the other apes and monkeys? Did man descend, let us ask from the Mandrill baboon, nature's caricature? or from the fiendish gorilla? No evolutionist today is foolish enough to claim man's direct descent from any one of the living apes or monkeys; but he cleverly evades the issue by fabricating an imaginary "proconsul" that is as unreal as his "pro-avis" (the missing link that is still missing, between reptiles and birds). "There just ain't so such animals" as the "pro-avis" and the "proconsul."

Since evolution teaches the GRADUAL change from proconsul to men, we ask—WHERE ARE THE INTERMEDIARY STAGES? No such "links" exist, either in the world of living animals or in the fossil world. The only place they exist is in the imagination of the evolutionists!

Many Authorities call attention to the STRIKING DIFFERENCES BETWEEN APES AND MEN

"Man with his highly developed brain, his upright posture, his sense of duty and his appreciation of beauty is VERY DIFFERENT from the lower primates" (Book of Knowledge, Vol. 1, p. 205).

"Apes have long arms; men short. Men have chins; apes have none. Apes have massive canine teeth; men do not. Apes can not oppose the thumb to the fingers; men can. Lacking this, no ape could be a competent tool-using animal." (Article on Anthropology, Pageant magazine, 10-'47).

"HOW TO HUMANIZE APE—Reshape his hands, develop pelvis, refine vocal organs, change the anatomy of his brain to produce man's ability for abstraction, symbolism and foresight; change the anatomy of his nervous system so he can develop skills basic to our culture." (Dr. Keith J. Hayes, Yerkes Laboratory of Primate Biology, Orange Park, Fla.)

"There is a subtle and misleading omission by scientists of facts opposing the supposed ape and man similarity. Why do they not tell us that female humans have a membrane (hymen) which female apes do not have, and that human males lack a bone (baculum) which apes have? Female apes have a poorly developed breast. From man's much longer head-hair to the obviously different foot, from buttocks to chest-rib shape—with its longer collar bone—the entire ape is APE, not human.

"The sperm and ovum differ. The diet is different. The human child, after birth, is greatly changed in leg length, and its skull will not calcify for some time. This is not so of the ape. The ape's brain is different, lacking the vital 'Broca (speech) area'.

"Our research has shown FULLY 150 VITAL DIFFERENCES between apes and men—not to speak of mental, moral and spiritual matters. And probably NO APE ORGAN OR STRUCTURE IS OR CAN BE QUITE LIKE A HUMAN'S." (Professor Leroy Victor Cleveland, Author, Anti-Evolution Compendium).

As a matter of fact, there are thousands of vital differences between apes and men, too numerous to list. But here are a few more of the more obvious:

Adapted to arboreal life, a chimpanzee before it is a month old can grasp a branch and suspend its own weight for more than a minute. (Dr. Keith Hayes). This is utterly impossible for a human baby of that age. Babies, as we all know, are born much more dependent on their parents and on society than apes or monkeys—and they stay dependent on society for a much longer time than any other form of life on earth.

Apes and monkeys spend at least part of their time in trees, and some of them rarely descend to the ground. Men live on the ground in houses. Apes sleep in crudely constructed nests.

Apes have fur; men must wear clothes.* Men light and use fires; Apes neither make a fire nor put sticks on a fire to keep it burning. Apes do not plant or keep gardens. Apes neither cry nor laugh. Apes live mostly on leaves, buds and fruit. Most monkeys have cheek pouches on the inside of their mouths where they can stuff extra food they gather. Men store their extra food in midriff bulges or double chins.

BONES

"Every bone of a gorilla," admitted Prof. Huxley, "bears marks by which it might be distinguished from the corresponding bone of man."

In half-erect monkeys the backbone is so curved that the weight of the body is bound to fall forward unless a very special effort is made; but in adult men the backbone is such that the weight of the body is well balanced and distributed in an upright position.

The pelvis of an ape is not designed for an upright posture. "Man stands alone because man alone STANDS."

The Foot

If anthropologists want to determine whether a skelton is that of a man or an ape, let them examine the FOOT, not the skull. If they could examine the brain inside the skull, they could easily decide if it be that of a man or an ape, for the brain of a man has certain lobes that the brain of an ape does not have. But the brain soon disintegrates after death. Mere skulls may be very

---

*In his DESCENT OF MAN Darwin explained this difference by saying that many females liked their monkey mates with less hair—and so there gradually evolved HAIRLESS MAN! This is totally unscientific, for monkeys live in hordes, and are promiscuous. They are either pregnant or nursing a baby, almost all the time; so they do not seem to have much concern about what male monkey is their mate, whether hairy or not.

misleading. The skulls of males, females and children ARE ALL OF DIFFERENT SIZES, with considerable variation in shapes —both for men and apes. Then too there are skulls deformed from diseases. BUT BY EXAMINING THE FOOT ONE CAN KNOW IMMEDIATELY IF IT BE THE "HIND HAND" OF AN APE OR THE FOOT OF A MAN. In the apes, the big toe is really a thumb—made to grab with. MAN'S TOES ARE UNIQUE: they are short and small and NOT like a thumb. The specialized foot of man is made to WALK on and not to grab things with.

"It is our FOOT, not our hand, that is the ONE PECULIAR FEATURE of our limb skeleton. Our ape ancestors were not four-footed types, but **four-handed** ones, and their 'hind hands' were better grasping organs than the front ones! (Man and the Vertebrates. Chapter on "The Human Body," pp. 373, 374).

All will agree with Sir Arthur Keith, "A child has never been seen with an anthropoid foot." "And what is more striking still," adds Prof. F. Wood Jones, "the human foot as soon as it is formed in the embyro is of characteristic HUMAN TYPE." (Problems of Man's Ancestry," P. 38).

Apes Are Dying Out

One more significant fact: "Apes are dying out. It seems probable that before long they will be extinct." On the other hand, the populations of mankind are rapidly increasing, the world over.

It is clear that the body of man, starting with two minute cells, is so complex, so intricate and involved in operation, that it **had** to be created by an all powerful Intelligence. It is equally clear that the brain and mind of man, with its amazing powers of speech, reason, memory and imagination, is so fantastically wonderful it **had** to be created by a Master Genius, infinite in His creative resources. To believe that such marvels as the eye and the ear, the brain and the hand, the liver and the blood, could develop by random mutations, or natural selection, is the height of absurdity.

The gulf between men and apes, or between men and some imagined predecessor of the apes, in every conceivable realm: physical, mental, moral, emotional, spiritual, is so vast—and with no "missing links" to span the gulf—that one is forced to the conclusion that man could NOT have evolved from the lower animals but was, as the Bible says, "created in the image of God."

Few prominent evolutionists in recent years have come so close to admitting that evolution is an unproven and unproveable theory as Prof. Eiseley. Listen to his picturesque portrayal of man's dilemma; this is what happens when man rejects the Biblical account and tries to solve the problem of man's origin.

"There are great gaps of millions of years from which WE DO NOT POSSESS A SINGLE COMPLETE MONKEY SKELETON, let alone the skeleton of a human fore-runner. . . . For the whole Tertiary Period which involves something like 60 to 80 million years, we have to read the story of Primate evolution from a few handfuls of broken bones and teeth. . . .

"In the end we may shake our heads, baffled. . . . It is as though we stood at the heart of a maze and no longer remembered how we came there. . . . Until further discoveries accumulate, each student will perhaps inevitably read a little of his own temperament into the record. . . . They will catch glimpses of an elfin human figure which will mock us from a remote glade in the forest of time. OTHERS JUST AS COMPETENT WILL SAY THAT THIS ELUSIVE. . . . ELF IS A DREAM, SPUN FROM OUR DISGUISED HUMAN LONGING FOR AN ANCESTOR LIKE OURSELVES." (Caps ours. Prof. Loren C. Eiseley, Prof. of Anthropology, University of Pa., in the June, '56 "Scientific American.")

Chapter 12

MYSTERIES, MIRACLES AND MISSING LINKS

MYSTERIES AND MIRACLES

THERE ARE A MILLION—and more—mysteries and miracles in nature that defy explanation. No man apart from Divine Revelation can explain the origin of matter, nor the secrets of the atom. Nor can anyone account for the origin of motion, in this vast universe, and the miracle of sustained, controlled motion.

We do not know what LIFE is, nor why or how a new unit of life can be started with the union of two small cells. Neither do we know why plants and animals grow—nor why, when they reach a certain stage, they **stop** growing—nor why they grow old and eventually die.

"It is one of the ironies of modern science that the most elementary questions are still the hardest to answer. How, for instance, is a new animal created? We know that an egg and a spermatozoon unite to form a single cell, and this union somehow sets in motion a chain of events that gives rise to a new being. But what trigger, what spark, starts the process? What, in short, is the secret of fertilization? We do not yet know, although many eminent biologists have searched many long years for the answer." (Albert Monroy, in "Scientific American.")

Scientists do not know how life arose from dead matter in the beginning; the gap between the inorganic elements and the simplest forms of life is infinitely great. No one knows why life is divided into the two major kingdoms: plant and animal. Why not all one kingdom (say, animal); or, why not more than two— plant, animal and some other radically different form of life? No one can explain the origin of sex; why male and female? Nor can any one explain the secrets of heredity, or instinct—that amazing property of animals which acts like intelligence but is not intelligence as we know it, but which enables certain creatures to do what man with all his intelligence can not do!

No person on earth can explain fully what electricity is, nor gravitation, magnetism, light, heat, sound or color! We know how these natural forces act—but WHY and exactly how they do so, no man knows.

How are we to explain this phenomenon—one can send an electric current through a copper wire at 10° below zero, and at the other end of the wire heat a heating coil to thousands of degrees. Where was the "heat" while going through the wire? We all know there are "laws" governing heat and electricity that explain HOW these phenomena work—but who knows WHY they work that way?

No one can explain the origin of chlorophyll, or the fascinating, involved process of photosynthesis.

"Photosynthesis — the amazing process of the synthesis of organic compounds from carbon dioxide and water by plants in light remains one of the great unsolved problems of biology." (Eugene I. Rabinowitch, "Scientific American," 11-'53).

WHY does a comparatively minor change in the number of protons, neutrons and electrons in an atom produce an entirely different element? Essentially, all the atoms of all the elements are built on the same general plan: a central nucleus, made up of protons and neutrons, with an equal number of electrons revolving around them, with unbelievable speed. WHY, by merely changing the number of the protons, neutrons and electrons, does one get the different elements—so vastly unlike? From hydrogen to uranium and beyond, ALL atoms are built on the same general principle, but they differ in the **number** of particles in the atom. What magic legerdemain did the Creator use to accomplish such uncanny results?

NO ONE YET HAS THOUGHT OF A SATISFACTORY THEORY FOR THE ORIGIN OF THE EARTH, IF GOD IS LEFT OUT OF THE PICTURE. Every theory advanced so far has been riddled full of holes by FACTS that disprove the theories. As an example of how "theories" have been made and blasted into bits, we quote from Fred Hoyle.

"Nearly all of the planets LIE VERY FAR OUT (far away from the sun)." This simple fact is "the death blow" of every theory that seeks for an origin of the planets in the sun; for, "how could the material have been flung out so far?" (See Fred Hoyle's argument, p. 85, in "The Nature of the Universe.")

Moreover, if all the planets were originally part of the sun, **Why do some of the moons of some of the planets revolve in retrograde motion—in the opposite direction to the others?**

If the earth were originally thrown off the sun, as a boiling mass of red-hot matter, WHERE DID THE WATER ON THE EARTH COME FROM? In a universe that is practically **without** water, why is there so much water on our earth?

A thousand mysteries envelop our earth! Who understands the mysteries of cosmic rays—and other forms of stellar radiation?

"In cosmic radiation we are dealing with a universal phenomenon that is energetic, basic and mysterious" (Shapley).

Who can explain why, except by admitting an act of God, there is an "ozone belt" about twenty miles up in the atmosphere, that filters out "killer rays" from the sun. Without that ozone belt no life on earth would be possible. WHO PUT IT UP THERE?

The whole gamut of life on earth, from bacteria to man, is so involved, so interdependent, so filled with specialized organs that perform functions needed by the SOCIETY OF LIFE, as well as for their own benefit, that it presents one grand miracle of integrated achievement. Life on earth in any organism is really a LIVING MACHINE, with a million parts, ALL necessary for the successful operation of the whole.

Bateson sensed this, though he professed to be an evolutionist. We quote:

"To supply themselves with food, to find it, to seize and digest it, to protect themselves from predatory enemies whether by offence or defence, to counter-balance the changes of temperature, or pressure, to provide for mechanical strains, to obtain immunity from poison and from invading organisms, to bring the sexual elements into contact, to insure the distribution of type; all of these and many more are accomplished by organisms in a thousand most diverse and alternative methods. These are the things that are hard to imagine as produced by any concatenation of natural events." (Quoted from "Problems of Genetics," in "Evolution, the Unproven Hypothesis").

Mysteries of the Microscopic World

Many people are familiar with the so-called plant carnivores—plants that trap and eat insects. Of such are the pitcher plant, with its reservoir of digestive fluid to drown—then digest—hapless insect victims; and the sundew plant, with its flypaper-like leaves that trap insects; and the Venus flytrap, with its snapping jaws.

"There are MICROSCOPIC PLANTS in nature as unique and cleverly designed as any of these (pitcher plant, sundew plant, and the Venus flytrap).

"These microscopic predators are fungi, or molds. . . . Some of them are equipped with traps and snares which are marvels of genetic resourcefulness. HOW THEY EVOLVED THEIR PREDATORY HABITS AND ORGANS REMAINS AN EVOLUTIONARY MYSTERY."

Yes, Prof. Maio, it is a very deep mystery to all evolutionists, and we can assure you, it will remain so—until they accept the fact of an all-wise, all-powerful Creator; then their problem will be solved, as it has for those of us who believe in Divine creation.

Let us listen further to Prof. Maio, as he unfolds some of the marvels and mysteries of microscopic predators, so necessary in the "balance of nature."

"One of these molds is **Arthrobotrys oligospora.** It develops networks of loops, fused together to form an elaborate nematode (a minute worm) trap. An extremely sticky fluid secreted by the mold . . . dooms the nematode. In its frenzied struggles to escape the worm only becomes further entangled in the loops, and finally after a few hours exertion it weakens and dies.

"Even more artfully contrived are the 'rabbit snares' employed by some molds. They are rings, made of three cells, having an inside diameter JUST ABOUT EQUAL TO THE THICKNESS OF A NEMATODE (the victim). When a nematode in the soil sticks its head into one of these rings, the three cells SUDDENLY INFLATE LIKE A PNEUMATIC TIRE, gripping the nematode in a stranglehold from which there is no escape. The rings respond ALMOST INSTANTLY to the touch of a nematode; in less than one-tenth of a second the three cells expand to two or three times their former volume, obliterating the opening of the ring. . . . We are not yet sure what cellular mechanisms activate these deadly nooses. If the nematode touches the **outer** surface of the ring, it will NOT trigger the mechanism, but if the worm passes **inside** the ring, its doom is certain." (Joseph J. Maio, Research Dep't., University of Washington; in an article, "Predatory Fungi," July, '58, "Scientific American").

Remember—these clever snares, that work so quickly and with such precision, are MICROSCOPIC—so small they cannot be seen by the eye of man! Scientists have not yet discovered the secret mechanism that triggers the snares, nor the "steps" that led to the development of such an intricate system, on so small a scale: BUT THERE IT IS: at work, all the time, in nature. WHO DESIGNED THIS CLEVER, INVOLVED EQUIPMENT? The brainless mold? No one is thoughtless enough to suggest that. Man had nothing to do with it; these molds were snaring nematodes long before man even knew about them.

Why are there thousands of animals and plants—both microscopic and visible to the eye—that have **unique characteristics**, utterly unlike other animals or plants?

The Mystery of the "Limitation of Hazards"

We present a phenomenon, not uncommon in nature, that might be termed the "limitation of hazards."

"Nature has a 'connectedness' that is sometimes astounding. Kenneth D. Poeder (Tufts University) and Asher Treat (City College of New York) made the discovery that the high-pitched beeps that bats emit also act as a warning to moths on which bats prey."

The investigators found that the moth's ear is beautifully adapted to hearing and locating high-pitched sounds (the bat's cries are above the frequency range of the human ear and cannot be heard by man). Nature, as it were, gives these moths "a fighting chance."

"That is not all. Nature's wheels within wheels go on without limit.

Treat found there is a parasitic mite which lives on this moth, and finds the moth's EAR especially nourishing. But, IT NEVER EATS BOTH OF A MOTH'S EARS — only one! If it deafened the moth, the moth would become easy prey for the bats, so the remarkably adapted parasite leaves the moth with some hearing for its own protection." (Condensed from an article in the Scientific American).

Who restrains that tiny parasite and keeps it from destroying both of the moth's ears? **Does God care for the life of a moth? He does!** The Creator has placed thousands of similar "hedges" around parts of His creation, lest the "balance of nature" be upset. "He doeth all things well." "Evolution" is mindless and meaningless. How can one account for such marvels as the habits of a parasite that consistently protects moths? A Super-intelligence planned it so!

The Miracle of Bioluminescence—"Cold Light"

In the flash of fireflies on a warm summer evening, in the greenish-white "phosphorescence" in the wake of an ocean-going vessel, in the glow of luminous bacteria on a piece of old meat, in the weird lights on the railroad worm* and the cucujo beetle, one witnesses the miracle of bioluminescence—a "cold light" that man can not duplicate.

"Cucujo is a West Indies firefly having three luminous organs: one on the under side of the abdomen, and two on the rear of the first segment of the thorax. It produces one of the brightest natural lights in the world." (Science Digest). WHO DESIGNED THE CUCUJO?

The miracle of luminescence is found at all levels of the ocean. "Bioluminescence," comments H. M. Andrews **(When Nature Lights Up)**, "is all around us, yet remains something of a scientific mystery."

The Miracle of Instinct in the Amazing Hunting Wasp

How can the **Ammophila** (Hunting wasp) detect a caterpillar underground? Who teaches the hunting wasp how and where to

---

*The railroad worm gets its name from the 11 pairs of greenish lights along its sides and a single red light on its head. It is about two inches long. A disturbance will cause it to turn on its red headlamp, while further annoyance leads it to turn on some or all of the green sidelights. A return to normal is signalled by fading out of the green lights and the eventual extinguishing of the red headlight! Could such an arrangement "just happen" through "random mutations"?

The unique railroad worm is another of the thousands of oddities in nature that witnesses for GOD and Creation. Brainless evolution could never produce such a marvelous **creature!**

sting its prey (a caterpillar) to paralyze it, but not kill it? In that way the hunting wasp provides **fresh meat** for its larvae.

Professors can teach medical students how to paralyze the brain of a frog so that it can go on living and yet be insensible to pain when dissection is performed on it. Skunks, to feed their young, catch and instinctively bite frogs and toads through the brain in such a way as to paralyze the animal but not kill it—thus preserving it as fresh meat to be used when needed!

The Miracle of the Metamorphosis of a Caterpillar

Inside the cocoon, "NOTHING REMAINS UNCHANGED: jaws, claws, claspers, pro-legs, digestive system, even the very shape—ALL disappears. Yet if we were to watch patiently inside the horny case of the chrysalis of a butterfly (made of tough chitin), we would see something wonderful happen before our very eyes. The shapes of the head, legs and the thorax gradually appear upon the chrysalis case. . . . the first rough draft of nature's work is dimly seen on the horny case of chitin. (Presently) out of the husk comes a trembling . . . creature. The ugly grub has vanished: in its place is a lovely butterly, as colorful as a flower."

The encased caterpillar seems gradually to MELT INTO A JELLIED, SHAPELESS MASS; and before long, out of this blob of "melted caterpillar" comes a gorgeous butterfly, having large, dainty, colored wings, instead of the crawling form of an ugly, hairy caterpillar!

It is a miracle—an unbelievable and inexplicable transformation—that can not be explained satisfactorily by evolution, no not in a million years. The miracle of the metamorphosis of the caterpillar into the butterfly demands a wonder-working GOD!

The Generation of Frogs is Another Unbelieveable Wonder

We have already discussed the miracle of the life cycle of bees. Such a phenomenon is an incontrovertible proof of the fact of Creation—for "random mutations" could never evolve such an intricate system of generation.

Consider the miracle of frog generation.

"A frog lays its eggs, yet no frog hatches, but something quite different; not a fish, yet in certain respects resembling one, for it has gills, and is entirely aquatic. Soon the tadpole begins to sprout legs, and in a matter of days it is a different creature! Its gills disappear, and lungs and other organs are formed—and presently the tadpole is transformed into a frog!" Its legs are perfected **before** leaving the water. It then can go on log or land without having to "evolve" legs and lungs.

So God bypasses "the millions of years" required by evolution, and produces EVERY SUMMER, BY THE MILLIONS, land animals (frogs), with lungs, from water animals (tadpoles) having gills. But the strange thing is, year after year, this strange procedure goes on and on, and never changes, always working in the identical cycle that all of us are familiar with. Only a work of GOD could do that!

The Miracle of Instinct in the Humble Fiddler Crab

It is a well-known fact among naturalists that the fiddler crab can fortell cyclonic storms. But as yet, "we don't seem to have any inkling of the field or fields in which any mechanism they have for doing so "functions," confesses one scientist.

Fiddler crabs, which live in shallow, water-filled holes just above the normal tide level, WILL LEAVE THEIR HOLES SEVERAL HOURS BEFORE A HURRICANE STRIKES, and will travel inland; and thus they escape the destruction that results when the sea rises and floods their holes and keeps them inundated for hours. It is an inexplicable phenomenon, but it is factual. Something seems to warn the fiddler crabs of the approach of a hurricane or a cyclone several hours **before** it strikes—and they get out of the danger zone! DOES GOD CARE FOR THE FIDDLER CRAB?

Miracles of "Regeneration"

Why can a salamander regrow an amputated limb, a lizard develop a new tail that has been bitten off, a crab regenerate a new claw that has been snapped off and a lobster grow a new eye? Higher animals can do no such thing!

Cut off a lobster's eye, and he will grow a new one—but not man! When a dog or a cat loses a leg, it does not grow another. When a man's finger is cut off, another does not grow on. Certainly, the power to regenerate lost parts, such as a crab or a salamander has, is a distinct advantage; why did "evolution" withdraw this unique ability from higher forms of life—if, as is claimed, evolution retains the qualities best suited to the "survival of the fittest"?

This fact, so damaging to the theory of evolution, is observed in scores of realms.

When, where, why and how did man lose the covering of fur, if he descended from the lower animals, all of which have a substantial covering?

Why is it that a cow can digest the tough cellulose of plants like alfalfa, and man has not this ability?

Why give replaceable teeth to fish and reptiles, but not to man? Did evolution bungle things here?

Why did not the high-pitched sonar system of bats pass on to higher animals and man? It would seem to be of great value.

Cockroaches and some water insects have auxiliary "booster hearts" to insure better circulation in their legs; why does not man have a similar "booster heart" to help keep his feet warm?

Man has lost the "WONDER NET"—a special arrangement of blood vessels that some animals have to conserve heat.

"A man standing barefoot in a tub of ice water would not survive very long. But a wading bird may stand about in cold water all day, and the whale and the seal swim in the arctic with naked fins and flippers continually bathed in freezing water. These are warm-blooded animals like man and have to maintain a steady body temperature. How do they avoid losing their body heat through thinly insulated extremities? The question brings to light a truly remarkable piece of biological engineering. . . . The principle is known as 'counter-current exchange.' . . . It is a method of heat exchange commonly used in industry. . . . In animals (such a system of heat exchange) is called **'rete mirabile.'** The blood in one vessel flows in the opposite direction to the adjacent vessel, and in that manner the warm stream passes its heat on to the cold stream." (See article, "THE WONDER NET," by P. F. Scholander, in the April, '57 "Scientific American").

MAN HAS LOST THIS AMAZING "WONDER NET" **(rete mirabile)**—an arrangement of blood vessels by which some animals can conserve heat and oxygen pressure by applying the principle of "counter-current exchange."

When it comes to the sense of smell, man is "found to be degenerate." A dog can smell better than a man.

Man has not the strength of the lion, the speed of the gazelle, the hide of the hippopotamus, the eye of the eagle for distant vision, the stomach of a cow—and a thousand other features in which the lower animals are man's superior.

These HANDICAPS that are found in all higher forms of life, especially in man, "utterly demolish the theory of evolution." Instead of the "survival of the fittest" in nature we see that the all-wise Creator made each type of life perfectly fitted to its environment and equipped it to perform its function in the overall economy of nature, as HE PLANNED IT. He placed HANDICAPS where needed, lest any form of life overrun the earth and destroy all the rest.

Instead of the instinct of the bee and the wasp, the bird and the eel, God has given man HANDS and a MIND with which to achieve and dominate his environment. Somewhere along the line, if evolution be true, evolution "pulled a thousand boners" and LOST for mankind many superior assets that animals enjoy but that are denied to man. But GOD makes no mistakes—and HE created man as HE saw best and HE created birds as HE saw best and HE created eels and fish and wasps and bees as HE saw best!

The Miracle of Distinctiveness

Not only are the platypus and the pangolin, the railroad worm and the praying mantis, the sea horse and the sea mouse, extraordinarily odd creatures, but also there are scores of plants and animals that have DISTINCTIVE FEATURES that are both mysterious and unaccountable.

The COLLAR-FLAGELLATES—protozoa—have a "delicate, transparent protoplasmic collar from the center of which emerges a single flagellum. . . . These collar-flagellates are of special interest because a similar type of cell occurs nowhere else in the animal kingdom, except in the sponges." **(Animals Without Backbones).** It is hard to see what the odd collar-flagellates "evolved" from.

In the open ocean is a transparent, slender animal (from one to three inches long) that looks like a cellophane arrow. These "arrow worms" are members of the phylum, **Chaetognatha.** These strange arrow worms are hermaphroditic: that is, both male and female sex cells arise from the lining of the coelom.

"Their body plan is SO DIFFERENT from that of other groups that it is difficult to say what relationships they have to other invertebrates." (Animals Without Backbones).

If evolution be true, we ask, from what did this weird creature descend?

THE MIRACLE OF "CAMOUFLAGE," "MIMICRY," "MASKS" AND PROTECTIVE COLORING

There are animals, like the striped tiger and spotted leopard, whose coloring gives concealment by matching the background of their environment. The leopard frog, which lives in the moist grass along the edge of ponds, not only wears a green coat to blend with the grass, but also has many irregular blotches of brown on his back which perfectly simulate the small spots of shadow among the green grass! Horned "toads" have a color so much like the desert sand in our southwest that the little animals so blend with their surroundings that it is hard to see them. The arctic fox and the polar bear have white coats that can scarcely be seen against the snow. The snowshoe rabbit and the ptarmigan have three suits—a brown suit for summer; white for the snow of winter; and for autumn and spring they have a brown-and-white mottled ensemble.

The chameleon, a small lizard, is able to change its color to that of the article on which it rests.

"Impersonation of other living creatures represents a subtler form of masquerade than imitation of a leaf or flower. . . . Moths masquerade variously as a spider, a weevil, a beetle and a scorpion. . . . Transcending mimicry, some insects have evolved features dramatically designed to inspire fear, like the EYE SPOTS of the Headlamp Click Beetle, whose pseudo eyes are luminous at night. . . . The ultimate in horror defense is worn by the Lantern Bug. At rest, it is inconspicuous; in flight it exposes **huge eye marks on its hind wings and its hideous head**—as a huge mask bearing the likeness of a tiny alligator complete with ravening teeth. As a final defense, its body is coated with distasteful wax." (Life Magazine).

Who can believe that evolution produced this miracle?

The miracle of protective coloring and camouflage is hard for "evolution" to explain.

"If we accept the idea of protective coloration, IT IS NOT EASY TO UNDERSTAND HOW THE PROCESSES OF EVOLUTION COULD HAVE PRODUCED THE EXQUISITELY PRECISE PATTERNS OF MIMICRY THAT SOME OF THESE ANIMALS DISPLAY. . . . We have overwhelming evidence that the protective colors and markings of animals cannot be mere accidents. An enormous number of animal forms have (such protective) patterns." ("Defense by Color," by N. Tinbergen, in the Oct., '57 "Scientific American.")

There is mimicry in plants as well as animals. "There are plants in South Africa which look like pebbles, while others are colored to resemble the earth."

Mimicry is widespread in the insect world. The **dead-leaf butterfly** folds its brilliantly-colored wings and is invisible among the dry leaves! Did this butterfly create its own camouflage?

The **owl's-head butterfly** of South America has "eyes" on its wings almost exactly the size, shape, and color of the true eyes of a small owl! Small insect-eating birds are afraid of owls, so the value of these markings to the insect is obvious. Whoever heard of "chance" mutations ending up with a realistic design like that?

The **walking-leaf** insect of Asia shows a remarkable resemblance to a green leaf. How can one explain such a phenomenon, save by admitting creation by a Master Intelligence? Not only are living leaves imitated but dead ones also. The **dead-leaf butterfly** of the East Indies looks remarkably like a dead leaf! How are we to account for that?

The brown **walking-stick** insect is shaped so much like a stick that it is difficult to see it on the trees where it feeds!

Fish, too, seem well versed in the art of camouflage. Some **butterfly fish** have faked eyespots on their rear sides; they swim backwards, so that if they are attacked they can dart rapidly in the direction least expected by the enemy!

Some of the arts of camouflage seem to be unusually clever. "The file-fish, who feeds among clumps of eel-grass, stands on its nose in times of danger, with fins gently waving to imitate a clump of grass! its mottled green color matches the flora perfectly."

The leaf-fish, about four inches long, is shaped like an elm leaf. His head is like the bottom part of a leaf, and he even has a stringy part hanging from his bottom lip to represent the stalk of the leaf.

The **wrasse** is the master undersea "quick-change" artist: it can change its brilliant colors to that of any fish with which it comes in contact! Explain **that** miracle, and you can do more than any living naturalist. It also can completely vanish from sight by taking on the color of any underwater object. Evolution could not produce ability like that in a billion years!

There are a thousand more illustrations of "camouflage" and protective "coloring" in the sea—and ten thousand more among the insects of the world.

Of the countless examples of acted camouflage in nature, the **spider crab** has one of the most astonishing tricks. "Taking the cuttings of seaweed, it chews the ends to give them better purchase, then affixes them among the hooked bristles that grow on top of its shell, **where they take root and effectively conceal the wily crab."** (From "Total Defense in Nature," by Alan Devoe). It is not the crab, however, that is "wily," it is a planned instinct given to the crab, for its protection, by the all-wise, benevolent Creator.

Books have been written on this fascinating theme. It is a miracle in nature that points unerringly to the Master Mind who works in and through all creation to protect certain vulnerable forms of life from extinction. It gives evidence of the fact that a Supreme Creator so designed ALL LIFE that each kind has either the advantages or the handicaps it needs to make a well-balanced economy in nature.

The Marvel of Hibernation

Why the spermophile (a ground squirrel) and the woodchuck—and some other mammals—spend the winter in "an extraordinarily deep sleep, which in some cases appears to be only slightly removed from death," is a mystery.

Mammalian hibernation has been the subject of sporadic research for at least 100 years, yet "the fundamental causes of the condition are still a mystery."

Myriads of other Mysteries and Miracles

As we have said before, nature is full of marvels and wonders, miracles and mysteries that no man can fully understand or account for.

How account for the "Death March of the Lemmings"? Lemmings are short-tailed relatives of meadow mice, that live on the bare tops of mountains in northern Europe or on the Arctic tundra. Every so often the number of lemmings grows far beyond the supply of food in their natives haunts—so they start their "death march" to the sea. Hordes of them will swim across rivers, travel across plains, and over mountains, until they reach the sea—and then plunge into the sea and swim out with all their might, until they drown from sheer exhaustion. WHAT IS THE EXPLANATION OF THIS STRANGE PHENOMENON?

We have called attention before to the miracle of migration. We know that salmon travel from mountain streams where they are hatched, down to the sea, and far out into the sea; then, when they are grown and have lived most of their lives, they go back to the home of their infancy, lay eggs—and die. WHY DOES THE EEL ACT IN JUST THE OPPOSITE WAY? Eels are born in the salt water of the Sargasso Sea in the Atlantic. They then go through thousands of miles of the trackless depths of the Atlantic ocean, until they finally emigrate to the rivers of Europe, where they live in fresh water from five to twenty years! WHY? No one knows why, except the Creator who made them so.

For over 200 years botanists have been puzzling over the problem of how water rises from the roots to, say, the highest branches of a 400-foot fir tree.

"Even today we do not know the complete story of how it does this (for) merely to raise water 450 feet requires a pressure or tension of about

210 pounds per square inch! And in some hardwood trees water rises at the rate of almost 150 feet per hour. A date palm in a desert oasis may need to raise as much as 100 gallons of water a day to make up its losses from evaporation from the leaves." (Victor A. Grenlach, in "The Rise of Water in Plants," Scientific American Magazine).

No one knows why, about every four years, there is a sudden and mysterious increase in the mouse population. So astonishingly large is this increase that some scientists say the mouse population "explodes". . . . Usually some kind of mouse sickness slows down this strange increase — and presently the summer fields are once again peaceful, with just enough meadow mice to feed their natural enemies.

Naturalists and travelers have observed the same species of plants in widely separate areas. Rutherford Platt and Francis Smythe both reported the existence of a peculiar type of saxifrage in widely separated and remote areas of the world. Rutherford Platt comments on this mystery.

"How account for the existence of precisely the same peculiar plant at two points thousands of miles apart, separated by oceans and continents?" (Scientific American).

Bible believers have an answer: the seed may have been scattered at the time of the flood!

No scientist can explain the amazing engineering accomplishments of the prairie dog. The rodent's "plunge hole" is a vertical chimney as much as 16 feet deep!

"Apparently well-counseled by instinct, the prairie dog also builds a 'flood-control dam' around his burrow entrance. Inattention to this flood-control work might be costly since midsummer cloudbursts can create lakes two to three inches deep."

And so the story goes — one miracle after another, one mystery added to another. It would take volumes to write the whole account, and then the half would not have been told, for man does not as yet know all the mysteries and miracles of creation, much less understand or explain them.

Let us now turn our attention to the subject of

MISSING LINKS

The public has been misled into thinking there are a "few" missing links in the chain of evolutionary descent. Far from being almost complete, the so-called chain of evolution is broken by millions of "missing links," many of which are so great as to be veritable gulfs and chasms. Let us note some of these "missing links".

Between Empty Space and the Creation of Matter

Since the advent of the atomic age we know that matter is NOT eternal; matter is a form of energy; it had a beginning and may be destroyed. Sir Ambrose Fleming once said,

"Between space, absolutely empty space, and space filled with even the most rarified matter there is a GULF which no theory of evolution has been able to pass or explain."

"Nothing" can not create matter; therefore we KNOW that the eternal and uncreated God made matter "in the beginning" as the first verse of Genesis says.

The Origin of Motion

Between the creation of matter and the beginning of MOTION is another "gulf" that cannot be bridged except by admitting GOD — for it takes POWER to put inert matter into motion. WHERE DID THAT POWER COME FROM, if not from the eternal One? No theory of evolution we ever have heard about even attempts to explain the origin of both matter and motion—without admitting an Original Cause: God.

The Origin of Life

Sir Ambrose Fleming, in an address to the members of the Victoria Institute, not only spoke of the origin of matter, but also of the origin of life.

"We (as scientists) have not the smallest knowledge of how empty space first became occupied with the most rudimentary form of matter. Neither have we any conception of how life originated. WE CANNOT IN ANY WAY BRING IT INTO EXISTENCE APART FROM PREVIOUS LIFE. . . . "

Irwin Schroedinger, "Nobel laureate in Physics," and leading atomic scientist, says,

"Where are we when presented with the mystery of life? **We find ourselves facing a granite wall which we have not even chipped.** . . . We know virtually nothing of growth, nothing of life." (Quoted in the New York Times, in "The Greatest Mystery of All—the Secret of Life," by Waldemar Kaempffert).

"MISSING LINKS" BETWEEN THE PHYLA

In 1898 the International Congress of Zoology organized an International Commission on Zoological Nomenclature to establish rules for the naming of plants and animals; and these rules have been adopted throughout the world. Linnaeus' "System of Nature" (1758) was taken as a basis for scientific classification. In our modern system of classification we start with TWO **KINGDOMS:** the Plant and the animal. The Plant Kingdom is divided into **DIVISIONS,** while the Animal Kingdom is divided into **PHYLA.** Each Division and Phylum is divided into **CLASSES.** Each class is divided into **ORDERS.** Each order is divided into **FAMILIES.** Each family is divided into **GENERA.** Each genus is divided into **SPECIES.** Each species is divided into "breeds," "varieties," or "races."

Science has divided the animal kingdom into 14 or more PHYLA; likewise, the plant kingdom is separated into many DIVISIONS.

The animal kingdom starts with Phylum PROTOZOA (single-celled animals, mostly microscopic), and ends with Phylum CHORDATA.

Having given a little sketch of what scientists mean by the terms "kingdom," "phylum," "family," "genus," "species," etc., we are ready to prove from their own writings that ALL THE PHYLA ARE SEPARATED BY "MISSING LINKS." We quote.

"If we could find an animal clearly intermediate in structure between two modern phyla, we would have good evidence that the two phyla are closely related. . . . Such an animal has never been found . . ." . . . "We have fossil records to show that certain species have remained unchanged for very long periods of time, but none are so old that they trace back to the time before all of the modern phyla evolved. Therefore we often speak of these MISSING ANCESTRAL FORMS as 'missing links'." (P. 235, Animals Without Backbones).

There we have the frank confession that the LINKS between the Phyla are ALL MISSING!

Lacking positive evidence, evolutionists IMAGINE what the "missing links" were! Note well this language that suggests GUESSWORK, THEORY AND IMAGINATION:

"It is HIGHLY PROBABLE that the capacity for photosynthesis was a characteristic of the ancestors of primitive organisms. FROM A HYPOTHETICAL (imaginary) ancestral type of 'plant-animal,' THE EXACT NATURE OF WHICH IS UNKNOWN, came at least two main lines of descent, the animal kingdom and the plant kingdom." (P. 338, ibid.)

"Considering the remoteness of the events with which we are dealing, and the INCONCLUSIVE EVIDENCE, it is clear that any 'invertebrate tree' (showing their concept of the steps in evolution) MUST BE CONSIDERED HIGHLY SPECULATIVE." (Pp. 337, 338, ibid.)

MISSING LINKS BETWEEN SINGLE-CELLED ANIMALS AND MULTI-CELLED ANIMALS

"The most primitive animals are **single cells.** . . . The exact manner in which **multicellularity** (animals having many cells) arose CANNOT NOW BE DETERMINED." (P. 338, ibid.) "The stage beyond the first multicellular organisms which led to the higher phyla CAN ONLY BE IMAGINED." (Ibid).

"By the passage of some of the cells from the surface into the interior, a two-layered animal was formed. This HYPOTHETICAL (imaginary) two-layered ancestor PROBABLY evolved from a different group of protozoa than that which gave rise to sponges. . . . Just what it looked like WE DO NOT KNOW . . .." (P. 339). (Caps ours).

"It is NOT KNOWN just how the radial gastrula-like ancestor became bilateral . . ." (P. 340, ibid). (Caps ours)

Read again the above-quoted paragraphs (from "Animals Without Backbones") and see that all that evolutionists have to base their theory on is IMAGINATION. They have absolutely NO PROOF of any evolutionary process or steps between the phyla.

Let us now take a closer look at some of the radical DIFFERENCES in body structure that exist in animals in different phyla. The ARTHROPODA are "jointed-legged" animals. A rigid cuticle furnishes a supporting framework for the tissues within, and provides a surface for the attachment of muscles. Such a supporting structure, as is found in spiders, beetles, ants, etc., is called an **exoskeleton.** In other words, the insect's skeleton that supports it, is ON THE OUTSIDE OF ITS BODY. In sharp contrast to this type of framework is the **endoskeleton** of vertebrates—animals with backbones—which lies on the INSIDE and is surrounded by the soft fleshly parts.

Now here is our question: by what possible route could an animal with its skeleton on the OUTSIDE, like a beetle, be transformed into an animal with its skeleton on the INSIDE, like a dog or cat? The method of construction is so vastly different, a step-by-step connecting route is out of the question. THERE IS NO EVIDENCE ANYWHERE IN NATURE TODAY OR IN THE FOSSIL WORLD OF YESTERDAY OF ANY SUCH GRADUAL CONNECTING STEPS, except in the imagination of evolutionists—and they cannot even IMAGINE the steps!

Another contrast: insects HAVE NO LUNGS. An insect gets air through little tubes called tracheae, which branch through the whole body. There are sacs at the ends of the tracheae which can be filled with air; the air in these sacs makes the body very light and buoyant, making it easier to fly. Vertebrates have LUNGS that are localized, and do NOT extend throughout the entire body. These breathing organs are so radically different, it is impossible to conceive of connecting links—that is, impossible except to the ardent evolutionist! To show how vastly different the breathing apparatus of an insect is, let us quote:

"Flying insects require more oxygen, ounce for ounce, than larger animals do. Insect evolution has met this demand by designing a respiratory system totally different from that of higher animals. Our 'rhythmic sipping of the air' supplies oxygen to our body cells by the roundabout route of lungs and bloodstream. The insect respiratory system by-passes the blood and delivers oxygen directly to each and every one of the millions of cells buried deep in the various tissues and organs of its body. Each insect cell, in short, has its own private lung to keep the fire of its metabolism burning. . . . The tracheal system embodies a refinement of biological engineering almost past belief." (See, "Insect Breathing." By Carroll M. Williams in the "Scientific American")

How could this amazing system of breathing, so marvelously designed and executed as "to be almost past belief" in its ingenuity and practicability, be the work of "random mutations"?

No involved invention of man—such as the telephone, TV, or radio—has ever happened of itself by mere "chance," but has ALWAYS been the result of thought, design and persistent effort. "CHANCE" never produced anything of an involved character that works with precision. Any involved mechanism, whether found in nature or in man's world of invention, HAD TO BE THOUGHT OUT AND MADE BY ADEQUATE INTELLIGENCE.

It is easy to see that the two systems of getting oxygen are so utterly different that it is a case of **using one or the other.** TO CHANGE OVER FROM THE INSECT'S SYSTEM TO THE VERTEBRATE'S SYSTEM by a slow, gradual change—by "chance mutations" and "random changes"—is utterly impossible; in fact it is absurd to believe it could be done. The "tracheae" system works well; the "lung" system works well; but a system in the process of changing from one to the other would be a monstrous impossibility. THIS ONE FACT IS ENOUGH TO SILENCE FOREVER THE EVOLUTIONIST. He cannot explain the change from "tracheae" to the lung system of breathing. Nor can be describe any hypothetical ancestral breathing apparatus which could have produced both!

There are scores of similar RADICAL DIFFERENCES between the phyla that prevent any step-by-step connection! For example, what possible steps could there be between a functioning "warm-blooded" animal and a "cold-blooded" animal. A mammal or a bird, being warm-blooded, can go about its business in a normal manner in a wide range of climates; but a snake, being cold-blooded, becomes almost completely incapable of any movement when it is chilled.

And what possible connection—or missing links—could there be between the scales of a reptile and the feathers of a bird? Scales are scales and feathers are feathers, and never the two shall merge! Scientists admit that "pro-avis" exists only in the imagination of evolutionists.

"So far the luck which paleontologists unfortunately need has failed to produce a specimen of 'pro-avis,' the name used by Heilmann for the 'feathered reptile' (more or less what Professor Wagner thought Archaeopteryx to be) that preceded Archaeornis. Nor have we found a form between Archaeornis and the modern birds." (See "Salamander's and Other Wonders," by Willy Ley, p. 141. Pub. by Viking Press, 1955, N. Y.).

The simple fact is, GOD MADE REPTILES WITH SCALES and GOD MADE BIRDS WITH FEATHERS.

To believe that the one (birds) descended from the other (reptiles) is pure imagination. In any event, THERE IS NO PROOF FOR THIS FANTASTIC THEORY OF EVOLUTION.

Harry Rimmer stated the case well in his book, "The Theory of Evolution and the Facts of Science."

"There have been approximately one hundred and twenty-five million species on the face of this planet in the vital life period of the earth's history, and according to Wassman the transmutations (evolution) of a closely related species into its next higher or subsequent species would require at least 1700 mutants, or variants, commonly called 'LINKS'. One hundred twenty-five million species multiplied by 1700 mutants gives the stupendous number of 212,500,000,000 definite life forms that must have existed to transmute the ameba (or other protozoa) into man." (P. 70).

The extreme difficulty in creating a "family tree" and seeking to make a step-by-step arrangement of the descent of one animal from another is seen in the evolutionists' crude attempts. We are told that "the bat probably evolved from the mouse." Supposedly, the mouse's feet started webbing. WHAT GOOD ON EARTH ARE WEBBED FEET TO A MOUSE? Eventually, a monstrous creature was supposed to have evolved that was half-mouse and half bat—a crazy creature that could not swim, walk, run or fly! What a freak it would be! While there are all sorts of bats on exhibit in our museum, and many kinds of mice, there ARE NO "LINKS" BETWEEN MICE AND BATS.

"One of the great problems of evolution," wrote Jacques Millot, in the Scientific American, "has been to find anatomical links between the fishes and their land-invading descendants. . . . Comparative anatomists have speculated for half a century on how the fin of a fish evolved into the forelimb of the frog."

THINK OF IT! A THOUSAND BIOLOGISTS HAVE SCRATCHED THEIR HEADS FOR FIFTY YEARS IN A VAIN EFFORT TO COME UP WITH PLAUSIBLE "LINKS" BETWEEN FINS AND FORELIMBS—and they are not nearly as diverse as scales and feathers! The fact is, THERE IS NO POSSIBLE WAY TO MAKE A DIRECT LINK BETWEEN FINS AND FORELIMBS without doing violence to all laws of anatomy and all rules of reason. AND NOWHERE WILL ONE FIND ANY "MISSING LINKS" BETWEEN "FINS" AND "FORELIMBS."

Prof. W. Bell Dawson, Canadian geologist, said,

"When Darwin's books appeared, people at once said: 'If men came from monkeys, we ought to be able to find creatures that are half-way between the two.' It was very reasonable to think this; and so the search began for what was called the 'missing link' between man and the monkey. We must remember that this began 100 years ago; and all the remotest corners of the world have been searched and the most degraded men have been hunted for, in every continent; BUT NO MEN HALF-WAY BACK TO MONKEYS COULD BE FOUND. Then it was said: We must go further back. And so search was made for skulls of men who lived before any history was written in their part of the world; but with no better success"

A noted biologist of the Smithsonian Institute in Washington, Dr. Austin H. Clark, said of this: "THERE ARE NO SUCH THINGS AS MISSING LINKS. Missing links are misinterpretations." And Professor Virchow said in his day, "The ape-man has no existence and the missing link remains a phantom."

Darwin himself was keenly disappointed in not being able to find the MISSING LINKS his theory demanded. He wrote:

"I do not pretend that I should ever have suspected how poor was the record in the best preserved geological section, had not the absence of innumerable transitional forms between the species which lived at the commencement and close of each formation pressed so hardly on my theory."

Fairhurst produced a sound argument when he wrote,

"It will surely be admitted that conditions favorable to the preservation of species are equally favorable to the preservation of transitional forms."

And La Conte commented, "The transitional forms ought not to consist of species at all, but simply of individual forms shading insensibly into each other, like the colors of the spectrum; but, THIS IS NOT THE FACT."

THE "MISSING LINKS" ARE STILL "MISSING"—because they are non-existent. They exist only in the imagination of the evolutionists.

The fact that great, vast gulfs exist between all phyla and between all classes and orders of all phyla, and between all families and all genera and many species is proof that EACH "KIND" IN THE ENTIRE GAMUT OF CREATION WAS THE DISTINCTIVE WORK OF THE MASTER ARTISAN WHO GAVE TO

EACH "KIND" ITS OWN PECULIAR CHARACTERISTICS AND PERFECTLY ADAPTED IT TO ITS OWN ENVIRONMENT AND ITS OWN SPHERE AND NICHE IN THE ECONOMY OF LIFE ON **EARTH.**

There is a difficulty in the path of evolution afforded by SIZE. Robert E. D. Clark, Ph.D. (teacher of chemistry, Cambridge Technical College), calls attention to it.

Even more basic is the difficulty afforded by size. It is a principle in engineering that one cannot, simply, imitate a small machine on a much larger scale. There comes a time when mere modification will not do; a basic redesign is called for. This fact arises from the consideration that weight increases as the cube of dimensions. but surface area and forces, which can be transmitted by wires, tendons, or muscles, vary only as the square. For this reason a fly the size of a dog would break its legs and a dog the size of a fly would be unable to maintain its body heat. So if evolution started with very small organisms there would come a time when, as a result of size increase, small naturally-selected modifications would no longer prove useful. **Radically new designs would be necessary for survival.** But by its very nature, natural selection could not provide for such redesign.

From this and much more besides, it becomes increasingly clear that it would be easier to show by science that evolution is impossible than to explain how it happened." (Christianity Today, issue of 5-11-59).

Chapter 13

"EYES," "SEX," AND "SPECIALIZED ORGANS"

NEXT TO THE BRAIN, the eye is the most wonderful of all God's gifts to His creatures. Were all creatures doomed to live in perpetual darkness, life—if it were possible at all—would be a dismal and boring experience of prosaic emptiness. After a survey, even though limited, of the eyes of animals and man, one is impressed with this: each creature has been given eyes, by the Creator, that best suits its needs and station in life.

A hare's eyes are so placed that it can look backward as well as forward without turning its head. It needs eyes like that, for it is the victim of predators, and must be able to see them to have a chance to escape. It has a complete 360° circle of vision.

Hawks—day animals—can see the slightest movement of a tiny animal far below them in the grass. Hawks need keen sight to spot and catch their food.

Birds generally "have a sense of vision that enables them to see with greater precision than any other living creature. Sight is their dominant sense, helping them to catch the tiny darting insects that so often form their diet."

The visual acuity of some birds' eyes is from eight to ten times that of the human eye. The eyes of hawks, eagles and vultures that dive toward their prey, have a peculiar ability to change focus rapidly. This speedy change of focus enables birds to catch insects on the wing, and to keep a tiny rodent spotted in the few seconds of a rapid power dive. "Birds' eyes are the finest and most remarkable of all the eyes of earth, being often both telescopic and microscopic." (Thomas Shastid, ophthalmologist).

The very large eyes of the owl are admirably adapted to seeing in semi-darkness, so that the owl can catch insects that fly after dark. "The sensitivity of the owl's eye in conditions of low light intensity has been shown experimentally to be about ten times that of the human eye."

The owl, as well as other birds, has an extra eyelid—a complete transparent membrane that sweeps down across the surface of the eye, starting from the inner corner. It not only moistens the eyeball (making it unnecessary for the large feathered eyelids to blink shut), but also protects the eye when its owner is forced to fly through such hazards as wind-blown dust and the closely laced branches of trees. This transparent nictitating membrane is drawn across the the eyes of many birds whenever they are in flight, as it was **designed** to give protection.

This transparent membrane that serves such a useful purpose MUST of necessity be a complete and entire piece of equipment to serve its intended purpose. IT IS UTTERLY IMPOSSIBLE FOR SUCH A PRACTICAL ORGAN TO DEVELOP GRADUALLY. Hence, the only solution to the problem of how, why and when this "transparent membrane" originated, is in the fact of instantaneous creation.

The MUD-SKIPPER of the tropics, which spends part of its life in water and part on land, "has movable, bulbous eyes which are adjustable to vision in the air as well as in the water. The fish has a special muscle which enables it to shift the lens close to the retina, so it can even produce a sharp image of distant objects."

Instead of having a thin transparent membrane, as birds have, to cover the eye while in flight, a whale has another highly specialized adaptation. Whales are able to dive to great depths in the ocean. Its whole body is adjusted to and adapted for this purpose. At a depth of 100 feet, the pressure of the water is 60 pounds to the square inch; but at 4,000 feet it is 1830 pounds! The result is, deep-water fish can not come to the surface: they will actually explode! Nor can most surface fish dive to great depths: they would be pressured to death! But a whale is "at home" in both surface waters and in great depths of the ocean. The eye would be the first organ to suffer from such exposure to terrific pressure; so the Creator equipped the eyes of whales with "a sclerotic coat, very thick and strong," to protect the eye when the whale goes into a deep dive.

When attacked by shark or sword-fish, the whale has only one effective defense: it dives to great depths—and if its pursuing enemies persist in accompanying the whale in its sudden plunge, they are killed by the great pressure.

QUESTION: Since observation proves that most other inhabitants of the sea dare not leave their depth element, but will perish, if they do, how can one explain the phenomenon of the whale's ability to plunge to great depths and live? IF IT DEVELOPED THIS ABILITY THROUGH LONG AGES IT WOULD HAVE PERISHED IN EVERY ATTEMPT! This ability **had** to be bestowed on the whale when it was created. And how is one to explain that extra strong, heavy coating for the eye? Obviously the eye of the whale was **made** to withstand the pressure of great depths—and at the same time function near the surface!

If this ability to dive to great depths was gradually "evolved" why did not sharks and swordfish obtain the same ability?

"The fish with the 'Built-in Bifocals' "—the **Anableps dowei**. This unusual fish lives in the quiet rivers and estuaries of the Caribbean. He feeds on tidbits which float on the surface of the water; therefore it is necessary for him to see in the air as well as in the water: and God made him with that marvelous ability. **Anableps dowei** has only two eyes (not four), but each of his eyes has two pupils. As **Anableps** swims along the surface, he can see clearly both above and below the surface—through air and through water at the same time! This highly complicated arrangement **must** have been so designed!

The **Anableps dowei** is not the only animal that has "built-in bifocals." The WHIRLIGIG BEETLE, which we may see on the surface of quiet water, is equipped to look up and down at the same time! Its eyes are divided so that the upper part sees the surface of the water and the lower part sees below the surface. There is a black layer of pigment between the two parts so that light from one does not affect the other. The entire arrangement is so highly complex, so practical for its intended purpose, so efficient. one must admit IT WAS DESIGNED AND MADE THAT WAY.

The eyes of the WOOD TURTLE are tipped downward to help it see what it is eating.

The camel—and other animals—have special built-in wind glasses! This desert animal, created for life and service on the desert, is equipped with a transparent third eyelid, which may be drawn at will over the eyeball without impairing the sight. This protects the eye from sharp bits of sand in desert dust storms.

Most DEEP-SEA CREATURES possess luminous organs which they flash on and off as occasion demands, and so, though they live in total darkness, they have eyes, and are able to make use of them! This special arrangement was manifestly DESIGNED for them by the Creator. Such miracles in nature do not "just happen" nor "evolve through chance mutations." What proof have we for that statement? This analogy: every complicated, working machine in the world today—such as the telephone, the radio, the camera, the typewriter, the automobile, the electric motor—**was designed and made by some man;** that is, by an outside intelligence. Complicated, complex working machines (whether living or inanimate) DO NOT "JUST HAPPEN" but in all instances are DESIGNED AND MADE BY AN INTELLIGENCE SUPERIOR TO THAT OF THE "MACHINE."

Most land vertebrates have perfectly adapted and well-functioning eyelids that blink many times each minute to keep the eyeball clear and moist. Of all land vertebrates, only snakes have no eyelids at all. Their eyes are fixed in a permanent glassy stare. To protect the delicate, lidless eyes of the snake, there is a transparent shield permanently in place over the entire eye opening.

The **position** of the eyes in the head of an animal may show great specialization. The **alligator** (a reptile) and the **hippopotamus** (a mammal) both have their eyes set in a raised position on the tops of their heads. With these "periscopic eyes" they can float in the water, almost entirely submerged, and still keep an eye on their surroundings. Clearly, this was so DESIGNED for them.

Different TYPES OF EYES

Not only do we observe in nature the phenomenon of perfect "adaptability" in the eyes of animals, but also we see great versatility in TYPES and KINDS of eyes.

Almost every one of the more than 38,000 species of vertebrate animals known to zoologists (4,000 mammals, 14,000 birds,

4,000 reptiles, 2,000 amphibians, and 14,000 fish) is born with functional camera-style eyes. There are a few of these animals, (mostly fish and salamanders) that live in total darkness (in the pools and streams of deep subterranean caves) that lose their sight as they mature.

Herein is a perfect argument against evolution: "Many thousands of years ago the ancestors of these blind cave species must have been carried in from the outer world by a surface stream." Even though thousands of generations have come and gone, THESE FISH STILL ARE BORN WITH EYES! The old Lamarckian idea of "acquired characters" has been exploded a thousand times. Through lack of use, the new-born fish soon lose their sight; but the essential nature of these fish to have eyes PERSISTS despite the fact that they have no use for their eyes.

The animal kingdom as a whole presents many different plans and styles of eyes. None however can compare with the wonderful "camera eye" possessed by birds and mammals.

"All vertebrate eyes are built much along the lines of a modern camera, but the all around precision and adaptability of the eye far surpasses our most modern and expensive cameras." (Nature's Wonders). (For a discussion of the marvels of the human eye, "camera" type—see Chapter 10, The Body of Man, in this book).

Beside the "camera" eye possessed by birds and mammals, there are other methods used by the Creator to give sight to His creatures. We list the following:

(1) The eye of the **chambered nautilus** lacks a lens, but functions well on the principle of the "pinhole camera," where a very tiny opening gives a universal focus.*

In the **arthropods** (which includes all the insects, spiders, crabs, centipedes and millipedes) by far the largest and "most successful" group of invertebrates, is a fascinating variety of both "simple" and "compound" eyes.

(2) The spider's eyes are known as "SIMPLE" eyes, because each has but ONE transparent lens to focus light rays on the sensitive nerve cells beneath it. If it be not as elaborate as the compound eye, the Creator

---

*It is interesting to observe that the **cuttlefish, squid,** and **octopus, closely related to the nautilus,** have eyes equipped with true lenses, the most specialized eyes of any of the invertebrates.

This strange fact presents a powerful argument against evolution, for WHY should these three marine mollusks have such highly developed eyes, similar to those of the higher vertebrates, when they are far below the organizational status of the vertebrates? Admitting that GOD made them so, the problem is solved; but to evolution it remains unsolved.

has compensated spiders by giving them eight of these simple eyes. They are placed strategically in two rows at the front of the head. Most millepedes have from one to several simple eyes on each side of the head.

The CYCLOPS, a fresh-water copepod, has one simple median eye.

(3) In the insects that have COMPOUND eyes, there is great variety. Some insects have enormous eyes that nearly encompass their heads. The common **housefly** has large compound eyes. The **dragonfly** represents the extreme, as it needs the best sight possible to capture flying insects on the wing. The compound eyes of the dragonfly have 30,000 facets! Each of these 30,000 units has its own light condensing apparatus! Next to the camera eye, the compound eye is most efficient. However, no man knows the exact nature of the image an insect gets from its elaborate compound eyes.

(4) Most insects have a combination of both simple and compound eyes. For example, the GRASSHOPPER has five separate and distinct eyes, three small simple eyes and two large compound eyes prominently placed at the sides of the head.

(5) Some animals (crustaceans) like the **crayfish** and the **lobster** have compound eyes **that are on stalks.** These eyes can be moved around for better vision.

(6) The **sphenodon** (lizard of Australia, about two feet long) has a **third eye,** or pineal eye, on the top of its head!

(7) There is also great variety in the way the pupils of eyes contract. In man's eyes there is a round pupil, and the opening automatically expands or contracts to let more or less light into the retina. In the domestic **cat** the enormous pupil opening will close to a vertical slit in the presence of bright light. On the other hand, the pupil opening in the eye of a **horse** takes the shape of a horizontal bar. **Lizards** and other **lower vertebrates** reduce their pupil openings "to a great variety of odd shapes and patterns."

Blind evolution would not be able to develop all of these various styles of eyes, all of which function **perfectly** for the needs of the animal to which they are given.

MORE MIRACLES—AND MORE PROBLEMS FOR THE EVOLUTIONIST

Let us contemplate a few more "miracles" about "eyes" in nature, and the problems they present to the evolutionist.

(1) Who can explain why the eyes of the Star-gazer (a fish of the Weever species) are placed horizontally on the upper part of its head, "in a position probably without parallel in nature," so that it is always looking up at the sky? Because of this, it is given its scientific name, **Uranoscopus.**

The Star-gazer completely buries itself in the sand on the ocean floor, so that only its eyes emerge.

Starting with the average fish with its eyes on the sides of its head, evolution has to explain **why** and **how** the peculiar eyes of the Star-gazer were evolved. With eyes normally placed, the Star-gazer would not be able to bury itself in the sand and see; in fact it would have no inclination to act as the Star-gazer now acts. And, one must conclude, the only reason the Star-gazer acts as it does, and buries itself, is because it has eyes on the top of its head! In other words, "evolution" didn't make the Star-gazer as it is; the Star-gazer acts as it does because it was made that way in the beginning!

(2) The Starfish is an oddity if ever there was one. WHY would any sea creature "evolve" into such an apparently absurd shape—as far as "sea-life" shapes go—as a Starfish?

The common species of Starfish have five arms or "rays," on the under side of which are hundreds of tube feet; and on the end of each arm is an eye! The Starfish is unable to swim but it walks along the bottom very slowly, over sand and shells, through a most ingenious system. It can go in any of the five directions its arms point to.

What did this strange creature evolve from? And WHY? The evolutionist has no logical answer. What was the starting point from which finally came the Starflish? Who could possibly trace the sequence of "chance mutations" that finally brought to pass the Starfish with an eye at the end of each of its arms? It is far more reasonable and logical to believe the Starfish was made as it is, and has always been as it is!

(3) Many snails have eyes at the ends of tentacles, which they can extend or compress, much as a telescope is lengthened or shortened. With these eyes at the ends of tentacles a snail can "look around a corner" without exposing its body. (Nature Magazine).

This unusual ability to see with eyes at the ends of stalks is no doubt a great advantage to the snail. But how could such an ingenious device be brought into existence in such a lowly animal by "chance mutations"? It is as easy to believe that the 100-inch telescope on Mt. Wilson "just happened" as to believe that such miracles, and such well-planned devices in nature, are the result of blind chance. Any one who has focused a telescope knows how careful the adjustment must be before the image is clear. Who gave the lowly snail the uncanny ability to see with its adjustable eyes, that can be lengthened or shortened at will? The successful use of such eyes involves optical and engineering principles that can be solved only by the Master Workman!

(4) The eye of a Pigmy shrew is little larger than the head of a pin; but it has the same camera-like eye arrangement as the grapefruit-sized eye of a great blue whale! Such an eye (as the Pigmy shrew has) must have been made by a mechanical Genius!

(5) The sole, or turbot, like the Stargazer, has its eyes directed upward. The fish lies in the sand at the ocean bottom in the daytime.

"They have extraordinary eyes that move in all directions, as though mounted on a universal swivel. Their eyes, with a rotary movement, will follow the movements (of an enemy above them); and those movements very often betray their presence." (The Underwater Naturalist; p. 219).

"Only at night do these fish search the surface of the mud, looking for worms. As this takes place at night and as their eyes are directed upwards, vision plays no role in their search, which is conducted purely by a sense of smell and by a sense of touch, from . . . special filaments on the under surface of the head." (The Underwater Naturalist; p. 219).

Question: if evolution is responsible for this state of affairs, didn't it make a serious mistake in placing this fish's eyes where it could not see the food it needs? Until it developed those special "smell' and "touch" filaments, the poor fish would starve to death! Who designed the "swivel" eyes of the sole? And Who gave them the sensitive "smell" and "touch" filaments?

(6) A Chameleon's eye "is one of the most remarkable organs exhibited by any terrestrial animal."

"The Chameleon has large protuberant eyes, covered with thick granular lids, perforated only by minute apertures for the pupils. THE TWO EYES CAN BE MOVED INDEPENDENTLY OF EACH OTHER. One can look straight ahead, while the other looks backward or up. . . . Why this doubling of the field of vision should accompany such excessively minute openings to the lids is a mystery."

If evolution alone were responsible, it would NOT have put a handicap on the otherwise marvelous eyes of the chameleon! WHY WERE SUCH WONDERFUL EYES, "able to move independently of each other"—so securing for the owner two entirely different fields of observation—SO DRASTICALLY LIMITED BY PLACING THEM BEHIND HEAVY GRANDULAR LIDS WHICH HAVE ONLY "MINUTE OPENINGS"? God, who does all things well, designed both their wonderful eyes and their limiting lids. "Handicaps" like this are so designed by the Creator who works for the welfare of all nature—not just one animal. "Evolution" we are told works for "the survival of the fittest." Deliberate and well-placed HANDICAPS are perfect evidence that nature is God's handiwork.

(7) The "compound eyes" of the dragonfly occupy most of the head, and are composed of nearly 30,000 separate units! The Creator had reasons for enlarging the vision of the dragonfly—at the cost of its brain development. But if evolution alone were responsible for this, the size of the brain would be proportionate to the size of the eyes! Here again we see the fact of an imposed "handicap" on certain animals (limited space for brain, in this instance), which is proof of the work of an outside In-

telligence who had in mind the **entire realm of creation** as He alotted to each its sphere in life and its handicaps and limitations as well as its specal functions and abilities.

(8) The large eyes of the **honeybee** make use of the ultraviolet portion of the sun's spectrum to see with. Man's eyes are not so made. As this gives a greater vision, why did "evolution" drop this phenomenal ability from the eyes of man? As it is a **distinct advantage** "natural selection" would have clung to it! Evolution has no adequate explanation of such phenomena.

(9) The eye of the **horseshoe crab** seems to have a unique feature not found in other animals.

"The eye of the horseshoe crab is amazingly simple. It is a compound eye composed of individual units (ommatidia), similar in type to the eyes of insects. **But unlike any other known animal, the horseshoe crab has a separate nerve fiber proceeding from each of these units toward the brain."** (L. J. Milne, in "Scientific American.")

The eye of the horseshoe crab is DIFFERENT from all animals below and above it on the "evolutionary ladder." This presents a **real** problem for the evolutionist. From whence did the horseshoe crab get this unique system of vision? The fact is, **the Sovereign, Almighty Creator made it so!**

(10) The kingfisher. and some other birds have a special area in the retina called the "fovea," in which the cells that line that area each have a private nerve fiber to the brain. This gives maximum visual acuteness to that limited area, called the **fovea.** If a bird desires special visual acuity it turns its head or eyes until the image is focused in the fovea. Some creatures actually have **TWO of these fovea areas in each eye.** With this magnificent system, they can not only obtain more acute sight, in a limited area, but also they can actually (by using **both** fovea areas) get a "bifocal" effect, and gain an accurate impression of both distance and depth. Owls use their eyes binocularly at all times.

"The kingfisher is one of the strangest users of the **two-fovea** system. Its eyes can notice both an object in the air and the **exact position** of a fish below the water surface, and also it can follow the fish accurately after its sudden dive into the pond.

"Vision in the air and vision in water are entirely different. When water comes into contact with the clear cornea it takes away all visual functions of the cornea. . . . (Therefore) in water the lens must act alone. Hence an eye that has normal vision in air is very long-sighted in water; and an eye that has normal vision in water is pathetically near-sighted in air. . . . The kingfisher (can see well in air and under water) through possession of an egg-shaped lens. When the bird uses its eyes monocularly with one of the two fovea in each eye, any prospective prey is kept in sharp focus through one end of the peculiar lens.

"But when the kingfisher enters the water, and its 'cornea' disappears, the image of the fish is formed through another axis of the lens on the **second** fovea of each eye. The fish is seen binocularly, straight ahead of the beak, in good focus, and the bird is able to complete the catch!

"The kingfisher thus has two eye systems in one—an underwater visual arrangement . . . and an aerial survey system with high visual acuity . . .". (See the June, 1950 "Science Digest," pp. 16, 17).

Who can believe that this amazingly intricate and highly ingenious system of sight, granted to the kingfisher to enable it to catch fish for food, is the result of "random mutations"? Here is a highly complex system of sight, involving elements entirely lacking in human sight, that equips this bird for its particular station in life and enables it to keep its prey (moving swiftly in water) in sight and in focus as it dives toward it from the air above!

Drop a penny in a bath tub full of water. Look at it from an angle, then reach for it, and you will miss it! Man's eyes give a false impression of the exact location of objects under water!

It is clear, the Creator gave each creature eyes suited to its environment and manner of life; and in most instances, the eyes He gave are so complex and vision is obtained through such an involved, complicated mechanism, one must admit this is the work of God!

We are reminded of the observation of Theodosius Dobzhansky, an outstanding scientist:

"Perhaps the most troublesome problem in the theory of evolution today is the question of how the haphazard process of chance mutations and natural selection could have produced some of the wonderfully complicated adaptations in nature. Consider, for instance, the structure of the human eye—a most intricate system composed of a great number of exquisitely adjusted and coordinated parts. Could such a system have arisen merely by the gradual accumulation of hundreds or thousands of lucky, independent mutations?

"Some people believe that this is too much to ask natural selection to accomplish, and they have offered other explanations. . . . But these theories amount only to giving more or less fancy names to imaginary phenomena." ("Scientific American," in article on "Strangler Trees").

In all fairness to Dr. Dobzhansky, we must say that he still is an evolutionist).

Read again the description of the wonders of the human eye, found in Chapter 10; then consider the marvels of the eyesight of the kingfisher, the owl, the dragonfly, the horseshoe crab, the lobster and the grasshopper; then ask yourself if it is reasonable to believe that such marvels came about entirely by "chance mutations" and "natural selection." It is easier to believe that a Mergenthaler Linotype machine, with its thousands of parts and hundreds of delicate adjustments, "just happened" than to believe such a complex organ as an eye (either simple, compound or camera-type) was gradually developed through "chance mutations." Evolution is not only merely a **theory,** unproved and unproveable, but also it is a very **illogical** theory.

When once a person admits the presence in the Universe of an Almighty Supreme Being, who created all things, all such marvels and miracles as the eye, the brain, the wonders of the atom, the mystery of gravitation, etc., are readily accounted for.

THE FASCINATING MARVELS OF SEX

The primary method of multiplication of unicellular life (protozoa) is by simple division, called "binary fission" by biologists. For some unknown reason the cell of a protozoa like the ameba splits in two and makes two identical cells. If all higher forms of life evolved from unicellular forms of life, as evolution teaches, and these original unicellular forms of life were asexual, and cell division of these primordial protozoa invariably produced two duplicates of the original cell, **how could sexless forms of unicellular life ever give rise to the higher forms with sex?**

True, mutations in sexless forms of life do occur, but they are mutations **that stay within certain bounds. For sexless forms of life ever to evolve by mutations into sexual forms is utterly impossible.**

Mutations of organisms which do not reproduce sexually give rise to "clones," the descendants of a single individual. A gene which mutates in a sexless individual cannot pass outside of the "clone," and thus can not be as widely distributed as mutations occurring where there is sex.

"The spectacular evolution of plants and animals into myriads of diverse forms probably could not have taken place without the process of sexual reproduction. The living forms that do not reproduce sexually but that carry on the life of their species by dividing, budding, or other means MAINTAIN A FAIRLY CONSTANT HEREDITY." (The Mystery of Sex, in "Popular Science," p. 743).

Here then evolution faces an impasse. Evolutionists believe there was a time when there was no reproduction by sex, They teach that reproduction by means other than sex (even though there are mutations) maintains "a fairly constant heredity"—that is, succession of life with very few changes. HOW THEN DID "SEX" EVER GET STARTED?

Since sex usually is the coordination of **two** functions—male and female—both male and female elements had to get started **at the same time.** For a series of mutations to develop the "male" element without the "female," or vice verse, would be an absurdity, for one is incomplete without the other. And to believe that both "male" and "female" elements developed concurrently by "chance mutations" is likewise an absurdity. Obviously, sex—the coordination of two unlike elements—had to be PLANNED, and DESIGNED, and CREATED that way.

Most cell division in unicellular protozoa, like amebas, goes on generation after generation for endless millennia without any change whatever. In some asexual forms of life, like certain bacteria, we know there are mutants, and "varieties" develop, but the essential nature of the bacteria remains unchanged, generation after generation, and there is NO transformism from one genus into another.

We know there are one-celled organisms that reproduce sexually—but they obviously **were made that way** in the beginning. An asexual system could never, of itself, develop into a sexual system, WITHOUT THE INTERVENTION OF AN OUTSIDE GUIDING FORCE. The fact of the presence of "sex" in life demands the work of an Intelligent Creator.

All of life that shows "design" and "purpose" is the result of creative Intelligence. Consider an illustration: all will admit that if one were to find a bolt with machine threads on one end, in the iron fields of Michigan, that **some one had to mine the ore, smelt it, and then form that bolt and thread it.** And then to find near the bolt a threaded nut that exactly matched the size and thread size of the bolt, would give one a complete, useful product. For either a threaded bolt or a threaded nut to appear "spontaneously," without the intervention of outside intelligence, would be an unheard of thing. Every highly complicated device in nature that shows design and purpose must of necessity be the work of an Intelligent Designer. The miracle and marvel of sex in life proves the presence of an outside Intelligence who designed and created it all.

Variety in Methods of Reproduction

Host higher plants reproduce by fertilized seeds, involving the male and female elements. Some plants reproduce by "vegetative propagation," such as by bulbs, tubers, runners and cuttings. Actually, most plants can be raised from stem cuttings. Algae, fungi, mosses and ferns reproduce by means of small specialized bodies called **spores.**

Algae seem to excell in the variety of methods of reproduction. Some algae reproduce simply by cell division, others produce offspring by means of asexual spores, some others "reproduce by fusion of sexually undifferentiated gametes," and finally there are algae that produce true sex cells—eggs and sperm which unite to produce new offspring.

The ameba reproduces by the simple process of dividing into two identical amebas.

Yeast cells reproduce by budding. A small bud appears on the outside of the cell; this grows and finally separates from the parent cell as a new cell.

Reproduction in animals is either "asexual" or "sexual." But there is great variety in both realms.

A starfish will deliberately divide itself—and reproduce in that manner; but if it be torn limb from limb and cast into the sea, from each limb (or ray) another starfish will result provided that a fragment of the central disc adheres to each severed ray.

If a small bit of the base of a sea anemone becomes separated from the parent animal, a new sea anemone will grow from this remnant! By means of muscular contraction in the middle of its body, the sea anemone may divide itself into two parts, and each half will become a new individual. The anemone may also reproduce sexually. The eggs are fertilized in the sea water and develop directly into new anemones.

Corals reproduce by "budding." New polyps grow off the old ones. Sexual reproduction by means of egg and sperm also occurs among corals.

Sponges reproduce both asexually and sexually. A new sponge will grow from almost any piece which has been broken from a living sponge. Buds or branches may break off and grow into new individuals. Sponges may also develop sex cells (eggs and sperms). In some sponges both kinds of sex cells may arise in one individual; in others, they occur in different individuals, in which case the sperms are brought into the female sponges by the water currents. The fertilized egg then develops into a flagellated **larva** (the young, free-living stage in the development of animals) which escapes from the parent body, swims about for a while, then settles down, becomes firmly attached, and grows into a new sponge. Evidently God intended that sponges should survive and multiply—He gave them so many ways by which to reproduce!

There are many other ways of reproduction, such as the

hermaphroditic method used by earthworms, in which each individual has a complete male and female sexual apparatus. Though each earthworm is hermaphroditic, it does not fertilize itself.

Some forms of life reproduce by **parthenogenesis,** that is, having a mother but no father. This occurs in such animals as bees, some marine worms, aphids, etc.

All of this speaks to us of the fact that the Creator adapted the means of reproduction to the station in life of the creature.

Strange Methods of Hatching Eggs

The female of the Giant Water Bug (about 4 inches long) cements her eggs all over the back of her husband! They stay there until they hatch.

Frogs use "solar energy" quite regularly in hatching their eggs. Their hundreds of eggs are each enclosed in a transparent jelly, and the entire mass has a convex shape which acts like a magnifying glass and concentrates the sun's rays, focusing them on the enbryos in the eggs. The frog's "incubator" is run by solar heat! Who taught the humble frog this trick?

The female Sawfly has a highly specialized ovipositor (egg-laying organ at the end of her abdomen) with which she cuts a hole in a leaf and lays her eggs. When the eggs hatch into larvae, they have the leaf right there for food! This specialized organ had to be made at once, to be useful. Slow "evolution" in no wise accounts for it.

The **sea horse** is different in more ways than one. After the female lays the eggs the father incubates them! The eggs are attached to a special pouch on the under side of his body and are carried there until they hatch. This was so designed, and is not the result of evolution, for it is different from average procedure.

The eggs of spiders are "all put in one bag." The eggs are inclosed in a silken bag which is then hung from the web, or is carried about by the female. When the young spiders are born they emerge from the egg sac, and look like miniature adults. Generation after generation spiders follow this procedure: the eggs are laid and inclosed in a neat, substantial bag where they incubate. There is never any deviation from this among the species where it is the method used. No one claims any "evolution" of spiders for the past millions of years.

Golden-eye lacewings lay stalked eggs! The eggs are attached to short stalks, and the ends of the stalks are securely fastened to leaves. After they emerge, the larvae spin silky cocoons from which they emerge as delicate, thin-winged adults. Why would blind evolution hit on such an unhandy plan? It is much easier just to lay eggs than to have to attach each egg to a long thin handle, and then fasten the end of the handle to a leaf. Remember, even if a female lacewing ages ago happened to put her eggs on stalks, the next generation would have gone back to the old method—for there is no such thing as "acquired characters." ALL CHANGES THAT COME ARE FROM MINOR MUTATIONS ORIGINATING IN THE SEX CELLS, and never come from any habits or abilities "acquired" during the life time by the parent. Such a radical change from just "eggs" to "stalked eggs" (eggs fastened at the end of poles) is a vast change, and such vast mutations DO NOT OCCUR IN NATURE. All observation indicates that viable mutations are all **minor,** only slight variations. So evolution is at a loss to account for the unique system of laying and suspending eggs from stalks, to hatch them!

The female gaff-topsail fish (**Felichthys felis**) lays her eggs and the male comes along and takes them into his capacious mouth, and keeps them there until they are hatched and the little fish—all 55 of them—are carried in his mouth until they are about 3 inches long. Knowing how voracious most fish are, this method of hatching eggs is a miracle! Why does the father want to burden himself with these eggs? How can he manage to **live** during the 65 days of his parental responsibility? And when the father gets **real** hungry about the time the little fish in his mouth are two inches long, why doesn't he just gulp them down? When this system was first inaugurated (let us say for the sake of argument) many millions of years ago, how did the mother persuade the father to keep the eggs in his mouth when all the neighborhood fathers were not burdened with such trying chores? THE ONLY POSSIBLE EXPLANATION FOR THIS PHENOMENON IS THAT GOD CREATED THESE FISH THAT WAY AND PUT INTO THEIR GENES THE INSTINCT TO CARRY OUT THIS PECULIAR CUSTOM OF RAISING THEIR YOUNG. Therefore, generation after generation, since they are born with a set instinct, these fish act as they do! **That instinct was put there by the Creator; evolution had nothing to do with it.**

Actually, there are literally **thousands** of unique methods of incubating eggs. Space does not permit listing any more, but we know that each different method is a witness for Divine Creation, showing the ingenuity and marvelous workmanship of the Master Workman.

We might mention this interesting fact: fish are generally prolific in laying eggs. The ling fish takes no chance in being left childless by laying 160 million eggs at one time! But the sunfish beats this by laying 300 million! The herring lays a mere 30,000 —but the eggs are coated with a glue-like substance so that they stick to rocks.

Devious Methods Used at Times in Sex

In addition to the orthodox methods of conjugation and propagation of species through fertilized eggs, spores, cell mitosis, etc., that we have mentioned, there are scores of "devious routes" followed by sex that we want to call to our reader's attention.

(1) The Strange Case of the Bedbug

"The male bedbug does not inject sperm into the female genital tract, but into an entirely separate structure known as **Ribaga's organ,** on the right side of the female's body. This organ has no connection with the ovaries. The difficulties encountered by the sperm are increased by the fact that Ribaga's organ contains cells that eat sperm. Nonetheless, some of the spermatozoa manage to survive and fertilize eggs. Passing between the cells in Ribaga's organ, they enter the body cavity, travel up the walls of the female's reproductive tract and ultimately reach the ovaries.

"Normal copulation is impossible because the large, inflexible sex organ of the male cannot fit into the female genital opening. Without the mutation responsible for the evolution of Ribaga's organ, bedbugs would have become extinct—to the advantage of the human race." ("Unorthodox Methods of Sperm Transfer," by Lord Rothschild, "Scientific American," 11-'56.)

How illogical can the evolutionist get? Remember, evolution teaches the "gradual" change by "random mutations" through "long periods of time." If normal copulation is impossible—and in the case of the bedbug it is—every bedbug in the world would have died childless long before "evolution" got around to establishing this devious sex route followed by the bedbug. Say it took a million years to develop "Ribaga's organ." Every bedbug in the world would have perished—and even their memory would have been long lost in the shadows of antiquity—while waiting, patiently waiting, for this organ to "evolve."

Say folks—since Evolution has taken to itself the credit for "evolving" Ribaga's organ, and since it is so anxious to achieve something, why don't we forget logic for the time being, and let it have the credit for "saving bedbugs for the world!"

Seriously, all logical thinkers can see that there has been imposed a HANDICAP on the pesky little bedbugs. Though God pronounced a judgment on the world—and this judgment is on nature also (see Genesis 3:17-19; Rom. 8:20-22)—apparently He in mercy set **limits** to that judgment, and, in the case of the bedbug, He made it difficult for them to propagate. HANDICAPS are to be seen everywhere in nature; and these "Handicaps" are a witness for the fact of Divine Intervention in life on earth. If "survival of the fittest" were the law of nature, the world would have been destroyed by pests or monsters ages ago; but the Creator so balanced all life that the evil that are strong would not prevail, and completely dominate and ruin His creation.

SPONGES, SPIDERS, LOBSTERS and LEECHES all use a roundabout method to achieve copulation. (See Lord Rothschild's article). And each is a distinctive witness for Creation and against evolution.

(2) The Curious Behavior of the Stickleback

Thousands of species of animals—birds, fish, mammals, reptiles and insects—go through a distinctive courtship routine, prompted and established in pattern by unchanging INSTINCT. As an example of this phenomenon we quote from an article by Prof. N. Tinbergen (zoologist).

"The sex life of the three-spined stickleback **(Gasterosteus aculeatus)** is a complicated pattern, purely instinctive and automatic, which can be observed . . . at will. The mating cycle follows **an unvarying ritual.**

"First each male leaves the school of fish and stakes out a territory for itself, from which it will drive any intruder. . . .

"Then it builds a nest. It digs a shallow pit, piles in a heap of weeds, coats the material with a sticky substance, and shapes the weedy mass into a mound with its snout. It then bores a tunnel in the mound by wriggling through it. The tunnel, slightly shorter than an adult fish, is the nest.

"Having finished the nest, the male suddenly changes color—from an inconspicuous gray to a bright red and a bluish white.

"In this colorful, conspicuous dress the male at once begins to court females. He performs a zigzag dance before them until a female takes notice. He then swims toward the nest, and she follows. She enters the nest . . . and lays her eggs . . . and slips out of the nest. He then glides in quickly to fertilize the clutch.

"One male may escort three, four or even five females through the nest, fertilizing each patch of eggs in turn. Then his mating impulse subsides, his color darkens. Now he guards the nest from predators and 'fans' water over the eggs to enrich their supply of oxygen. This he does daily until the eggs hatch. For a day or so after the young emerge the father keeps the brood together. But soon the young sticklebacks become independent and associate with the young of other broods." (Condensed from "The Curious Behavior of the Stickleback," by N. Tinbergen; "Scientific American," 12-'52).

It is safe to say that almost all life on earth—below the level of mankind—is guided largely by INSTINCT. Instinct creates behavior patterns like that given above; instinct teaches a bird how to build its particular type of nest; instinct teaches the hunting wasp how to paralyze but not kill the caterpillar; instinct teaches the bee how to make the honeycomb. Since instinct enables an animal to exhibit intelligence in actions, without having actual intelligence, instinct must be a gift of the Creator to His creatures. The fact that all life is largely guided by God-given INSTINCT is one of the most powerful of all arguments in favor of creation. **There is no proof anywhere for the evolution of instinct. It is unchanging. Instincts can not be evolved gradually.**

(3) Dandelions Have Said Good-by to Sex

Science Digest (May, 1957) had an interesting article on this theme, "Dandelions Have Said Good-by to Sex," by Joseph Wood Krutch. Dandelions are one of the "highest" of all plants, and as a race, dandelions are "prospering and inheriting the earth."

"But in this most recent . . . of all plant groups, the flower is . . . **devolving** rather than evolving. Some plants have returned to a more primitive form of sexuality . . . but the dandelion is one of the very few plants that has gone these other (plants) one better (or worse); it has abandoned sex entirely. Its ovaries are not fertilized by pollen from a stamen in the same flower. **They are not fertilized at all.** No sexual process takes place. Every seed and therefore every new generation is the product of a virgin birth. For good or ill, dandelions have said good-by to sex.

"Sexuality made the dandelion what it is. The abandonment of sexuality will keep it almost precisely that. . . . If it lasts for ANOTHER HUNDRED MILLION YEARS IT WILL . . . 'IMPROVE' NOT AT ALL."

And so God has chosen the humble dandelion—as well as the bedbug—to be one of His witnesses! Read the above paragraphs again, and see this amazing confession: **Without sex, involving the interchange of genes from both parents, there is little prospect of any change.** Having "abandoned sex," they say, "the dandelion will remain unchanged for the next hundred million years." But since sexual reproduction is NOT the original method of reproduction, say the evolutionists, but simple cell division is, we know that evolution is a vain theory, for if there was a time when there was no sex, that time gave little hope of transmutation from one genus into another, even though minor mutations developed. It is **"sex"** that gave the terrific drive to mutations! A SEXLESS WORLD, we are told, is a world that has little hope of ever changing. Evolutionists admit that in the world that existed before "sex" there was little possibility of "evolution"—so slight, in fact, that a "sexless dandelion" has no chance to improve or evolve in the next 100 million years! HOW THEN DID THAT PRIMEVAL WORLD EVER GET OUT OF ITS SEXLESS RUT?

This is not the only instance in which evolutionists are thoroughly confused and have to resort to such statements as "they are **devolving,** not evolving." We have read similar confessions many times. The truth is, the facts of nature when viewed as a whole and **in detail** confute the theories of evolution.

(4) It is Sex that made the Midwife Toad Famous!

"Most frogs and toads which live on land return to the water when mating time comes, for the eggs have to hatch in water for the sake of the tadpole stage. But the midwife toad does not return to the water for mating. This takes place on land, and while it goes on the male, using his hind legs, literally pulls the eggs out of the body of the female. The eggs form a long string, about 30 inches in length, consisting of a jelly-like substance in which the eggs are imbedded at regular intervals. The male loops this string of eggs around his hind legs. . . . He then digs himself a hole in moist sand, or soil, which he does with great skill and very fast. There he sits with the egg string, waiting patiently while the eggs incubate. After waiting a few weeks, the male finds water, jumps in and starts swimming very energetically. This breaks the egg membranes and the tiny tadpoles scatter in all directions." (Salamanders and Other Wonders, pp. 29, 30).

Darwin's Frog (**Shinoderma darwinii**) presents an even more striking oddity. The male has modified vocal sacs which he converts into receptacles for the eggs of his mate! The pouch, which becomes an extensive chamber under the body is entered by two channels on the floor of the mouth, and into this—"the most curious of all nurseries in terrestrial animal life"—the eggs are received. There the dozen or so young are born and **they stay there until they pass through the larval stage!** This father is taking no chances in having his youngsters (tadpoles) eaten! Here again evolution is dumb and helpless; it has no adequate explanation for this phenomenon. The only possible answer is, GOD MADE IT SO; and He assured the continuance of this odd life cycle by impressing on the hereditary genes the stamp of this odd body and weird manner of rearing young. What croaking frog would wish such a task as this on himself?

It is our conviction that the Creator injected such reversals of the general trend in nature to demonstrate the fact of His handiwork. Established by inflexible instinct, the midwife male frog goes through this trying procedure generation after generation, while his wife enjoys herself! The poor male would be the first one to grab onto some stray, "chance mutation," to get out of his slavery to the maternity ward—but no luck. In this instance, as in thousands like it, evolution explains nothing; all it does is give us a confused babble of meaningless words.

(5) The Curious Life Process of the Alpine Salamander

"The Alpine Salamander, living from 3,000 to 10,000 feet up the slopes of the Alps, produces her young alive, and that by the most curious process yet observed. Of 50 eggs which the oviducts may contain only two are fertile. When the two tadpoles emerge from the eggs, they are not extruded from the parent body, but are nourished upon the substance of the remaining 48 eggs," so there in the mother's body the twins undergo their metamorphosis, protected and with abundant food, and emerge, like their parents, only smaller!

Here is an amazing adaptation to climate that permits the Alpine salamander to live and reproduce in conditions normally adverse to salamanders. Instead of spending their tadpole stage out of doors, as other salamanders do, these little ones are fed from a wellstocked pantry, and are brought through their tadpole stage right in the protection and coziness of the mother's body! To believe that this adaptation happened through a chance mutation, or "random changes," millions of years ago, and through a period of millions of years, is more than we can swallow — and we'll tell you why: In the course of those millions of years, so we are told by geologists, the alpine areas went through several radical climatic changes from ice age to a warm climate and back again to another ice age, and then a return to a warm climate! **When would evolution have time enough, with so many climatic changes, to perform its wonders?**

(6) The Butterfly: A Witness Against Evolution

"Human genius has never invented anything lovelier than a butterfly, nor anything so wonderful."

In all nature one can scarcely find anything more beautiful than the butterfies! But before a butterfly becomes an adult, it must go through a complete metamorphosis in four stages: egg, larva (worm or caterpillar stage), pupa (or chrysalis), and then the adult butterfly. Why such a roundabout path to produce a butterfly? If unguided nature or evolution were doing it, according to Darwin's theory of "natural selection" and "survival of the fittest" the impractical devious route would not have a chance, but the butterfly would hatch directly from the egg, as would seem to be the normal route. Unguided evolution in a billion years could not even think up such an involved plan as "complete metamorphosis"—much less put it into a workable plan!

Could it be that God, the Master Teacher, so designed the life cycle of the butterfly **to teach us a lesson?** Undoubtedly, spiritual and moral truths are illustrated in nature—and the metamorphosis of the caterpillar into the butterfly is an obvious lesson. If the grovelling, repulsive, greedy, earth-bound caterpillar pictures man in his lowly, fallen estate, then the transformation into the butterfly is a lesson in the need and reality of the new birth (see John 3:3, 5, 7). And the butterfly, released from its cocoon, flying heavenward, is a picturesque display of glory, speaking of the glory of the coming resurrection for the saved of earth (1 Cor. 15:42-44; Phil. 3:21).

One student, writing of this miracle of metamorphosis, says,

"The metamorphosis of the butterfly cannot be reasonably explained by any mechanical theory of evolution. The idea that this mysterious process, by which a certain form of animal is changed comparatively suddenly into something entirely different, and which goes on with undeviating regularity generation after generation, could have come about by the selection of chance variations or mutations, without plan and without directing force, is so contrary to intelligence and so basically unscientific that it cannot be supported. It is manifestly absurd, and the more one considers the process of metamorphosis the more obvious it becomes that no theory of 'fortuitous variations' large or small can explain it." (Evolution: the Unproven Hypothesis; p. 48.)

The life story of the butterfly begins with the tiny egg which the butterfly deposits upon a branch. And, mysterious miracle —each kind of butterfly seems to prefer its own special kind of plant or tree. No one knows why. Another miracle: the eggs "are as exquisitely beautiful as gems—lustrous as pearls, more delicate than hand-wrought jewels. They are fluted, ribbed, patterned in a score of different ways—perfect as works of art, yet contrived with marvelous skill for the admission of the fertilizing substance. The material of which the eggs are made also provides the larvae with their first meal after they have hatched from the shells." Most anyone can see that the "design" and beauty in these eggs is the work of the Master Artist, the Creator whose works are perfect!

"The grown-up insect goes back to the plant or tree trunk on which it was nurtured in its early life. This is wonderful. No moth or butterfly eats solid food (though some butterflies drink nectar); some can not even take moisture. Yet all lay their eggs on a substance which will be cradle and larder to the caterpillars into which those eggs will hatch! . . . Generally (with a few exceptions) there is one food, and one only, for a species. If that fails, the caterpillar will die in the midst of abundance, starving while caterpillars of other species are flourishing. The parents, to which solid food is not necessary, find it without fail for their offspring which the parents may never live to see. . . . Yet Nature, by some magic, guides the parent to the right tree, bush or weed. There, on the very substance essential to the creature yet unborn, the egg is laid. **There is no more perfect example in the world of unerring instinct."**

Evolution is not equal to a feat like that! To put such ability in an insect is the work of Infinite Intelligence.

Another feature about the egg laying is its great variety.

"The eggs may be laid singly, in cluters or in masses. . . . In some species they hatch in a number of days, in others the egg is buried underground or covered with a coat of varnish, and survives the winter, hatching the following spring."

Surely, this speaks of the Great Designer, who loves variety in His creative handiwork.

If the eggs are interesting, the career of the caterpillar till it becomes an adult butterfly is even more so. Having eaten the shells of the eggs from which they emerge, caterpillars begin a "campaign of gorging" and almost burst with food!

The larva of **Polyphemus** (the American silkworm) in its two-months' career actually consumes 86,000 times its own weight when first hatched! The caterpillar of the Goat-moth reaches a weight 72,000 times as great as its weight when first hatched!

The larva of the monarch butterfly is about an eighth of an inch long when it is first born. Soon it sheds its skin in the first of four molts. In about two weeks it is full grown, and it then begins preparation for a major change in its way of life.

Seeking a convenient leaf or stem he proceeds to spin a tough, flat button of silk. This amazing feat is done by means of a liquid secretion of glands in his head; the secretion hardens into a thread when it is squeezed out into the air from an opening on the lower lip. How can one account for the fact that he not only possesses a chemical factory, but he also is an

"architect and designs a house, though he has never lived in one before and has never even seen one? His first attempt follows a pattern that is standard dwelling for all caterpillars of his variety, and is perfect for its purpose!" He is not only an architect and a builder; he is an interior decorator and a water-proofer as well. And he builds for himself a habitation "that the genius of man cannot duplicate."*

Lazy days pass in the caterpillar's pupal house. There the chrysalis takes its final shape, and the outer skin hardens. Within this dry shell the organs of the caterpillar are dissolved; special cells are generated (in the apparently lifeless body) whose function is to devour the organs which once worked for the caterpillar, and reduce them to a pulp—a seemingly formless glob, "a kind of soup." A miracle then takes place!

"Nothing remains unchanged, save perhaps its system of breathing. Jaws, claws, claspers, pro-legs, digestive system, even the very shape—all disappear. Then the shapes of the head, legs and thorax of the butterfly gradually appear upon the chrysalis case, and the first rough draft of the coming butterfly is dimly seen on the horny case of the chitin."

The hour arrives for the insect to wake up and come out of its chrysalis. At this time it voids a quantity of a rather corrosive liquid which softens and partly dissolves the silk at one end of the chrysalis. Through the opening thus formed the butterfly emerges. The ugly grub has vanished; and in its place is a lovely winged butterfly as colorful as a flower, and in the case of the Monarch butterfly, capable of winging its way across an ocean! When it emerges, it is "resurrected"—full grown—and does not have to grow up like a baby chick. There is no growth thereafter for either moths or butterflies, whether it be the tiny moths of the leaf-mining group, or the giant Atlas moths of Africa, which have a wing span of nearly a foot!

We must call attention to one more miracle: **the subtle beauty of the butterfly's wings.** Their beauty is proverbial.

---

*There are many varieties of moths and butterflies and many varieties of patterns of life followed by them. For example, the "Leaf-rollers" (moths) cement together two edges of one leaf, or two different leaves, and "in the little room thus formed make their home, snugly furnished with a couch of web spun from their silk gland."

There is a poisonous species of caterpillar (the Puss-moth larva), that spit out their poison a considerable distance, reminding one of a little spitting Cobra.

The Wooly Bears (larvae of moths) weave their own (but now useless) hairs into their silken cocoons; the caterpillars of the **Dicranura** chew such hard materials as wood and even sandstone, and mix that with their silk! Most wonderful of all, an African moth (**Nyctemera** group) wraps itself merely in a cloak of bubbles that it blows up and then goes to sleep! Practical thinkers know there is an Intelligent Creator behind such marvels.

One writer says, "Most butterflies appear like animated pieces of art with an amazing combination of small-patch color schemes." Another says, "The most striking thing is the way the colors shift and vary with every change in the angle of the light or of the eye of the viewer." (Nature's Wonders; p. 99.)

Another writer, describing a certain gorgeous butterfly, says, "It shows a play of **iridescent colors** that can hardly be matched in jewel-like tones by any other one thing." We will explain why the colors of many butterflies are iridescent.

Under a grant from the Radio Corporation of America, Drs. T. F. Anderson and A. Glenn Richards, Jr., studied the brilliant blue tropical butterfly (**Morpho cypris**), which is prized as a decoration for coffee trays and the like. They used the electron microscope which can "see" objects smaller than a wavelength of light. They discovered this incredible phenomenon:

"There is a three-dimensional architecture on each wing scale. The wings are covered with these minute wing scales that overlap much as a roof is covered with shingles. The surfaces of these scales are covered with perfect structures—rows upon rows of them that look like long narrow skyscrapers on arching supports! Imagine each 'skyscraper' to be made of a transparent material like glass and the distance between reflecting floors to be half a wavelength of blue light! Each of these scales reflects blue light and no other. These 'skyscrapers' have 'floors' only 1/100,000 of an inch apart. This is NOT guesswork; pictures taken with the electron microscope are so sharp that details as small as three-ten millionths of an inch can be seen in the 'walls' of the 'skyscrapers'. (And remember, these 'skyscrapers' are but minute portions of the infinitesimal scales on the wings of a butterfly!) These 'details' may well be the molecule-sized 'bricks' of which these 'skyscraper' structures on the wing scales are built. But how these 'bricks' make the 'skyscrapers' is still an unsolved mystery."

Can you beat that for miracle in the unseen world of nature! The color in a butterfly's wings is not from pigment, but from the reflected light from these transparent wing scales, **made to reflect different colors according to the size of the "skyscraper" arrangement on each wing, or on each part of each wing.** This is amazing beyond description, and is the fitting work of an infinitely wise, all powerful Creator. Such a wing could no more develop through the so-called evolutionary processes, by "random changes," than that a dog could jump to the moon. Let us be honest; give the Creator the credit due Him.

The Testimony of a Great Scientist

A German biologist, Richard Goldschmidt, "set himself the task of proving by laboratory research what Charles Darwin had **assumed** to be true." He became an authority on the gypsy moth **(Lymantria)**, following about the same line of work on it that others have done with the fruit fly **(Drosophila).** Goldschmidt became director of the famous Kaiser Wilhelm Institute in Berlin, and recently he has been head of the department of Zoology in the University of California. In his book, **The Material Basis of Evolution** (Yale University Press), he tells of his disappointment **in not being able to verify the theory**.

He argues that there are "large species . . . which are distinct from one another, and separated from one another by 'bridgeless gaps' with no transitions from one to another (p. 29). Within each of these groups varied changes may occur, but such changes never amount to enough to form a distinct different kind, **and we have no scientific knowledge that new species have been formed in this way."** (Boldface type ours).

This witness is all the more valuable for Professor Goldschmidt still believes in organic evolution **"somehow."** But being an honest man, he gives as his considered judgment that there IS NO SCIENTIFIC EVIDENCE for belief in evolution.

MORE ABOUT "SPECIALIZED ORGANS"

Many times in this book we have pointed out the miracle of "specialized organs," organs created for a special function. Again and again we have called attention to the fact that ALL SPECIALIZED ORGANS HAVE TO BE PERFECT TO WORK— and that a partly developed or partly formed "special organ" is useless; hence we KNOW that all specialized organs came into existence AT ONCE, IN THEIR COMPLETED, PERFECT FORM; otherwise they would not function properly.

Actually, the world around us is full of wonders, including many strange and fascinating "specialized organs." One author says,

"Artists in the Middle Ages painted snakes—mostly the well-known 'sea-monsters'—as some kind of monstrous (animal). Today we smile at these grotesque conceptions. . . . Science has done away with all these fantasies and replaced them by real creatures, the complexity of whose body structure FAR EXCEEDS THAT OF THE STRANGE CREATURES OF IMAGINATION." (Reptile Life).

Let us examine some of these strange creatures.

The **Lionfish (Pterois radiata)** is a queer fish that lurks on coral shelves, 130 feet below the surface of the ocean. It has long, singular bristling spines that inject a potent poison into any living thing that touches them. The human victim experiences excruciating pain, if his arm or leg is pierced by one or more of these spines. No antidote is known. Whatever possessed evolution to turn out such a frightful creature? The evolutionist might counter with, why accuse a benevolent Creator of making such a repelling, poison-inflicting creature? The answer is: Nature is full of symbols of evil as well as good. There are poisonous snakes as well as milk cows. As nature is a reflection of a fallen, sin-cursed world, much that is evil and injurious is in evidence—to teach men moral lessons. Poison in nature, with its lethal consequences, is a picture of sin and its deadly consequences. It is to be avoided.

Aside from the reason **why** there is such a creature as the death-dealing Lionfish, we know of no link of intermediary forms that lead up to the Lionfish.

The **Lungfish** has been described as "the strangest fish in the world." In South America and in Africa they live in stagnant pools that dry up in the rainless season. In such a situation, fish that breathe air would die. But the Creator made a fish for just such a situation. When dry weather dries up the pool, the lungfish digs into the ground, curls up comfortably, and goes to sleep after enveloping itself in a sort of a mucilage cocoon! It gets its air through a hole that extends to the surface of the ground. And so it sleeps on through the dry season; and when the spring rains fall and fill the pool again, the water melts the cocoon, and releases the lungfish to swim around in the pond! The Darwinian theory of "natural selection" falls down completely here, for the first dry season for a normal fish would KILL IT when the pond dried. Natural selection cannot reach into the execution chamber of a dry pond and save even one fish that is not FROM THE BEGINNING equipped for such an emergency as is the lungfish. CREATION is the only logical answer to such a strange creature as the lungfish.

The **South American sloth** is a singular quadruped.

"Though all other quadrupeds rest on the ground, this singular animal is destined by nature to be produced, to live, and to die in the trees. He has no soles to his feet, and he is ill at ease when he tries to move on the ground. He spends most of his life **hanging upside down from the limb of a tree!** In fact, he spends his whole life in trees and never leaves them but through force or by accident. And what is more extraordinary, the sloth rests not UPON the branches or limbs, like the squirrel or monkey, but UNDER them! He moves suspended from the branch, he rests suspended from it, he sleeps suspended from it. To enable him to do this HE MUST HAVE A VERY DIFFERENT (PHYSICAL) FORMATION FROM THAT OF ANY OTHER KNOWN QUADRUPED. . . . When his form and anatomy are attentively considered, it is evident that the sloth cannot be at ease in any situation where his body is not suspended, as from a limb of a tree." (Charles Waterton).

How can even the most imaginative evolutionist possibly come up with **an explanation of how the sloth got its entirely different body for its radically different manner of life?** He (the Creator) "hath given it a body as it hath pleased Him."

The characteristic animals of Madagascar are the **lemurs** (related to monkeys). They live in trees, and are all night prowlers. One of the wierdest of the lemurs is the **Aye-aye** that has large protruding ears designed to catch the faintest sound made by insects; and one of its fingers is more than twice as long as the others, "as skinny as a living limb can be and equipped with a curved hooklike nail for dragging insects out from under the bark of trees." (Salamanders and Other Wonders, p. 171).

With ears designed to **hear** the unsuspecting insect, and finger clearly designed to drag insects out from under the bark, this strange animal needs its ears and slim finger to make its living! Ears and finger work together for an intended end. In this instance, the coordination of TWO unusual developments was essential to accomplish what the Aye-aye lemur has. To believe that **one** series of "chance mutations" produced the weird finger is asking a lot; but to have to believe that TWO series of "chance mutations" came at the same time and developed simultaneously into a cooperating pair is more than we can believe. To us the case is clear: the Almighty Creator gave that little creature both the unusual ears and the more unusual finger, to enable it to make its living in the trees!

Man and the bats are not the only creatures in God's world who use the secrets of sonar. There is a fish that lives in the Nile **(Gymnarchus niloticus),** shaped like a compressed eel, that has the ability of storing electricity in its stubby tail and of discharging it into the water in controlled bursts. "What is more, it can pick up or receive these impulses as they bound back from solid objects. Thus it uses its electromagnetic energy for an efficient form of underwater radar—and it manages to interpret those reflections, just as bats do with air-borne waves, in time to alter its course and so avoid running into things when darting backward, even in muddy water at night!" This is such a highly developed "specialized organ" that its development by chance is ruled out. Common sense says, God made it so!

Even a cursory examination of nature reveals literally hundreds of thousands of "specialized organs" in all forms of life. Who designed the unique mouth strainer for the **boleen whale**—"ingenious horny plates with fringed edges"—that permits the small plant and animal plankton to sift back into the ocean, but keeps in the krill for food when the whale gulps in great mouthfuls of sea water?

Who gave the brainless **starfish** an extraordinary stomach that "turns itself inside out" to envelope its food? And what Engineer devised the unusual means of locomotion for the five-rayed starfish, so that it can move about by means of a most amazing "hydraulic pressure mechanism," known as a "water vascular system"?

"Water enters by minute openings on the upper surface of the starfish and is drawn down a tube to a **ring canal,** encircling the disk. From this central ring canal go five radial canals one for each arm. Each of these connects by short branches to hundreds of pairs of tube feet—hollow cylinders that end in suckers. On each tube foot is a muscular sac. When this sac contracts, the water, prevented by a valve from flowing back into the radial canal, is forced into the tube foot; this extends the tube foot, which attaches to the sub-stratum by its sucher. Then the tube feet contract—shorten—and draw the animal forward a tiny bit. This process is repeated and the starfish slowly moves forward." (Animals Without Backbones," pp. 300, 301). This method of locomotion is obviously DESIGNED as a working mechanism for this odd creature; though we must confess, it seems to us quite an engineering problem to devise a system of locomotion for such a strangely shaped creature!

Obviously, a brainless starfish could not devise such an intricate system, using "valves, water pressure, canals, tube feet, muscular contraction, suckers, etc.", all finely coordinated to give this humble creature controlled locomotion! This interesting creature is so different from all other animals, we ask the evolutionist, What could this singular animal have evolved from?

We next mention the **sponge** as a highly specialized creature and a witness for Creation. Get a mental picture of a sponge. Can you conceive of an animal more unusual than a sponge? It is a real puzzle to evolutionists, too. Read carefully this statement:

"**The sponge body plan is unique.** No other many-celled animals use the principal opening as an exhalant opening instead of a mouth, or have the peculiar collar cells, or show so low a degree of co-ordination between the various cells. Hence, it is thought that the sponges have evolved from a group of protozoa different from the ones that gave rise to all the other many-celled animals. And the phylum **Porifera** has sometimes been set aside as a separate sub-kingdom of animals. THERE IS NO EVIDENCE THAT THE SPONGES HAVE EVER GIVEN RISE TO ANY HIGHER GROUP. This does not mean that the sponges have been a failure, for they are an abundant and widespread phylum. . . . But in the general trend of animal evolution the sponges are little more than a side issue." (Animals Without Backbones, p. 68; caps ours).

"There is NO EVIDENCE" that sponges have ever given rise to any higher group." And there is NO EVIDENCE that sponges ever evolved from any lower group! And so the humble sponges become—along with the equally humble bedbugs, dandelions and starfish, as well as many more—witnesses for God and His amazing creative work.

We call on another witness for God and Creation: the amazingly complex SPIDER, with its highly specialized organs.

This little creature and its habits are so wonderful, we ought to give a whole chapter to it; but space is limited. We believe, after considerable research, that a study of spiders is as compensatory and as interesting as that of either bees or ants.

For ages past spiders have been "ballooning through the upper atmosphere, diving under water with oxygen tanks, and spinning filaments so fine that even modern science can't duplicate them." These wonder workers are called **"Araneida."** But the so-called "common spiders" are some of the most uncommon creatures on earth! They are also among the most numerous. Naturalist W. S. Bristowe estimated that there were 2,265,000 spiders **per acre** on a certain grassy plot in England.

There is a tremendous versatility among the 30,000 odd species of spiders. Not all spiders spin webs.

(1) The non-spinners. Among the non-spinners are the RUNNING SPIDERS—hairy, speedy spiders that can be located under logs; the JUMPING SPIDERS—chubby little fellows that jump around like bucking broncos.

Among spiders that spin are:

(2) TRAP-DOOR SPIDERS that build tunnels for their permanent homes. They cement the walls with glue to keep them dry and prevent cave-ins; then they line them with silk to make them warm and attractive. Next, then fit them with a real, hinged trap door! And every new generation of these spiders builds the same type of home, with the same type of trap door—even though they have never seen such a building before, nor ever made one!

(3) The CRAB SPIDER. When she lives in the yellow plumes of the goldenrod, she too is yellow in color! When an innocent bee arrives this "villian" jumps out from her ambush and actually lassoes the bee with

silken thread hurled speedily over her wings; and this is quickly followed by more silken strands over her legs to stop her thrashing and to hogtie her. The spider then injects a chemical into the bee to paralyze her—and soon she begins her tasty meal.

Certain Crab spiders hurl their silken strands across the gap between a flower or branch on which they are sitting and an adjacent one. After the far end has been successfully snagged they walk over their new suspension bridge!

(4) The GRASS SPIDER weaves a "blanket" on top of grass (or other plants) and then she strings a series of sticky lines above the blanket to stop flying insects. The snare works, and when the insects land on the blanket below, the spider runs out and captures them.

(5) Some spiders construct "balloons" or "kites" by means of which they float around, sometimes going many miles.

"Some spiderlings," says Dr. Willis Gertsch (Curator of spiders at the American Museum of Natural History,) "climb up on threads like little acrobats, and in this way control the ship they are flying!"

(6) Another spider binds together dead leaves with its silk so it can sail downstream on its own canoe. When prey is spotted, it leaps from its craft, strides on the surface of the water (easily done, since its feet are constructed like little snowshoes) and soon returns with its victim to its floating dining room.

(7) There is also a European WATER SPIDER that uses its silk to construct an undersea house. The female spreads a silken sheet between underwater plants and then makes repeated trips to the surface to collect air bubbles, which enable her to survive under water. At mating time, the male builds a smaller house alongside the female's and joins the two with a silken tunnel! Peter Farb, writing on "NATURE'S WONDERFUL WEAVERS" says of these wonderful creatures:

"Spiders have achieved all this without a glimmer of intelligence. **They are creatures of blind instinct, locked into patterns of behavior that go back a hundred million years. And, through all the countless generations since, they have methodically continued to weave their individual webs, WITH NARY A VARIATION."**

And so Mr. Farb says exactly what Bible believers have been saying: God created all things to reproduce "after their kind" (Gen. 1), and ALL the different genera God has made have "methodically continued" their original manner of life "with nary a variation." WHERE DOES THAT LEAVE EVOLUTION?

Speaking of the amazing water spider, the late Philip Mauro in his book, "EVOLUTION AT THE BAR," fitly remarks: "It is manifest that its extraordinary manner of life, and the highly specialized organs, which are vital to it, could not possibly be the outcome of a long and slow process of evolution. Before the life of a water spider could begin, it must be equipped, first, with the means of secreting a waterproof material; second, means for spinning that material into a watertight cell; third, protective hairs to keep it from getting wet; fourth, the peculiar apparatus for filling its under-water 'house' with air; fifth, the instincts which prompt the doing and give the 'know how' for the doing of these things."

(8) Let us now consider the WEB-SPINNING types of spiders.

"Every species of spider MAKES ITS OWN KIND OF WEB, and builds it by instinct. When a baby spider spins its first web, even if it has never seen a web before, it makes one just like its forebears, except on a smaller scale." **(Spiders' Webs,** Peter Wilt, in "Scientific American.")

"The orb web is one of the most marvelous of natural objects—a truly marvelous engineering work. Each circle and spoke is laid with geometric accuracy, to a degree or two; and the whole web, consisting of thousands of separate parts, takes the spider less than an hour to complete. It consists of a framework of DRY lines that bridge an open space and lines that radiate outward from a central hub. On this are laid down many spiral turns of a STICKY silk. Insects that walk or fly into this trap struggle helplessly in the seemingly flimsy, elastic lines. But the wily spider, who hangs away from the web, touches **only** the DRY lines with the tips of his legs when he walks out to further ensnare his prey, and so he manages not to become entangled."

A spider's thread sometimes is only a millionth of an inch thick, and is then invisible to the naked eye.

From the moment of birth a spider starts spinning, and thereafter, for the rest of his life, it never loses the ability. A spider as it walks, climbs or jumps, lays behind it a silken lifeline which guards against falls. The silken strand also serves as the telegraph line to announce when prey has arrived at the trap, and then the victim is promptly handcuffed and strait-jacketed with it.

With their ingenious traps, spiders have snared objects hundreds of times their own weight. One observer saw a mouse trapped in a spider's web; and within 12 hours the mouse had actually been hoisted a couple of feet off the floor "by the soundest engineering principles of block-and-tackle lifting."

"The spider's silk is undoubtedly the most versatile substance in nature. It is also the strongest, for its size. Some of these silks can stretch a third of their length before snapping.

"The average spider has six minute spinnerets on its belly, each shaped much like the nozzle of a watering can. They can be manipulated as easily as we move our fingers.

"Each nozzzle is made up of roughly 100 tubes and each tube is connected to its own silk-making gland. But that is not all. The glands manufacture a VARIETY of silks, usually three or four. The spider can use as many of the tubes as she wishes, combining them in a well-nigh infinite assortment, to cope with every possible need." ("Nature's Wonderful Weavers," by Peter Farb).

Consider then the little SPIDER as a unique witness for God and creation. According to scientists, it has not changed its ways for millions of years, and gives NO EVIDENCE now of "evolving" or having ever "evolved." Spiders always have been spiders, since the day they were created.

These spinnerets are essential in the makeup of the web-building spider. A complicated machine only works when **completed.** A partly assembled typewriter is useless. Logic assures us that these highly specialized organs—spider's spinnerets—were completed when the spider itself was made: and that of course means CREATION.

Chapter 14

FOSSILS, FRAUDS AND FABLES

Prof. Thomas Huxley, one of the greatest exponents of evolution of all time, said frankly, "Evolution, if consistently accepted, makes it impossible to believe the Bible." So many present day geologists have accepted the theory of evolution, that to them "the Creation, Fall and Flood" are mythological. Basing everything on the theory of evolution, these geologists have closed their minds, and refuse even to consider the staggering amount of evidence that refutes the theory of evolution. One geologist says plainly,

"Everything contrary to geological Uniformity is impossible, therefore no amount of evidence can ever prove any past world conditions which would be contrary to Uniformity"* (or Continuity).

So he plainly denies the catastrophic changes that took place during the primeval judgment on the earth indicated in Genesis 1:2, Genesis 3 (the Fall of Man, and consequent judgment on the earth), and Genesis 6-8 (the universal Flood). Much that seems impossible to the believer in Uniformitarianism is perfectly clear and logical to the Bible believer who accepts the historical facts of TWO OR MORE OVERWHELMING DELUGES in the history of our earth. But the fact is, neither the geological strata nor fossils are found in the orderly CONTINUITY the evolutionist desires.

"There is not a single spot on earth where the whole series of the different strata appears; no cases where more than three, or, at most four ages are found one on top of another and these three or four ages may be any three or four of the numerous ages that are said to exist. Though the bottom, or earliest age, is, as would be expected, at the bottom, those ages above are not always in the same genealogical order."

Even Sir Charles Lyell (an evolutionist) admitted, "Violations of continuity are so common as to constitute even in districts of considerable area, the rule rather than the exception." (The Case For Creation, pp. 37, 38).

"Many geologists, compelled by facts that have been accumulating for a century, are now doubting whether fossil remains can be graded in a life-

---

*"Uniformitarianism" in geology has been defined as "the doctrine that all things and all forces continue as they were from the beginning"—and this of course rules out sudden catastrophic changes in the earth's surface due to the tremendous upheavals of such cataclysmic events as described in Genesis 1:2 and the Flood (Genesis 6-8).

succession at all. 'In the present condition of our knowledge,' **admitted** Prof. Huxley, 'one verdict—NOT PROVEN AND NOT PROVABLE—must be recorded against all grand hypotheses of the paleontologist respecting the general succession of life on the globe'."

In the last few years there has arisen a considerable weight of scientific opinion that "has challenged the fundamental principle of the system established by the nineteenth-century geologist, Charles Lyell. He supposed that geological processes of the past always proceeded at their present rates: processes such as rainfall, snowfall, erosion and the deposition of sediment. However, in 1955, Leland Horberg (geologist) showed that unless the radiocarbon method was entirely fallacious, **there was a very marked acceleration of the rate of these geological processes during the last part of the ice age. Some factor must, therefore, have been operating that is not operating now.** . . . By use of 'Carbon 14' dating (using the radioactive isotope of carbon) . . . scientists revised the date of the end of the last ice age, making it only 10,000 years ago, instead of 30,000 years. . . . The importance of all these problems compel us to admit that **we do not now have an integrated,** effective **theory of the earth we live on.** . . . (In the last 100 years) at least fifty theories have been produced to explain the 'ice ages' but none of them has been satisfactory." (Charles H. Hapgood, "The Earth's Shifting Crust," in Jan. 10, '59, The Saturday Evening Post).

FOSSILS—A Witness for Creation

"Any evidence in the materials or (sedimentary) rocks of the earth's crust that gives some idea of the size, shape, or structure of the whole or any part of a plant or animal that once lived is called a **fossil**. Fossils can be formed (or preserved) in a variety of ways. . . . Most fossils are formed . . . when the skeletal structures are slowly dissolved by water and are gradually replaced by minerals such as calcite, silicon dioxide, or iron sulphide, which are deposited in the cavities left by the slow dissolution of original materials." **(Records from the Invertebrate Past, pp. 231, 322, in** "Animals Without Backbones").

"GEOLOGICAL TIME" has been divided into six **eras**: (1) **Azoic** ("no life") era marks the origin of the earth and the formation of rocks: no life was present. (3 billions years ago). (2) **Archeozoic** ("primitive life") era. "If life had evolved, the rocks show little evidence of it." (2 billion years ago). (3) **Proterozoic** ("first life") era. (1,200,000,000 years ago). "Rocks of this era have only rarely yielded a recognizable fossil; yet this era must certainly have been a time of great evolutionary development, for by the **Cambrian** period (first of the Paleozoic eras), the animal kingdom is already highly diversified." (4) **Paleozoic** ("ancient life") era (550 million years ago). (5) **Mesozoic** ("middle life") era (200 million years ago). (6) **Cenozoic** ("recent life") era (60 million years ago). (See Croneis and Krumbein, **Down to Earth).**

According to the evolutionary theory primitive life should have "evolved" in the **Archeozoic** era and shown rapid and widespread development in the **Proterozoic** era. But all geologists note this strange phenomenon: THERE IS LITTLE IF ANY FOSSIL RECORD BEFORE THE CAMBRIAN PERIOD (first of the **Paleozoic** era)—long after the fossil record SHOULD have appeared, if evolution be correct!

"Nearly all Phyla which leave any kind of a fossil record are well represented in Cambrian rocks—many of them by several groups, which already show the distinctive characters of modern classes. WHY PRE-CAMBRIAN FOSSILS ARE SO RARE IS NOT YET UNDERSTOOD." (Animals Without Backbones, pp. 324, 325).

The fact that there are few if any pre-Cambrian fossils is **fatal to the theory of evolution,** as the following quotations show. Evolutionists explain this lack of pre-Cambrian fossils by saying, "The records have been obliterated."* The only proof they have for this assertion is, "Since evolution **must** be true—'for there is no alternative'—therefore we know that the earlier living forms **must** have existed!" (Evolution, the Unproven Hypothesis). The natural inference from the fact that there are no pre-Cambrian fossils is, **life began on earth suddenly and in great variety.** All thinkers should be able to see this, except those who have been blinded and prejudiced by the evolutionary hypothesis.

To help us realize the full force of the fossil evidence, let us present this summary (based on a similar summary in "Creation's Amazing Architect," pp. 49, 50).

1. "The era of ancient life arrived abruptly and without warning" (Wells and Huxley). "'The fossils, instead of appearing slowly and sporadically, **suddenly appear in their thousands in the Cambrian strata,** whereas in the pre-Cambrian strata the fossils cited are very few, very far from being intermediates, and they all have been disputed. It has also been demonstrated that these alleged pre-Cambrian remains may well have been produced by inorganic means."

2. "Every species that ever occupied the earth throughout the vast reaches of geological time, when it appears in the record of the rocks (as a fossil) for the first time, appears complete and fully organized. "'There is no evidence in the history of the rocks that any 'half and half' form ever existed."

---

*"In passing from the Permian to the Mesozoic, we are conscious of entering a new world in the succcession of life. . . . A WHOLE VOLUME OF RECORDS IS MISSING. When we next gather up the threads of the story we find that the organic world had made extraordinary advances during the age **of which we have no available records."** (Prof. Howchin. See p. 89, "Creation's Amazing Architect").

3. The life that lived in the waters of the Cambrian period was "highly organized and differentiated" (Howchin) . . . "diversified and not so simple as the evolutionist would hope to find it" (Percy Raymond). "They have not the simplicity of structure that would naturally be looked for (if evolution is correct)" (Dana). "They are perfect of their kind, and highly specialized structures" (Dana).

4. The fauna of the Cambrian is "In essentially the same form as that in which we now know it" (Clark). "The majority of the fundamental types of the animal kingdom come before us without any links between them from a paleontological point of view" (Deperet).

Darwin himself admitted the failure of geology to support his views. He freely admitted that "all but one of the greatest geologists and paleontologists of his day were against him". (See end of chapter 9, "Origin of Species").

"Geology," said the disappointed Darwin, "assuredly does not reveal any such finely graded organic chain; and this perhaps is the most obvious and gravest objection which can be urged against my theory."

Many modern scientists (some of them evolutionists) admit the failure of fossils to support the theory of evolution.

"LACKING THE MORE CONVINCING EVIDENCE OF A FOSSIL RECORD, and basing our ideas on the principles of homology and recapitulation, we are able to construct animal trees . . . which attempt to show the order of evolution . . .". "The fossil record . . . IS OF PRACTICALLY NO USE IN RELATING THE PHYLA TO EACH OTHER. For, as we dig deeper and deeper into the rocks, expecting to find a level at which the most recently evolved phyla no longer appear, WE FIND INSTEAD THAT THE FOSSIL RECORD IS OBLITERATED." ("Animals Without Backbones," p. 335, etc.; caps ours).

Lacking FACTS they use their imagination to develop their theories.

Douglas Dewar, British naturalist, at one time a believer in evolution, turned from it as the result of his own scientific research. He said,

"Paleontology (study of fossils) cannot be regarded other than as a hostile witness (against evolution).

"It is not possible to draw up a pedigree showing the descent of any species, living or extinct, from an ancestor belonging to a different order. The earliest known fossils of each class and order are not half-made or half-developed forms, but exhibit, fully developed, all the essential characteristics of their class or order. . . . It is not possible to arrange a genealogical series of fossils proving that any series has in the past undergone sufficient change to transform it into a member of another family.

All the changes proved by fossils to have taken place in animals are within the limits of the family".

In the book, "IS EVOLUTION PROVED?" Douglas Dewar quoted Sir J. William Dawson, F. R. S., of McGill University (Toronto), a trained geologist. Prof. Dawson said in his day:

"The evolutionist doctrine is itself one of the strangest phenomena of humanity, but that in our day a system destitute of any shadow of proof . . . should be accepted as a philosophy, and should enable adherents to string upon its thread of hypotheses our vast and weighty stores of knowledge is surpassing strange

In a truly monumental work (published in 1954) by Dr. Heribert-Nilsson, Professor of Botany at the University of Lund, Sweden, he gives the results of his life's studies in genetics and other subjects. Speaking of fossil flora, he says,

"If we look at the peculiar main groups of the fossil flora, it is quite striking that at definite intervals of geological time they are ALL AT ONCE and QUITE SUDDENLY THERE; and, moreover, in full bloom in all their manifold forms. . . . Furthermore, at the end of their existence (if they are now extinct) they do not change into forms which are transitional towards the main types of the next period: such are entirely lacking. This all stands in **as crass a contradiction to the evolutionary interpretation as could possibly be imagined.** There is not even a caricature of an evolution."

His general conclusion is: "The final result of all my investigations and study, namely, that the idea of evolution, tested by experiments in speciation and allied sciences, always leads to incredible contradictions and confusing consequences on account of which the theory of evolution ought to be entirely abandoned, will no doubt enrage many; and even more so my conclusion that the theory of evolution can by no means be regarded as an innocuous natural philosophy, but that it is a serious obstruction to biological research. It obstructs—as has been repeatedly shown—the attainment of consistent results, even from uniform experimental material. For (to the evolutionist) everything must ultimately be forced to fit this speculative theory. An exact biology cannot therefore be built up."

Let us now consider the efforts of evolutionists to trace man's descent from the lower primates through

FOSSIL MEN . . . "MISSING LINKS"*

We reject the hypothesis that man is descended from the lower animals. Anthropologists, who accept the theory of evolution, believe men are "not direct descendants of apes . . . but both apes and men descended from the same ancestor." Consider these truths:

---

*Athropologists themselves do not use the term "missing links"—though their theories and conclusions justify the popular use of the term; hence we use it in our discussion.

1. Consider well this fact: the brain volumes of living men **vary from 790 cc. to 2,350 cc.** Then too, there are microcephalic idiots **with brain volumes of 500 cc. and less.** These unfortunate individuals are found in every human race. Living apes' brains vary from 87 cc. to 685 cc.* Consider too the variation in sizes of skulls from infants to adults, from male to female, from seven foot giants to four foot pygmies—all human skulls. Some diseased human skulls are actually **smaller** than the skulls of the larger apes!

2. Consider also this fact: Researchers have scoured every continent and every major island in the world, during the last 100 years, in a frantic search for "missing links" (skulls), and in the course of their searching **they have found and discarded tens of thousands of skulls**—and kept out a few bushel baskets full as the "missing links." Some that they prize most highly **are but small portions of a skull!** We are told they discard the rejected specimens usually "because there is no stratigraphical proof of their age." But the methods of arriving at the age of either bones or strata are highly uncertain—subject to vastly different interpretations.

3. Consider this fact: There has been bitter controversy **over every so-called "missing link."** Some experts will label a bone "human," while others will say most emphatically it is from an ape. Some will say the creature that possessed the bone walked upright; others will say that he most certainly walked on all fours.

We are of the opinion that the entire effort to find and reconstruct MISSING LINKS between apes and men is a pathetic farce, entirely beside the point. Their whole desire to find such "missing links" is based on a theory—the theory of evolution—a vain, misleading theory. If they do find an ancient bone fragment, they can not tell with certainty what it is. So we suggest to the evolutionists a much better way: Instead of searching for skull bones, that the experts can not and will not agree on, why not concentrate on

FOOT BONES?

There is a radical difference between the hind hand-foot of an ape and the foot of a man. **The hind hand-foot of the ape has a long thumb, that is opposable, enabling him to grasp the limb of a tree;** the foot of man has toes that enable him to walk upright, but do not enable him to grasp the limb of a tree! So let all searchers give up this vain search for mission skull bones—for they can not prove anything when they find them. Let them rather search for **foot bones** that show gradations from THE HAND-FOOT OF AN APE TO THE FOOT OF A MAN. If such a series of complete foot

---

*These figures are taken from Raymond A. Dart's article on "South African Man-Apes," in the 1955 Annual Report of the Smithsonian Institution.

bones are found, in intermediary stages from the hind hand-foot of an ape to the foot of man, they will have a most positive argument!

4. Consider this fact: Since the foot of an ape is so radically different from the foot of man, and his method of walking upright is "awkward," we are told that

"Rather late in history, there ventured a queer, somewhat old-fashioned mammal which had evolved, for reasons still not clearly understood, A FANTASTICALLY AWKWARD MODE OF PROGRESSION. It walked on its hind feet. . . . It was venturing late into a world dominated by fleet runners and swift killers. BY ALL THE BIOLOGICAL LAWS THIS GANGLING, ILL-ARMED BEAST SHOULD HAVE PERISHED, but you who read these lines are its descendants" (Loren C. Eiseley, in **"Fossil Man,"** Scientific American. Caps ours.)

One of the nation's leading anthropologists tells the world that man's upright posture and awkward way of walking put him at such a disadvantage that he had practically no chance of surviving the swift, deadly animal predators—but survive he did! If evolution were factual, and if Dr. Eiseley's judgment is correct, man would probably NOT have survived his "evolutionary" experiment—and you and I would not be here; but as we **are** here, we must give the credit to creation by the All-wise and All-powerful Creator!

5. The Biblical account of the creation of man (see Genesis 2:7) leaves no room for the theory of "Theistic evolutionists" who believe that "at some point in the evolution from the ape-like ancestors to man God put a human 'soul' in the creature and called it a man." But the Biblical account of the creation of **both Adam and Eve** completely negates the vain theory of Theistic evolution. (See gen. 1:26-27; 2:7, 18-25).

6. Consider this fact, even if the body of man evolved from the lower animals, **one has yet to explain the amazing mind of man.** (See our discussion of this in Ch. 12). Quoting Dr. Eiseley again:

"A student of man's evolution on earth is confronted today with an odd paradox. From a wealth of skulls and bones unearthed in the last few decades we can now piece together a reasonably convincing account of how and from what forebears man first came into existence more than a million years ago. But there the story trails into mystery. How the primeval human creature evolved into **Homo sapiens,** WHAT FORCES PRECIPITATED THE ENORMOUS EXPANSION OF THE HUMAN BRAIN—these problems ironically still baffle us." (Op. cit.)

Since there is no scientific evidence whatever of the possibility of the transformism of one genus into another, the "wealth of skulls and bones" that have convinced Dr. Eiseley of the evolution of men leave us uninfluenced — especially since we know that evolutionists do not derive their theory from facts. It is clear, evolution can NOT account for the marvelous mind of man. Some other explanation is necessary, and the only explanation that really solves all problems is, man was created by God in His own image and likeness, as the Bible says (Genesis 1:26, 27).

All scientists do not have Dr. Eiseley's faith in "skulls and bones." Dr. Austin H. Clark, noted biologist of the Smithsonian Institute, said:

"Man is NOT an ape and in spite of the similarity between them there is not the slightest evidence that man is descended from an ape. . . .

"While man's bodily structure is most nearly like that of the man-like apes, yet all the early remains of prehistoric man so far discovered are distinctly those of man. or are the misinterpreted fragments of apes. NO MISSING LINK HAS EVER BEEN FOUND."

"There is no fossil evidence whatever that the most ancient man was not a man. There are no such things as missing links. Missing links are misinterpretations. Fossil skulls which have been dug up and advanced as missing links, showing connections between man and monkey, have all been shown as misinterpretations."

The theories of modern anthropologists are in a state of flux and uncertainty. This can be seen readily from these statements from Dr. Eiseley's article:

"Is **Pithecanthropus erectus** (Dubois' 'missing link' of 1892) safe from the heretical hands of the modern generation of anthropologists? . . .

"In the 1890s all that was needed to tell the story of human evolution was to arrange on a classroom desk the skull of a chimpanzee, the skull cap of Pithecanthropus and the skull of Neanderthal. If the instructor placed his own head at the end of the line, a student could comprehend in a glance the full course of human evolution. . . . Today this state of affairs is vastly changed. We have a series of low-vaulted massive skulls with jutting brow ridges . . . a fairly comprehensive gallery of 'cave men.' Pithecanthropus belongs in this gallery. Though at various levels of development in brain size, we can say with assurance THAT THEY ARE ALL MEN. They represent the true human plateau. . . . They ranged from Java and China to the Middle East, Africa and Europe. . . .

"(Now) here is a point where tempers rise and staid investigators jab excited fingers at one another. In our gallery of beetle-browed ancestors **there are three or four specimens that throw the whole sequence out of order.** They are the well-known **Piltdown** skull, the **Swanscombe** skull and the **Fontchevade** cranium. . . . These three well-documented finds suggest NOT beetle-browed cave men, but true **Homo sapiens** or something approaching him." (Op. cit.)

Here we find these amazing inferences: Modern anthropologists are warring amongst themselves, and their theories about the descent of man are in a state of great confusion. Furthermore, we are frankly told that the prize of exhibits of yesterday ("Pithecanthropus," "Piltdown," "Swanscombe," "Fontechevade," etc.) are NOT "missing links"—intermediates between apes and men —but are ALL ACTUALLY MEN! And that is exactly what many of us who are Bible believers have been saying for years!

And then Dr. Eiseley further states (speaking of the **Australopithecines** of Africa),

"In South Africa we have a variable assemblage of walking APES with many human anatomical characters." (Op. cit.) (He tells also of some lemurs of Madagascar that stand on their hind legs "like little men.")

And that, too, is what many scientists, as well as Bible believers, have said for years—the South African skulls are from APES, not men, nor intermediates.

Listen again to Dr. Eiseley:

"We are no longer sure that the human precursor first arose in Asia. The great tablelands of Tibet and the neighboring regions HAVE YIELDED NO TRACE OF THIS EARLY STAGE. One of our greatest authorities upon fossil man in Asia, Pierre Teilhard de Chardin . . . is now convinced that Africa is the original homeland of the human race." (Op. cit.)

And so our leading evolutionists today are leading the race back to the apes of Africa—not a very exalted origin for man, who walks upright and who "has eternity in his heart".

But Dr. Eiseley is not sure of himself, nor of his belief. He says,

"Two facets of the 'mystery of man' deserve our particular attention: (1) How did man achieve his upright posture, and (2) how did the human brain arise, and what has carried it to its present peak of achievement? Neither of these questions has, in my opinion, been satisfactorily answered."

He also mentions a third problem he has. "It is difficult to see precisely why ONLY ONE GROUP (of primates) TOOK TO BIPEDAL HABITS. Other primates, notably the Baboons, have taken to a ground existence, but in spite of a considerable manual dexterity, they have retained a four-footed posture." (Op. cit.)

Yes, Dr. Eiseley, those questions present real problems to you, for which evolution HAS NO SOLUTION. We urge you to forsake the bickering uncertainties, the confusion, the heated arguments, and the "finger pointing" of the stymied evolutionists, and **admit the fact of God's direct creative work**—then all these problems have a satisfactory solution and answer.

FACTS ABOUT FAMOUS "MISSING LINKS"

As we have been told by Dr. Eiseley that anthropologists have classified as "men" the prize exhibits of evolutionists of yesterday, it seems almost pointless to review these so-called "Missing links." However, it is good to remind ourselves of what evolutionists have done and believed to support their theory. Let us take a hurried look at some of the better known.

(1) **Pithecanthropus erectus**—"Java Man." Starting with ONE BONE from the top of the skull, discovered by Dr. Eugene Dubois in 1891, in Java, and a leg bone and two molars, a plaster-of-Paris "reconstruction" was made, from the waist up, showing "a flat nose, short chin, with a bull neck." In the same general area were found "three adults skulls . . . in fair shape . . . and parts of the upper and lower jaws with a number of teeth." So *"Pithecanthropus"* is a RECONSTRUCTION, made from the imagination of what an evolutionist **thought** he should look like! Many thousands of pictures of **Pithecanthropus** have been taken and published in school books.

(2) "The **Heidelberg Man.**" In this instance they had only a **jawbone,** discovered by two workmen in a sandpit near Heidelberg, in Germany, in 1907. Dr. H. F. Osborn made a "reconstruction," starting with this jawbone, of an ape-like creature, carrying the carcass of a wild boar over his shoulder. Anthropologist Hrdlicka said that the teeth of this jaw "'are unquestionably human teeth."

Many living people "have the same type of receding chin indicated by the Heidelberg jaw."

(3) **"Sinanthropus Pekinensis"—"Pekin Man."** Many skulls and skull fragments were found in cave deposits near Peking, China, in 1929. All of these skulls "fit within the range of human skulls of today." A "reconstruction" of Sinanthropus was made by Dr. Franz Weidenreich, which makes one think of an intermediate creature, between apes and men. But again, the RECONSTRUCTION was conceived in the mind of an evolutionist, and represents what he SUPPOSED and WANTED it to look like! Dr. Davidson Black, of the Rockefeller Foundation, said that "all of these skulls (found near Peking and Choukoutien, China) were skulls of men."

(4) **Eoanthropus.** Charles Dawson, of Piltdown, England, announced in 1911 that workmen had found parts of a cranium that apparently were from a primitive type of man. Later a lower jaw and some teeth were found. The English paleontologist Sir Arthur S. Woodward decided that the bones were of a now-extinct type of man, which he called **Eoanthropus** (dawn man) and which also became known as Piltdown man. Today we all know that **Eoanthropus** was a deliberately planned hoax—a hoax that deceived the world's leading anthropologists for forty years!

(5) The **Swanscombe** skull, found in a gravel bed at Swanscombe, England, in 1935, consisted of the back and one side of a woman's skull. "It was only pieces of a skull cap, and not the whole skull or face" (**Early Modern Men,** p. 26). There is not enough of the skull to prove anything though it is usually classified as "Neanderthal."

(6) The **Fontechevade** skulls, found in France. One author laments, "If we only had enough of this find to be sure the brow ridge belonged to an adult, male individual, we'd be well off." (**Early Modern Men**). "But that brow ridge," he continues, **"might** have come from a youngish female Neanderthaler." Here again, there is not enough of the skulls to make positive identification.

(7) The first **Neanderthal** skull cap was discovered in 1857 in a limestone cave, near Dusseldorf, Germany. Virchow, the great German pathologist, declared it was the cranium of an idiot!

Anthropologists tell us that since the original find, "there have been over 100 'Neanderthal' skulls found; 20 were in good condition." Admitting there was a type or race of men that anthropologists call "Neanderthal," that is now extinct, one has proven no more than that a distinct "race (such as the Biblical "Canaanites" or "Hittites") that once lived is **now extinct.** It is impossible to prove the great ages credited to these races (Neanderthal, Heidelberg, etc.). They are in no sense in an "evolutionary chain."

Outline of Science (p. 80) makes this interesting observation, "At the same time there lived (in the same community as "Neanderthal" man) a race which resembles very much our present Negro. It is difficult to believe that such vast changes (as asserted by evolutionists) should have taken place in the Neanderthal man and left the progenitor of the Negro, the Grimadi race, untouched."

So much has been said and written about Neanderthal man, we want to quote an important statement from J. E. Weckler, in his article on "NEANDERTHAL MAN," in the December, 1957, **"Scientific American."**

"The fossils (of man) leave us mystified about his beginnings. Long study of the skulls **has failed to give any conclusive picture of man's early evolution;** in fact, many of the theories have not stood the test of new fossil finds. Among all the enigmas, Neanderthal man, . . . is still perhaps the most puzzling."

To show the confusion and uncertainty that dwells in the camp of the anthropologists, we quote again from Mr. Weckler:

"The oldest (of the fossils found in Europe) is the jaw of Heidelberg man, believed to date from about 500,000 years ago. Heidelberg man was once thought to be an ancestor of Neanderthal, BUT HIS TEETH TURN OUT TO BE MORE ADVANCED THAN NEANDERTHAL'S, and **like those of Homo sapiens."**

Even though evolutionists attached great ages to all these skulls, it can be said with assurance, backed by the testimony of qualified anthropologists, that ALL of the foregoing skulls and types are HUMAN.

(8) **Australopithecus africanus,** i.e., the South African ape. Dr. Robert Broom and J. T. Robinson (and others) found the remains of "about 100 different infantile, adolescent, and adult . . . specimens of these australopithecines from Taungs, 80 miles north of Kimberley, S. Africa (and from other sites in South Africa)". Undoubtedly, these are the skulls of APES.

"Unfortunately (for the evolutionist) it has been difficult to link the man-apes to the tools (found in the vicinity). The man-ape's bones have been found only in caves, whereas the tools generally have turned up in open river valleys, where bones are not preserved." REMEMBER, APES DO NOT MAKE OR USE TOOLS.

Here is a touch of unthinking inconsistency often found in

the statements of evolutionists. "Some stone tools found associated with fossil bones of australopithecines." (Reported in **Nature,** a British journal, by Dr. J. T. Robinson and R. J. Mason, of the Archeological Survey of the Union of South Africa). We quote from "Science Digest":

"Finding the tools with the fossilized bones of the man-apes does not necessarily mean that Australopithecus either made or used the tools, Dr. Robinson points out." "Although the Australopithecines," adds Mr. Mason, "may not have been capable of inventing and making tools, it seems very possible they were intelligent enough to make use of them." If they weren't intelligent enough to INVENT them, **no tools would be there for them to use,** since they were man's ancestors and man had not been "evolved" yet!

All reputable anthropologists agree that the Australopithecines were APES, though of course most of them who are evolutionists will say they were "man-apes." Some who believe in evolution see in them ancestors of man. But this has been challenged. In 1951 Dr. Montague Francis Ashley Montagu (Prof. of Anthropology) wrote, "It is quite possible that the Australopithecines pursued a parallel evolution with early man."*

All evolution can EVER do is to advance a THEORY that modern man descended from "some ape-like ancestor." The Bible believer has this three-fold assurance that man was CREATED in the image of God:

(1) The Bible that gives full evidence that it is a revelation from God, clearly teaches that man was created in God's image (Gen. 1:26, 27).

(2) Christ, the Son of God—the One who demonstrated the truthfulness of His claims to deity by His resurrection from the dead (see Rom. 1:4)—clearly tells us that God **created** mankind: male and female (see Matt. 19:4). If Christ be God—and all true Christians believe He is—then He knows these facts, and His Word is to be believed.

(3) No facts in nature or in the realm of science have ever upheld the theory of evolution that teaches that the higher genera evolved from the lower. THERE IS NO SUCH THING IN NATURE AS TRANSFORMISM (transmutation) FROM ONE GENUS TO ANOTHER. So, the evolutionists are promulgating a theory that is scientifically undemonstrable.

The issue finally devolves into this: Whom are you going to believe—Christ or the modern evolutionist? Will you rely on FACTS or THEORIES?

FOSSIL FRAUDS, FAKES AND FABLES

We already have called attention to the colossal fraud perpetrated by, or in the name of Charles Dawson anthropologist of

---

*An Introduction to Physical Anthropology, (p. 120.)

England in foisting off on the public (and the world of science as well) the "'Piltdown Man."* True science cannot be blamed for such a forgery, but the whole affair shows that MANY SCIENTISTS CAN EASILY BE DECEIVED.

Many people think that it takes many thousands of years to produce a fossil. This is not necessarily so. There are fossil men in the ruins of Pompeii, overwhelmed by an eruption of Mt. Vesuvius, in 70 A.D.

A fossilized Mexican sombrero was found not many years ago; it couldn't have been over 200 years old!

True, the fossilized bones of man have been found in great abundance—but such fossils need not necessarily be over a few thousand years old. Some will say that, through the modern "fluorine" test, it has been proven that some human fossils go back to at least the Middle Pleistocene (500,000 years ago). But here again modern anthropologists **completely ignore** the Flood, with its universal spread of highly mineralized **sea water** all over the face of the earth. Naturally, bones that soak for months in sea water and that are covered with earth soaked with sea water for several years will absorb fluorine far quicker than bones that lie in ground with very little fluorine content.** All authorities agree that the "rate of the accumulation of fluorine depends on the fluorine content of the soil in each particular area"—hence "fluorine dating" is **relative,** depending on the fluorine content of the soil where the fossil is buried. Then too, bones, through local floods and washouts, may and do change their environment—hence change their rate of fluorine absorption. All of these possibilities make the "fluorine test" highly unreliable as far as dates of origin are concerned.

It has not been long since William Jennings Bryan was publicly ridiculed for refusing to believe that **Hesperopithecus Haroldcookii,** the so-called "Nebraska Man," was a million years old. "Hes-

---

*See "The Great Piltdown Hoax," Pub. by Smithsonian Institute.

It would be an interesting experiment to soak human bones in sea water for a few years and compare the rate of fluorine absorption with bones (1) buried in a damp soil, and (2) bones bleaching on top of the soil. Even with such information, since no one knows positively the history of any particular fossil, during the past thousands of years, it would be IMPOSSIBLE to give any scientific reliability to the fluorine test. It is also quite possible that the rate of fluorine absorption **decreases as fluorine saturation approaches.

peropithecus" was a reconstruction—this "man-like ape" was built from head to foot **from a single tooth!** Later it was found that this tooth belonged to an extinct species of pig! (The Theory of Evolution and the Facts of Science, by Dr. Harry Rimmer).

Another example of how anthropologists have been misled is in how they interpreted the main example of the classical Neanderthals, "the Old Man of La Chapelle-Aux-Saints." Almost the complete skeleton was found, including twenty-one vertebrae. The structure of the vertebrae seemed to suggest that there was no cervical curve in the vertebral column. Those doing the RECONSTRUCTION decided that Neanderthal Man walked with a bent-knee gait and that he was stooped with his head hanging forward. "This 'RECONSTRUCTION' was generally accepted until December, 1956, when William L. Straus, Jr., in a paper presented to the anthropology section of the American Association for the Advancement of Science declared that Neanderthal Man walked as erect as we. Straus had made an intensive investigation of the original skeleton and discovered that the 'old man' was suffering from the advanced stages of **osteo-arthritis of the spine!** That of course was the reason for the lack of a cervical curve in the spine." (From the paper, **"Paleoanthropic vs. Neanthropic Fossil Men,"** by Claude Stipes, anthropology professor).

See what we mean when we say even scientists are easily misled? It is extremely hazardous to make RECONSTRUCTIONS from ancient bones—parts of skeletons.

In July of 1923, Dr. Ales Hrdlicka, perhaps the greatest anthropologist of his day, was invited to inspect, personally, the original bones of **Pithecanthropus erectus.** As a result of his examination, Dr. Hrdlicka said, as quoted by Harry Rimmer:

"None of the published illustrations or casts now in various museums are accurate. The jaw fragment was from ANOTHER and later type man."

Dr. Wilhelm Gieseler, of the University of Munich, is reported to have made a study of **Pithecanthropus erectus,** and then to have given his considered opinion. He stated that he believed that Pithecanthropus was "human," largely on the basis that "his eye-sockets are man-like rather than ape-like." Now recall these eye-sockets were part of the RECONSTRUCTION job and were made out of plaster of Paris!

Prof. Ernst Haeckel, German zoologist, was tried and convicted by a university court at Jena on the charge that his charts, attempting to prove the "recapitulation" theory of evolution, were deliberately **altered** from the facts of embryology, to bolster his theory. He finally admitted that he had made certain "alterations" that were NOT true to physiology; but he excused himself by saying that many other embryologists and zoologists of his time were guilty of the same procedure!

Consider the "logic" of a prominent evolutionist, who said,

"Evolution is more than proved by the bare consideration that no alternative remains . . . except creation. . . . 'Spontaneous generation is quite the ordinary thing; we have only to remember that matter was from the very first ready to produce life. . . . At some time or other spontaneous generation MUST have taken place. If the hypothesis of evolution be true, living matter MUST have arisen from non-living matter; therefore life DID originate thus, and the truth of evolution is established." (See, The Case for Creation, p. 29).

Refusing to believe in creation, this "thinker" (?) argues, that since evolution **must** be true, and since life is here, it HAD to come by spontaneous generation!

SOME THOUGHTS ON THE "AGE OF MAN"

Modern evolutionists believe that "the earliest fossil bones of men yet found are about half a million years old." **(Prehistoric Men,** by Robert J. Braidwood, Dep't of Anthropology, University of Chicago). He continues, "It is sure that mankind is older than a half million years, but no fortunate accident of discovery has yet given us evidence to prove it." (Op. cit.). To a paleontologist, a half million years is "recent" in the history of the earth—so all biologists and anthropologists state that man's advent on earth is comparatively "recent", even though they mean by that from one to twenty million years ago.

"'Prehistoric men did not appear until long after the great dinosaurs had disappeared. . . . Paleontologists know that fossils of men and of dinosaurs are not found in the same geological period. The fossils of dinosaurs come in early periods, the fossils of men much later." (Op. cit.)

As we seek to consider the subject of the Age of Mankind objectively, we are struck with a fact of outstanding importance:

(1) Man's "Written history"—when man began to write—DID NOT BEGIN UNTIL ABOUT 5,000 years ago. (Op. cit.) That would be about 3,000 B.C.

This remarkable fact—that "written history" begins from 3000 to 3500 B.C. conforms closely to the Bible chronology of mankind!

Dr. Eiseley says that in the Second Interglacial period MAN'S

BRAIN HAD REACHED ITS PRESENT SIZE. That is about 500,000 years ago. We would like to ask Dr. Eiseley, IF mankind has had its present brain capacity for a half million years, why did not men learn to write and read **much sooner?** He admits that

"Stone age savages of today are capable of learning to fly airplanes, play chess and take on the virtues and vices of advanced society." (Fossil Man).

(2) How can one explain the phenomenon of hundreds of highly developed languages, the world over, even among primitive tribes, IF language evolved slowly through the ages? The evidence is, all tribes and races obtained the use of language SUDDENLY; and that fits in with the Bible record of Divine creation.

IF men, with brains the size of man's brain, can learn "to fly airplanes and play chess" in ONE GENERATION—why did it take mankind 500,000 years to learn to read and write? It does not make sense.

Lay aside men's theories of evolution — that man slowly evolved from the lower animals—and the facts in the case indicate that **mankind has been on earth for only about 6,000 to 8,000 years**—a few thousand rather than a few hundred thousand years.

(3) In addition to the evidence of "writing" we have another corroborative fact: Archeological evidence indicates that **"Food production** probably began in the Near East somewhere between 6,000 and 5,000 B.C." (Op. cit.) "The earliest village-culture materials now known start about 5,000 B.C." (Op. cit.)

By "food producing" economy, paleontologists mean, "men began **producing** their food, instead of simply **collecting** it." This they describe as "a revolution, just as important as the Industrial Revolution. In it men learned to domesticate plants and animals."

"See the picture of man's life after **food-production** had begun. He stored his meat 'on the hoof,' his grain he stored in silos or pottery jars. He lived in a house; it was worth his while to build one, because he couldn't move too far from his fields or flocks. . . . They all lived close to their flocks and fields in a village." (Op. cit.)

Now, we ask again—IF MANKIND HAD ITS PRESENT BRAIN FOR THE PAST 500,000 years, why is it they did not plant crops and herd sheep until a few thousand years ago? It just does not make sense.

Leaving out the theory of evolution with its long ages for mankind, the actual facts of history clearly indicate JUST WHAT THE BIBLE INDICATES THAT GOD MADE ADAM AND EVE ABOUT 6,000 to 8,000 YEARS AGO!

(4) Furthermore, it can be demonstrated, by taking the rate of population increase, per century, and working back from our present world population of 2,500,000, that mankind started with two people not very long ago!* The late Sir Ambrose Fleming, F.R.S., pointed out,

"If mankind had existed on earth in the vast periods of time invented by evolutionary speculation, the world . . . could not possibly have accommodated the human race—a fact to which vital statistics give increasing point."

In conclusion let us say, A THOUSAND FACTS, TEN THOUSAND VOICES, A MILLION SPECIALIZED PLANTS AND ANIMALS bear witness to DIVINE CREATION! And what speaks forth for evolution? Only "unproven and unprovable theories!"

Let man have the dignity that GOD gave him, by creating him in His own image and likeness! Teach man that he is descended from the beasts, and he soon will act like a beast. Teach man that he is a mere animal—a cog in the social machine—and **communism will enslave him.** Teach man that he is created by, and has a responsibility to, God—and the dignity of human life, and the importance thereof, will be brought into government.

Teach man, as the Bible does, that though he is created in the image of God, the race is fallen, and needs a Saviour, and that men must turn to God, through Christ, for **His** salvation, **His** righteousness, **His** love, and **His** holiness.

Let our young people **know** that evolution is a baseless, groundless theory, and that the **facts** are all on the side of creationism, and they then will not be swept off their feet by the **ipse dixit** of biased professors and the propaganda of blinded evolutionists.

One reason so may young people are confused and misled by the teaching of evolution is, practically **all** biology teachers today teach evolution and tell their classes, "Every educated person today believes in evolution." And so the young people are misled—and deceived. THE TIME HAS COME TO GIVE FACTS, GIVE LIGHT, GIVE THE TRUTH TO OUR YOUNG PEOPLE. Let them know that evolution is NOT true—that it is a fable, a myth, an unproven and unprovable theory. The

---

*"On the first Christmas Day the population of our planet was about 250,000,000. Sixteen centuries later human numbers had climbed to a little more than 500,000,000. Today there are 2,500,000,000 of us." (Pp. 8, 9, **Brave New World Revisited,** by Aldous Huxley). Other figures are available that point back to the beginning of mankind a few millennia before Christ.

Word of God is true—and its teachings can be defended! Arise, in this day of battle, and stand by the eternal truth of God's Word! Honor God as the Creator, and Christ as the Redeemer and the Bible as God's Word!

Encourage research,* let the light of true knowledge flood our classrooms, but do not lock God outside the door! For after all, the God who made us demands an accounting!

"God . . now commandeth all men everywhere to repent (and believe the Gospel): because He hath appointed a day, in the which He will judge the world in righteousness by that man (Christ Jesus) whom He hath ordained; whereof He hath given assurance unto all men (that these facts are so), in that He hath raised Him (Christ) from the dead" (Acts 17:30, 31).

---

*Evolutionists often accuse creationists of believing in "magic" when they give credit to the Creator for His wonderful works. It is not accepting a "magical" explanation to believe that a Supreme Being created the endless marvels in our world and universe—it is just good common sense; for "creation demands a Creator" as much as a house demands a builder.

Those who credit blind, senseless evolution with the innumerable miracles of life all around us are the ones who really resort to "magic"— for they ascribe the most wonderful works to an imaginary, theoretical force that exists only in the imagination of evolutionists.

BIBLIOGRAPHY

A PARTIAL LIST OF THE BOOKS, PAMPHLETS AND MAGAZINES used for reference material in the preparation of this book:
Many articles from the SCIENTIFIC AMERICAN; Science Digest; Science Monthly; Book of Popular Science, Encyclopedias, Textbooks, etc.
Twenty or more recent Releases from The Smithsonian Institute, Washington, including: **"The Flight of Animals,"** by James Gray, Professor of Zoology, Cambridge University; **"The Migration of Mammals,"** by L. Harrison Matthews, Director, Zoological Society, London; **"Some Observations on the Functional Organization of the Human Brain,"** by Wilder Penfield, Professor of Neurology and Neurosurgery, McGill University; **"Genetics and the World Today,"** by Curt Stern, Dep't of Zoology, University of California; **"Genetics in the Service of Man,"** by Bentley Glass, Professor of Biology, The Johns Hopkins University; **"Barro Colorado—Tropical Island Laboratory,"** by Lloyd Glenn Ingles, Dep't of Biology, Fresno State College, Calif.; **"Phosphorus and Life,"** by D. P. Hopkins, Smithsonian Publication 4116; **"The Geological History and Evolution of Insects,"** by F. M. Carpenter, Harvard University; **"The Time Scale of Our Universe,"** by E. J. Opik, Armagh Observatory, Northern Ireland; etc., etc.
"Animals Without Backbones," by Ralph Buchsbaum, Dep't of Zoology, University of Chicago; The University of Chicago Press, Chicago, Ill.
"Reptile Life," by Zdenek Vogel; Spring Books, London.
"Prehistoric Animals," by Joseph Augusta, Professor of Paleontology; Spring Books, London.
"The Hunting Wasp," by John Crompton; Houghton Mifflin Co., Boston (1955).
"Evolution and Human Destiny," by Fred Kohler; Philosophical Library, New York.
"The Living Sea," by John Crompton; Doubleday and Co., Garden City, N.Y.
"The Underwater Naturalist," by Pierre de Latil; Houghton Mifflin Co., Boston (1955).
"Salamanders and Other Wonders," by Willy Ley; The Viking Press, New York; 1955.
"Evolution in the Genus Drosophila," by J. T. Patterson and W. S. Stone, Professors of Zoology. University of Texas; The Macmillan Co., New York.
"Animal Wonder World," by Frank W. Lane; Sheridan House, New York.
"Nature Parade," Frank W. Lane; Sheridan House, New York.
"Man and the Vertebrates" (Vol. Two), by Alfred Sherwood Romer, Prof. of Zoology, Harvard University; C. Nicholls and Company Ltd., Harmondsworth, Middlesex, England.
"Nature's Wonders," by Charles L. Sherman; Hanover House, Garden City, N.Y.
"Mammals," "Reptiles and Amphibians." "Birds," "Insects," "Flowers," "Seashores,"—Nature Guide books—by Herbert S. Zim, Ph.D., and Hobart M. Smith, Ph.D., Dep't of Zoology, University of Illinois; Simon and Schuster, New York.
"The Strange World of Nature," by Bernard Gooch; Lutterworth Press, London. Etc., etc.

BIBLIOGRAPHY
(For further study against Evolution)

"Evolution." by a medical student; International Christian Crusade, 14 Park Road, Toronto, Ont., Canada; 25c.

The Bible and Modern Science, by Henry M. Morris, Ph.D.; Moody Press, Chicago, Ill.

Evolution: Fact or Theory? by Cora A. Reno; Moody Press, Chicago, Ill; 35c.

The Evidences of GOD in an Expanding Universe—"Forty American Scientists declare their affirmative views on Religion"—edited by John Clover Monsma; G. P. Putnam's Sons, New York; $3.75.

God and the Cosmos, by Theodore Graebner; Eerdman's Pub. House; Grand Rapids, Mich.

Rastus Augustus Explains Evolution, by the late B. H. Shadduck, Ph.D.; The Higley Press. Butler, Ind; 25c. Six other pamphlets against evolution, by the same author may be had from the same publisher; 25c each.

The Theory of Evolution and the Facts of Science, by Dr. Harry Rimmer; Berne Witness Co., Berne, Ind. Several other books and pamphlets against evolution, by Dr. Rimmer, may be had from the same publishers.

Did Man Just Happen, by W. A. Criswell; Zondervan Pub. House, Grand Rapids, Mich. $2.00.

Creation's Amazing Architect, by Walter J. Beasley, F. R. G. S.; Australian Institute of Archaeology (Offices: 174 Collins St., Melbourne, C. 1, Australia). Ask for a catalogue of their other releases on the subject. "Modern Science and the Bible."

The Case for Creation. by E. C. Wren; Christian Evidence League, P. O. Box 277, Malvern, N. Y.; 50c. (Ask for cataloge of their other books and pamphlets).

Publications of the **"EVOLUTION PROTEST MOVEMENT,"** of Great Britain. Write to Prof. L. V. Cleveland, Canterbury, Conn., U.S.A., for a list of their publications.

INDEX

A

"Adaptations" and "Design," 99-107
"Acquired Characters," 199
Age of Man, 338
Aims of Evolutionists, 9
Air Plants, 135
Algerian Locust, 23
Alpine Salamander, 311
Ameba, 86, 87, 137
Anteater, 113
Ants, 129, 214-224
Apes are Dying out, 271
Archaeopteryx. 158
Archer Fish, 155
Army Ants, 222
Astronomy, Limitations of, 66
Atom, 71-74
Atmosphere, 34-37
Australopithecines, 334-335
Aye-aye Lemurs, 318

B

Bacteria, 84-85, 134
"Balance" in Nature, 109-113
Bats, 196
Beauty, 155, 189, 190, 313
Bedbugs, 307
Bees, 201-213
Beetles, 124-126
Bibliography, 342, 343
Bills (Beaks) of Birds, 166-170
Birds, 111, 158-196
Birds' Eggs, 171-173
Birds' Nests, 173-178
Bladderwort, 135
Blood, 233-236
Bombardier Beetle, 23, 125
Bones of Birds, 161
Bones of Man, 247
Bower Birds, 184
Brain, 250-253
Bubble-nest Builders, 157
Built-in Bifocals, 294
Butterfly, 311-316
Butterfly's Wings, 314-316

C

Cactus, 102, 110
Camouflage, 281-284
Cell Mitosis, 90
Cells, 88-93, 230
Cerebellum, 252
Cerebrum, 250, 252
Chino-Mark Moth, 23
Chromosomes and Genes, 90-95
Chromosomes and Genes—effects of Radiation on, 93
Cicadas, 138
Circulation in Seas, 50
Clark, Austin H.—His Testimony, 16
Classification System, 286
Climbing Perch, 157
Clouds, 41
Clover in England, 116
Collar-Flagellates, 281
Combinations on Earth that Prove Creation, 28
Commensalism (see Symbiosis), 20
Community Instinct, 211, 216-217
"Comparative Anatomy" Argument, 267
Complexity of Proteins, 78, 79
Conception, 226
Crabs, 150-152
Creation of Life in a Test Tube? 81
Creation of Matter, 286
Cross-pollination, 117, 118, 206
Curlew, Bird that gets a new Stomach, 24

D

Dandelions, 309
Darwin's Frog, 310
Dead-leaf Butterfly, 282
Death March of Lemmings, 284
Deceptions that Deceived Professionals, 7
Deep Sea Fish, 152, 153
"Design," a Proof of Creation, 99-108
Digestive System, 241

Differences between Apes and Men, 269
Drosophilia, 199
Ductless Glands, 243
Dust, a Witness for Creation, 37

E

Eanthropus, 333
Ear, 246
Earth, Theories of Origin of, 25
Echidnas, 131
Eels, 284
Eggs, 172-174, 305-307
Einstein's Theory of Curved Space, 65
Electric Eel, 148
Electric Fish, 148, 149
Elements, Stable, 44, 45
Embryo, Embryology, 227-228
Evolution and Creation, mutually Exclusive, 8
Evolution Defined, 12
Evolution is Gloating, 9
Eunice Gigantea (Sea Worm), Miracle of Complexity, 24
"Expanding Universe," 64
Exploding Stars, 69
Eyes, 293-301
Eyes of Bees, 204
Eyes of Man, 237-240

F

Falcon, 184
Feathers of Birds, 161
Feet of Birds, 165-166
Fidler Crab, 279
Fish Hatched in Father's Mouth, 149
Fish, Strange, 139-157
Fixity of Species, 127-129
Flea Examined under Microscope, 104
Flies, 114, 118
Fluorine Dating, 336
Fontechevade Skulls, 333
Food Chain in Seas, 109
Fossil Men, 328-335
Fossil Frauds, 335-338
Fossils, 324-328
Fresh Water—a Miracle, 49

G

Genes and Chromosomes, 90-95
Geological Time, 325
German Warblers, 187
Globefish, 156
Glory in the Stars, 62
Goldschmidt's Testimony, 315-316
Grasshoppers, 105, 129, 136
Growth of Body, 229
Grunion, Dance of, 146

H

Hair, 243
Hand, 244
Handicaps and Safeguards, 110
Heidelberg Man, 333
Henpecked Hornbill, 177
Heredity and Bees, 207
Honey-guide of Africa, 179
Horseshoe Crab, 152
Human Cell, 230
Humming Bird, 186
Hunting Wasp, 106-107, 124

I

Instinct, 110, 211, 212, 219, 279

K

Killer Plants, 135
Kiwi, 179

L

Ladder of Creation, 8
Language of Ants, 219
Language of Bees, 210
Lantern Bug, 282
Leaping Spawner Fish, 155
Leaves of Trees, show Design, 102
Life, Creation of, 75
Life on Other Planets? 27
Lightning-Nitrogen Cycle, 40
Limitation of Hazards, 276
Lionfish, 316
Liver, 245
Luminous Fish, 153
Lungfish, 317

M

Man Did not Descend from Monkeys or Apes, 266, 267
Man, Mystery of, 225-250
Man's Moral Emotional and Spiritual Nature, 262-266
Medulla Oblongata, 252
Metamorphosis of the Caterpillar, 278
Methods of Reproduction, 303
Microscopic Predators, 275, 276
Midwife Toad, 310
Migration in the Sea, 50
Migration of Birds, 187, 188, 194-196
Mimicry, 281, 283
Mind of Man, 253-262, 330
Miracles, 273-285
Missing Links, 285-292, 328, 332-335
Mitochondria, 89
Molds, 44
Moon and Tides, 33
Moons that Revolve in Retrograde Motion, 27, 274
Moral Lessons taught by Nature, 136
Mosquito's Tool-kit, 106
Mountains and the Rain Cycle, 42
"Moving Left Eye", 21
Mutations, 17, 93, 99, 199
"Mutations" but no "Transmutations," 14
Mysteries and Miracles, 273-285
Mysteries in the Universe, 68
Mysteries of Heredity in the Cell, 90
Mysteries of the Sea, 46, 49

N

"Natural Selection" ruled out as explanation of Evolution, 19
Neanderthal Man, 334
Nebraska Man, 336
Nervous System, 242
Nitrogen Cycle, 35
Nuclear Energy, 73, 74

O

Oceans, 46, 47
Ocean Cycles, 51
Oddities in Bird Life, 193

Organic Compounds, necessary before Life, 76
Origin of Motion, 286
Oriole, 176
Owl's-head Butterfly, 282
Oxygen, 36
Oyster, 144

P

Paddle Fish, 155
Palolo Worm, 147
Paramecia contrasted with Amebas, 87
Pekin Man, 353
Penguins, 180-182
Phoebe, Saturn's Moon that goes in retrograde Motion, 26
Pine Cone Fish, 155
Pithecanthropus, 331, 333, 337
Plankton, 141
Plants show special Design, 102
Pollination; see Cross-pollination
Portuguese "Man-of-War," a Witness for Creation, 20
Praying Mantis, 138
Pro-avis, 159, 290
Problem of Evil Discussed, 107
Pronuba Moth, 119
Protective Coloring, 281-283
Proteins, a Prerequisite for Life on Earth, 76, 79
Proteins, highly Complex, 78, 79
Protozoa, 86
Pulsating Stars, 69

R

Radiation: its effects on Genes and Chromosomes, 93, 94
Railroad Worm, 23
Rain Cycle, 41, 42
Random Mutations ruled out as cause of Evolution, 19
Recapitulation Theory, 228
Red Blood Cells, 234-236
Regeneration, 279
Rings of Saturn, 26
Rivers and the Rain Cycle, 41

S

Sea Cucumbers, Witness for Creation, 21
Sea Horse, 143
Seas, Mystery and Miracles of, 46-50
Sea Worm, Miracle of Complexity, 24
Second Law of Thermodynamics, 77
Sex, 302-312
Skin, 231
Sloth, 317
Snow, 42
Soil-replenishing Cycle, 42
Songs of Birds, 191, 192
Sound, 154
Species Defined, 13
Specialized Organs, 316-320
Spider Crab, 283
Spiders, 111, 320-322
Spittlebug, 137
Sponge, 319, 320
Spontaneous Generation is impossible, 95, 96
Stars, a Witness for God, 54-69
Stickleback Courting, 308
Sting Ray, 143
Sundew, 135
Sun in Relation to Earth, 30
Sun's Rays, 32
Sun's Stability, 31
Swanscombe Skull, 333
Symbiosis, 20, 117, 216

T

Tents made by Ants, 215
Termites, 23, 125
Theories of the Universe, 64-65
Tongue, 240
Tonsils, 240
Tracheal System of Breathing, 289
Types of Eyes, 295

U

Undersea Garden, 142
Undertakers, Nature's, 114
Uniformitarianism in Geology, 324
Universe: A Witness for God, 54-69
Universe, Law in, 58

V

Varieties of Species, 122
Viruses as Witness for God, 80-84
Viruses—Method of Reproduction, 81
Vultures, 115

W

Walking-leaf Insects, 282
Walking-stick Insects, 283
Walnut Shell, a Miracle of Design, 103
Water, a Miracle, 39, 40
Water Bug, 137
Water Ouzel, 178
Wealth of the Seas, 48
Webster vs. Choate (Court Case), 7
Woodpecker, 182-183
Wrasse, 283

Y

Yucca Plant and Pronuba Moth, 119

Lightning Source UK Ltd.
Milton Keynes UK
UKHW010808281021
392990UK00001B/15